S0-CKI-729

Progress in Physics
Vol. 5

Edited by
A. Jaffe and
D. Ruelle

Birkhäuser
Boston · Basel · Stuttgart

Gauge Theories: Fundamental Interactions and Rigorous Results

Lectures given at the 1981 International Summer School of Theoretical Physics, Poiana Brasov, Romania

P. Dita, V. Georgescu,
R. Purice, editors

1982

Birkhäuser
Boston • Basel • Stuttgart

Editors:

P. Dita
V. Georgescu
R. Purice

Department of Theoretical Physics
Central Institute of Physics
Bucharest, Romania

Library of Congress Cataloging in Publication Data

International Summer School of Theoretical
Physics (1981 : Poiana Brasov, Romania)
Gauge theories.

(Progress in physics ; v. 5)
1. Gauge fields (Physics)--Congresses.
2. Field theory (Physics)--Congresses.
I. Dita, P. (Petre), 1942- . II. Georgescu,
V. (Vladimir), 1947- . III. Purice,
R. (Radu), 1954- . IV. Title. V. Series:
Progress in physics (Boston, Mass.) ; v. 5.
QC793.3.F5I593 1981 530.1'43 82-12817
ISBN 3-7643-3095-3

CIP-Kurztitelaufnahme der Deutschen Bibliothek

Gauge theories: fundamental interactions and rigorous results:
lectures given at the 1981 Internat. Summer School of Theoret.
Physics, Poiana Brssov, Romania/P. Dita ..., ed.
Boston; Basel, Stuttgart:Birkhäuser, 1982
(Progress in physics; Vol. 5)
ISBN 3-7643-3095-3

NE: Dita, Petre (Hrsg.); International Summer School of
Theoretical Physics (1981, Poiana Brasov); GT

ISBN 3-7643-3095-3

Printed in USA

TABLE OF CONTENTS

---

* Lecture given by Professor Sternberg, whose main lecture, "Geometry and Physics of the Momentum Map," is not included in this volume.

** Professor J. Tarski also had a talk "Quantized Fields on Conformally Flat Spaces," not included in this volume.

PREFACE

This book contains the lectures given at the Poiana Brasov Summer School in Theoretical Physics, 1981, covering the following main subjects:

i) Physical aspects of gauge theories and models for fundamental interactions

ii) Geometrical methods and classical solutions of gauge fields

iii) Constructive field theory methods and lattice gauge fields.

These subjects are quite independent of one another, but all throw some light upon the as yet incompletely understood theory of quantum gauge fields; moreover, it was the organizers' conviction that a successful theory of fundamental interactions will emerge in the future by a fusion of ideas and methods of the above fields.

As we presently lack that theory, the phenomenological models are valuable in organizing the experimental data and in isolating useful concepts and ideas.

The latest developments suggest that it is useful and natural to think of gauge fields as geometric objects. In solving the nonlinear equations of classical gauge fields and in classifying their solutions, a very effective frame of concepts and methods has been provided by algebraic geometry and topology. It is now believed that algebraic geometry will give a systematic procedure for solving large classes of nonlinear equations.

On the other hand, the experience accumulated in constructing and studying relativistic quantum field models will certainly provide a frame for a future theory of fundamental interactions. Already, the new tools exhibited by constructive field theory are so strong that they may help one to discard some phenomenological models (including perhaps quantum electrodynamics) as being trivial. Further, a natural frame is provided for studying such phenomena as phase transitions which require a nonperturbative approach.

Therefore, the organizers thought it interesting to bring together people working in the above fields as an opportunity to learn of recent progress in other areas. Efforts were made to begin the lectures at an elementary level, reaching as smoothly as possible the current state of the art.

The book also contains an abstract of the lecture Professor Yu. I. Manin intended to give at this School. In spite of all our efforts it was regrettably impossible for Professor Manin to attend the School.

There are many to whom we owe a debt of thanks. In making any School a success, the importance of the invited professors is obvious. In this regard we were most fortunate, and our thanks are due first of all to them. The physics staff memebers at the Central Institute of Physics cooperated in many ways. We are especially grateful to Drs. M. Ivascu, V. Ceausescu, and G. Costache.

The running of the School was made smoother by the help of Dr. M. Beadea, Ms. Viviane Prager, Ms. Daniela Marin, and Messrs. Mirza, Chefneux, Damian, Banu, and Vasilescu.

Last but certainly not least, Ms. Geta Uglai, the technical secretary of the School, deserves special acknowledgement for her outstanding and compassionate handling of hard and sometimes tedious work.

P. Dita
V. Georgescu
R. Purice

Bucharest, 1982

LIST OF PARTICIPANTS

Angelescu, N.	Joint Institute for Nuclear Research, Dubna
Anghel, V.	Institute for Nuclear Power Reactors, Pitești
Athorne, Ch.	University of Durham, Durham
Atiyah, M.F.	University of Oxford, Oxford
Banica, C.	Dept. of Mathematics - INCREST, Bucharest
Banyai, L.	Institute for Physics and Technology of Materials,Bucharest
Baran, A.	Dept. of Mathematics - INCREST, Bucharest
Barth, W.	Universität Erlangen-Nürenberg, Nürenberg
Berceanu, S.	Institute for Physics and Nuclear Engineering, Bucharest
Berthier, Anne-Marie	Université Paris VI, Paris
Bogatu, N.	Institute for Physics and Technology of Radiation, Bucharest
Bovier, A.	Physikalisches Institut der Universität Bonn, Bonn
Brezuleanu, A.	University of Bucharest, Bucharest
Brydges, D.	University of Virginia, Charlottesville
Bulgac, A.	Institute for Physics and Nuclear Engineering, Bucharest
Caprini, Irène	Institute for Physics and Nuclear Engineering, Bucharest
Ceausescu, M.	University of Bucharest, Bucharest
Ceausescu, V.	Institute for Physics and Nuclear Engineering, Bucharest
Ciubotaru, Luminita	Institute for Physics and Technology of Radiation, Bucharest
Constantinescu, A.	Dept. of Mathematics - INCREST, Bucharest

Constantinescu, R.	"T. Vladimirescu" High School, Craiova
Corciovei, A.-director of the School	Institute for Physics and Nuclear Engineering, Bucharest
Costache, Gh.	Institute for Physics and Nuclear Engineering, Bucharest
Czyz, J.	Institute of Mathematics of the Polish Academy of Sciences,Warsaw
Deciu, Elena	Polytechnical Institute, Bucharest
Diţa, P.	Institute for Physics and Nuclear Engineering, Bucharest
Dittrich, J.	Nuclear Physics Institute, Prague
Domitian, V.	University of Bucharest, Bucharest
Dumitrescu, B.O.	Institute for Physics and Nuclear Engineering, Bucharest
Dumitru, D.R.	Institute for Physics and Nuclear Engineering, Bucharest
Efimov, G.	Joint Institute for Nuclear Research, Dubna
Ek, B.	Royal Institute of Technology, Stockholm
Fazakas, A.	Institute for Physics and Technology of Materials, Bucharest
Ferrara, S.	CERN - Génève
Fraser, Caroline	Laboratoire d'Annecy-le-Vieux de Physique des Particules, Annecy-le-Vieux
Georgescu, V.	Institute for Physics and Nuclear Engineering, Bucharest
Ghika, G.	Institute for Physics and Nuclear Engineering, Bucharest
Ghitoc, Marina	Centre for Astrophysics and Astronomy, Bucharest

Gologan, R.N.	Dept. of Mathematics - INCREST, Bucharest
Grigore, R.D.	Institute for Physics and Nuclear Engineering, Bucharest
Grosu, Corina	Computer Centre CIMIC, Bucharest
Grosu, Marta	High School no.31, Bucharest
Grünfeld, C.P.	Centre for Astrophysics and Astronomy, Bucharest
Guerra, F.	Istituto Matematico "Guido Castellnuovo", Roma
Gussi, G.	Dept. of Mathematics - INCREST, Bucharest
Helayel Neto, J.A.	International School for Advanced Studies, Trieste
Hernandez-Vozmediano, Angeles	Universidad Autonoma de Madrid, Madrid
Hill, A.	University of Groningen,Groningen
Horetzky, P.	Universität Wien, Wien
Horoi, M.	Institute for Physics and Nuclear Engineering, Bucharest
Hranitzky, N.	University of Timisoara, Timisoara
Hristea, M.R.	Institute of Medicine, Bucharest
Inozemtsev, V.	Joint Institute for Nuclear Research, Dubna
Ion, B.D.	Institute for Physics and Nuclear Engineering, Bucharest
Iosifescu, M.	Institute for Physics and Nuclear Engineering, Bucharest
Iou Tun Wu	Atomic Research Institute, Academia Sinica
Isar, A.	Institute for Physics and Nuclear Engineering, Bucharest
Jaffe, A.	Harvard University, Cambridge

Kämpfer, B.	Zentralinstitut für Kernvorschung Rossendorf, Dresden
Kimio, U.	Research Institute for Mathematical Sciences, Kyoto
Lazanu, I.	University of Bucharest, Bucharest
Manea, Mihaela	University of Bucharest, Bucharest
Manolache, N.	Dept. of Mathematics - INCREST, Bucharest
Marchesoni, F.	Universita Pisa, Pisa
Markowski, B.	Institute of Nuclear Energy and Nuclear Research, Sofia
Ma Sien Sin	Technical University, Academia Sinica
Marra, Rossana	University of Salerno, Salerno
Martinescu, A.	High School no. 6, Craiova
Mazon, M.	Instituto de Estructura de la Materia, Madrid
Messager, A.	Centre de Physique Théorique, Marseille
Mezincescu, A.	Institute for Physics and Technology of Materials, Bucharest
Mezincescu, L.	Institute for Physics and Nuclear Engineering, Bucharest
Mezincescu, Nadejda	Institute for Physics and Technology of Radiation, Bucharest
Micu, Liliana	Institute for Physics and Nuclear Engineering, Bucharest
Micu, M.	Institute for Physics and Nuclear Engineering, Bucharest
Mihai, A.	University of Bucharest, Bucharest
Mihalache, Gh.	Research Institute for Energetical Problems, Bucharest
Mihalache, N.	Dept. of Mathematics - INCREST, Bucharest

Mihul, Eleonora	Institute for Physics and Nuclear Engineering, Bucharest
Milea, T.	University of Bucharest, Bucharest
Minti, H.	Research Institute for Energetical Problems, Bucharest
Mot, P.	Institute of Medicine, Bucharest
Mostow, M.	North Carolina State University, Ralleigh
Müller-Preusker, M.	Joint Institute for Nuclear Research, Dubna
Nadiu, G.S.	University of Oradea, Oradea
Naidin, Andreea	University of Bucharest, Bucharest
Nanopoulos, D.V.	CERN - Génève
Nenciu, Gh.	Joint Institute for Nuclear Research, Dubna
Olive, D.	Imperial College, London
Oncica, A.	Centre for Astrophysics and Astronomy, Bucharest
Osterwalder, K.	ETH Zentrum, Zürich
Pantea, Alexandrina	Institute for Physics and Technology of Radiation, Bucharest
Pantea, D.	Institute for Physics and Nuclear Engineering, Bucharest
Percacci, R.	Universita di Trieste, Trieste
Pometescu, N.	High School Filiasi, Filiasi
Popp, O.	Institute for Physics and Nuclear Engineering, Bucharest
Purice, R.	Institute for Physics and Nuclear Engineering, Bucharest
Puta, M.	University of Timisoara, Timisoara
Putinar, M.	Dept. of Mathematics - INCREST, Bucharest
Quiros, M.	Instituto de Estructura de la Materia, Madrid

Radescu, E.	Joint Institute for Nuclear Research, Dubna
Radulescu, D.C.	Polytechnical Institute, Bucharest
Raina, A.	Université de Lausanne, Lausanne
Ramirez, C.	Physikalisches Institut der Universitat Karlsruhe, Karlsruhe
Rosca, M.	University of Bucharest, Bucharest
Rosu, Aurelia	University of Bucharest, Bucharest
Rosu, H.	Centre for Astrophysics and Astronomy, Bucharest
Rosu, R.	University of Bucharest, Bucharest
Ruiz, J.	Centre de Physique Théorique, Marseille
Saliu, L.	University of Craiova, Craiova
Sararu, A.	Institute for Physics and Nuclear Engineering, Bucharest
Sararu, Mihaela	Institute for Physics and Nuclear Engineering, Bucharest
Saru, D.	INCREST - Bucharest
Satir, A.	Middle East Technical University, Ankara
Schotsman, J.E.	Institute for Theoretical Physics, Utrecht
Scutaru, H.	Institute for Physics and Nuclear Engineering, Bucharest
Seiler, E.	Max-Planck Institut München
Silisteanu, I.	Institute for Physics and Nuclear Engineering, Bucharest
Simaciu, I.	Polytechnical Institute, Ploiesti
Simms, D.	Trinity College,Dublin
Skrypnik, V.	Institute for Theoretical Physics, Kiev
Soper, A.	University of Cambridge,Cambridge
Sparpaglione, M.	University of Pisa, Pisa
Steinbrecher, G.	University of Craiova, Craiova
Sternberg, S.	Tel Aviv University, Tel Aviv

Stihi, M.	Institute for Physics and Nuclear Engineering, Bucharest
Stoica, C.	Polytechnical Institute Ploiesti
Stoica, L.	Dept. of Mathematics - INCREST, Bucharest
Stratan, Gh.	Institute for Physics and Nuclear Engineering, Bucharest
Tarski, J.	Technische Universität Klausthal - Zellerfeld
Tataru, L.	University of Craiova, Craiova
Taylor, J.C.	DAMTP Cambridge
Timotin, D.	Dept. of Mathematics - INCREST, Bucharest
Trache, Maria	University of Bucharest, Bucharest
Trautmann, G.	Universität Kaiserslautern
Turcu, I.	Institute for Physics and Technology of Radiation, Bucharest
Turok, N.	Imperial College, London
Verhuyck, K.	Rijksuniversiteit Gent, Gent
Visinescu, Anca-Ilina	Institute for Physics and Nuclear Engineering, Bucharest
Visinescu, M.	Institute for Physics and Nuclear Engineering, Bucharest
Voiculescu, D.	Dept. of Mathematics - INCREST, Bucharest
Wallner, R.P.	Institute for Theoretical Physics, Viena
Wollenberg, M.	Institut für Mathematik, Berlin
Wünsch, R.	Zentralinstitut für Kernforschung Rossendorf, Dresden

CONTRIBUTED PAPERS

I. Gauge Theories of Fundamental Interactions

STANDARD ELECTROWEAK THEORY AND DECOUPLING THEOREMS

J.C. Taylor

1. The Standard Electroweak Theory

1.1 The weak group

The standard theory is defined first of all by the local invariance group of its Lagrangian

$$SU(2) \times U(1) , \tag{1}$$

with generators weak isospin $\underset{\sim}{t}$ of $SU(2)$ and weak hypercharge $\frac{1}{2}y$ of $U(1)$. This group contains the electromagnetic gauge group $U(1)$ whose generator is the electric charge operator Q , and

$$Q = t_3 + \tfrac{1}{2}y . \tag{2}$$

The invariance of the vacuum under the group (1) is spontaneously broken by the Higgs-Kibble mechanism; so the vacuum is not a zero eigenstate of $\underset{\sim}{t}$ and y , though it is of Q .

1.2 The particle representations

The fields of the first "generation" of fermions are assigned to representations of (1) according to table 1. Here $\alpha = 1,2,3$ is the colour index, and n' is the Cabibbo mixture of n and s , to be defined later. L and R refer to the left- and right- projections of the fields

$$e_{L,R} = \tfrac{1}{2}(1 \pm \gamma_5)e , \text{ etc.} \tag{3}$$

	t	t_3	y	Q
ν_L	$\frac{1}{2}$	$+\frac{1}{2}$	-1	0
e_L		$-\frac{1}{2}$	-1	-1
(ν_R)	0	0	0	0
e_R	0	0	-2	-1
p_L^{α}	$\frac{1}{2}$	$+\frac{1}{2}$	1/3	2/3
$n_L'^{\alpha}$		$-\frac{1}{2}$	1/3	-1/3
p_R^{α}	0	0	4/3	2/3
$n_R'^{\alpha}$	0	0	-2/3	-1/3

Table 1.

Because all right-handed fields are SU(2) singlets in this and all known generations, SU(2) is sometimes called $SU(2)_L$. But the left-handedness is not a property of the group but of the representations we give to the fields. There would be no mathematical contradiction if some new heavier fermions were discovered with right-handed (charged) weak currents. Note that the left-handed V - A structure of τ-decay

$$\tau^- \to e^- + \bar{\nu}_e + \nu_T$$

has been well confirmed [1].

Since the field ν_R has zero weak quantum numbers it has no interactions with the intermediate vector mesons, and so is never produced unless the neutrino has a non-zero mass (which it can acquire from an interaction with Higgs fields).

The eigenvalues of y in table 1 are chosen so that left- and right- handed fields have the same values of Q, that is the electric current is vector-like.

1.4 Glashow mixing

The local gauge-invariance under (1) demands the existence of vector fields $\underset{\sim}{W}^{\lambda}$, B^{λ} ($\underset{\sim}{W}^{\lambda}$ is a weak-iso-vector, and λ is a Lorentz index), coupled to the $\underset{\sim}{t}$ and y currents $\underset{\sim}{j}^{\lambda}$, j_y^{λ} by

$$g\underset{\sim}{j}^{\lambda} \cdot \underset{\sim}{W}^{\lambda} + \tfrac{1}{2}g' j_y^{\lambda} B_{\lambda} \ , \qquad (4)$$

where g and g' are coupling constants. In the case of g , all representations must be coupled with the same coupling constant g . Crudely we can see this must be so, because $\underset{\sim}{W}$ is a Yang-Mills field and is self-coupled. Once we fix the strength of the self-coupling to be g , all other couplings have to have the same strength for consistency. Consider, for example, a representation of SU(2) containing a charged and a neutral member R^+ , R^o . Since it must couple to $\underset{\sim}{W}$, the transition

$$W^+ \leftrightarrow R^+ + R_o$$

is possible. Both sides of this reaction are sources of the field W^o , and they must produce fields W^o of the same strength, otherwise the W^o field would have to change discontinuously.

On the other hand, there is no reason why different representations should not be coupled with different values of g' . The argument above does not go through. We just put all the couplings g' equal by hand. This is one of the scandals of the standard model. Of course, it is a scandal that we have lived with for a long time in electromagnetism. There is no reason why all electric couplings should have strength e . The electroweak theory removes the problem from e to g' .

Because of the spontaneous symmetry breaking, W_3 and B (each having zero Q) will not in general be mass eigenstates. Defining the true mass eigenstates to be A and Z , there will be an orthogonal transformation

$$\begin{aligned} B &= \cos\theta\, A - \sin\theta\, Z \\ W_3 &= \sin\theta\, A + \cos\theta\, Z \end{aligned} \qquad (5)$$

where θ is often called θ_W. Then (4) becomes (suppressing Lorentz indices)

$$g(j_1W_1 + j_2W_2) + (g\sin\theta j_3 + \tfrac{1}{2}g'\cos\theta j_y)A$$

$$+ (g\cos\theta j_3 - \tfrac{1}{2}g'\sin\theta j_y)Z \; . \tag{6}$$

Because we have assumed that local $U(1)_{e.m.}$ is unbroken, the field A (say) is a zero-mass eigenstate. Comparing (6) with (2), we see that

$$g\sin\theta = g'\cos\theta = e \; . \tag{7}$$

Then (6) may be rewritten

$$\frac{e}{\sqrt{2}\sin\theta}(jW^\dagger + j^\dagger W) + ej_QA$$

$$+ \frac{e}{\sin\theta\cos\theta}(\cos^2\theta j_3 - \tfrac{1}{2}\sin^2\theta j_y)Z \tag{8}$$

where

$$j = j_1 + ij_2 \; , \quad W = \frac{1}{\sqrt{2}}(W_1 + iW_2) \tag{9}$$

are the charged combinations.

1.5 The effective weak interaction at low energies

If the $W^\pm$ and Z have masses M_W and M_Z, the interaction (8) gives from the exchange of a W or a Z an effective weak interaction

$$(e^2/2\sin^2\theta M_W^2)jj^\dagger + (e^2/\sin\theta\cos\theta M_Z^2)(\cos^2\theta j_3 - \tfrac{1}{2}\sin^2\theta j_y)^2 \; , \tag{10}$$

where we have neglected the momentum transfer in the vector-meson propagator (and also contributions from $q_\mu q_\nu/M^2$, in the propagator, which are of order $(\text{fermion mass})^2/M^2$, the same order as Higgs couplings).

Comparing the charged current term in (10) with the conventional definition of the weak coupling G, we see that

$$\frac{g^2}{M_W^2} = e^2/(\sin^2\theta\ M_W^2) = 4\sqrt{2}\,G\ . \tag{11}$$

To understand how all the powers of 2 came in, consider a particular term

$$G\ 2\sqrt{2}\ [\ \bar{e}\ \gamma_\lambda\ \tfrac{1}{2}(1+\gamma_5)\nu_e\ ]\ [\ \bar{\mu}\ \gamma^\lambda\ \tfrac{1}{2}(1+\gamma_5)\nu_\mu\ ]^\dagger\ . \tag{12}$$

The $2\sqrt{2}$ here is put in because it gives the same muon decay rate as the old parity-conserving interaction

$$G\ \ (\bar{e}\ \gamma_\lambda\ \nu_e)\ (\bar{\mu}\ \gamma^\lambda\ \nu_\mu) + (\bar{e}\ \gamma_\lambda\ \gamma_5\nu_e)(\bar{\mu}\ \gamma^\lambda\ \gamma_5\ \nu_\mu)\ ]\ . \tag{13}$$

Inserting the experimental value of G into (11) gives

$$M_W = (37/\sin\theta)\ \text{GeV}\ . \tag{14}$$

Using (11) and the relation

$$j_Q = j_3 + \tfrac{1}{2} j_y\ , \tag{15}$$

the neutral current interaction may be written

$$2\sqrt{2}G\rho\ (j_3 - \sin^2\theta j_Q)^2 \tag{16}$$

where

$$\rho = M_W^2/(M_Z^2\ \cos^2\theta)\ . \tag{17}$$

Note that a factor $\frac{1}{2!}$ has appeared in (16) from second order perturbation theory, because the two currents are identical (as opposed to $jj^\dagger$ in the charged current case).

It has been pointed out [2] that the presence of the cross-term $-2\sin^2\theta j_3 j_Q$ in (16) is to be expected in any theory with a neutral weak current, because some mixing with the photon is inevitable, by reactions like

$$W_3 \leftrightarrow e^+e^- \leftrightarrow \gamma \tag{18}$$

(though from this point of view one might expect $\sin^2\theta$ to be of order e^2). What is characteristic of the standard electroweak theory is the j_Q^2 term in (16). To test this, one may replace (16) by the more general interaction

$$2\sqrt{2}\, G\rho\ [\ (j_3 - \sin^2\theta j_Q)^2 + c\ j_Q^2\ ] \tag{19}$$

and look for the experimental limit on the parameter c. Measurements [3] of Bhaba scattering $e^+e^- \to e^+e^-$ and of $e^+e^- \to \mu^+\mu^-$ at PETRA yield 95% confidence limits on c of about 0.03 (assuming $\sin^2\theta = 0.23$) .

Interactions of the form (19) arise in electroweak theories which have two Z bosons, because the group (1) is extended for instance to SU(2) × U(1) × SU(2) or SU(2) × U(1) × U(1) . In these models, one Z is lighter than the standard model Z and the other one is heavier.

Let us now assume that $c = 0$, and return to (16). The simple looking form $(j_3 - \sin^2\theta j_Q)$ has a variety of different matrix elements, because of all the different quantum numbers in Table 1. Some examples are given below:-

$\bar{e}e$	axial	$-\frac{1}{4}$
$\bar{e}e$	vector	$-\frac{1}{4} + \sin^2\theta$
$\bar{q}q$	axial t = 0	0
$\bar{q}q$	vector t = 0	$-\frac{1}{6}\sin^2\theta$
$\bar{q}q$	axial t = 1	$\frac{1}{4}$
$\bar{q}\bar{q}$	vector t = 1	$\frac{1}{4} - \frac{1}{2}\sin^2\theta$

where $t = 0,1$ means, of course, $\frac{1}{2}(\bar{u}u \pm \bar{d}'d')$.

Measurements of the ratios of these matrix elements (see Table 2.2 in ref.4) test the theory and determine

$$\sin^2\theta = 0.23 \pm 0.01\ . \tag{20}$$

Using (11), this predicts $M_W = 78$ GeV (21)

Finally, the strength of the neutral current interactions measures ρ in (19), defined by (17). It is found [4] that

$$\rho = 1.002 \pm 0.015 \quad (22)$$

1.6 The Higgs fields

To proceed further, and discuss the value of ρ in (17), we need a theory of the vector meson masses coming from the Higgs-Kibble mechanism. If there is a Higgs field ϕ, its gauge-invariant kinetic term in the Lagrangian is

$$|(\partial_\lambda + ig\underset{\sim}{t} \cdot \underset{\sim}{W}_\lambda + \tfrac{1}{2}ig'yB_\lambda)\phi|^2 \quad (23)$$

(where the symbols $\underset{\sim}{t}$ and y here stand for their representations in the ϕ basis). A non-zero vacuum-expectation value $\langle\phi\rangle \neq 0$ contributes a mass-matrix for the W and B. Substituting the Glashow mixtures (5) into (23), we see that (23) gives the ratio

$$M_Z^2 / M_W^2 = (\cos\theta g t_3 - \tfrac{1}{2}\sin\theta g' y)^2/(g t_1)^2 \ . \quad (24)$$

Since $\langle\phi\rangle$ has $Q = 0$, $\tfrac{1}{2}y = -t_3$ by (2), and using (7) we get

$$\rho^{-1} \equiv M_Z^2 \cos^2\theta / M_W^2 = \tfrac{1}{2}y^2/[t(t+1) - \tfrac{1}{4}y^2] \quad (25)$$

(having used $2t_1^2 = t(t+1) - t_3^2 = t(t+1) - \frac{1}{4}y^2$).

There must be a Higgs field to couple invariantly to $\bar{e}_L e_R$ to give the electron mass, and (from Table 1) this must have $t = \tfrac{1}{2}$. $\langle\phi\rangle$ must have $t_3 = \tfrac{1}{2}$, $y = 1$. It is easy to check that this single field is capable of giving mass to all the fermions in Table 1. So the minimal Higgs structure has this $t = \tfrac{1}{2}$ representation alone, and then (25) predicts $\rho = 1$, in agreement with (22).

With this value of ρ,

$$M_Z = (74/\sin 2\theta) \text{ GeV} = 89 \text{ GeV} \ . \quad (26)$$

It is of interest to find different forms of the condition for $\rho = 1$. Performing the Glashow rotation (5), we find, assuming $\rho = 1$, that

$$\begin{pmatrix} 1 & 0 & 0 & 0 \\ 0 & 1 & 0 & 0 \\ 0 & 0 & \cos\theta & -\sin\theta \\ 0 & 0 & \sin\theta & \cos\theta \end{pmatrix} \begin{pmatrix} M_W^2 & 0 & 0 & 0 \\ 0 & M_W^2 & 0 & 0 \\ 0 & 0 & M_Z^2 & 0 \\ 0 & 0 & 0 & 0 \end{pmatrix} \begin{pmatrix} 1 & 0 & 0 & 0 \\ 0 & 1 & 0 & 0 \\ 0 & 0 & \cos\theta & \sin\theta \\ 0 & 0 & -\sin\theta & \cos\theta \end{pmatrix}$$

$$= \begin{pmatrix} M_W^2 & 0 & 0 & 0 \\ 0 & M_W^2 & 0 & 0 \\ 0 & 0 & M_W^2 & M_W^2\tan\theta \\ 0 & 0 & M_W^2\tan\theta & M_W^2\tan^2\theta \end{pmatrix} \tag{27}$$

Thus a necessary and sufficient condition for $\rho = 1$ is that $M_{W_3} = M_{W^\pm}$.

1.7 The accidental global SU(2)

Let $t = \frac{1}{2}$ Higgs doublet be

$$\phi = \frac{1}{\sqrt{2}} \begin{pmatrix} \phi_o + i\phi_3 \\ i\phi_1 + -\phi_2 \end{pmatrix} \tag{28}$$

where ϕ_o and ϕ_3 are neutral, and without loss of generality, we can assume that $\langle\phi_o\rangle \neq 0$, $\langle\phi_3\rangle = 0$. We can also arrange the Higgs fields into a 2×2 matrix

$$\phi = \phi_o + i\underset{\sim}{\tau} \cdot \underset{\sim}{\phi} = \begin{pmatrix} \phi_o + i\phi_3 & i\phi_1 + \phi_2 \\ i\phi_1 - \phi_2 & \phi_o - i\phi_3 \end{pmatrix}$$

$$= (\phi , \quad i\tau_2\phi^*) \tag{29}$$

Since ϕ and $i\tau_2\phi^*$ transform the same way under SU(2) transformations (since $(i\tau_2)\,(i\underset{\sim}{\tau})^*\,(i\tau_2) = i\underset{\sim}{\tau}$, where * denotes complex conjugation), we have that under an SU(2) transformation $\phi \to U\phi$,

$$\Phi \to U\,\Phi\ . \tag{30}$$

On the other hand, a hypercharge transformation $\phi \to e^{i\alpha}\phi$ can be written as

$$\Phi \to \Phi\ e^{i\alpha\tau_3}\ . \tag{31}$$

We may think of this as a special case of another SU(2) transformation

$$\Phi \to \Phi\ V\ . \tag{32}$$

We will call the transformations (30) and (31) $SU(2)_L$ and $SU(2)_R$ respectively. Of course, $SU(2)_R$ is not a symmetry, local or global, of the weak Lagrangian.

Note that Φ transforms as a $(\frac{1}{2},\frac{1}{2})$ representation of $SU(2)_L \times SU(2)_R$, that is $(\phi_o, \phi_1, \phi_2, \phi_3)$ is a 4-vector representation of the SO_4 which has the same Lie algebra as the above group.

The Higgs part of the Lagrangian may be written, in the Φ notation, as

$$\frac{1}{4}\,Tr(D_\mu\Phi^\dagger D^\mu\Phi) + \frac{1}{4}\,\mu^2\ Tr(\Phi^\dagger\Phi) - \frac{1}{8}\,\lambda\,[Tr(\Phi^\dagger\Phi)]^2 \tag{26'}$$

(note that $2Tr(\Phi^\dagger\Phi\Phi^\dagger\Phi) = [Tr(\Phi^\dagger\Phi)]^2$), where

$$D_\mu\Phi = \partial_\mu\Phi + \tfrac{1}{2}ig\underset{\sim}{W}_\mu\cdot\underset{\sim}{\tau}\Phi + \tfrac{1}{2}ig'B_\mu\Phi\tau_3\ . \tag{27'}$$

Now consider the global transformation of the group $SU(2)_{L+R}$ defined by putting

$$U = V \tag{28'}$$

in (30), (32), and rotating $\underset{\sim}{W}$ according to this transformation, i.e.

$$\underset{\sim}{\tau}\cdot\underset{\sim}{W} \to U\underset{\sim}{\tau}\cdot\underset{\sim}{W}U^\dagger\ . \tag{29'}$$

_If_ $g' = 0$ (that is to say, $\theta = 0$) , (26') is invariant under this "accidental" global symmetry.

Next, we consider the invariance of the Higgs-fermion couplings under $SU(2)_{L+R}$. We will show that, if the members of a fermion doublet have _equal_ masses, then the coupling is invariant. Take any doublet $F_L = \begin{pmatrix} U \\ D \end{pmatrix}_L$, U_R , D_R . Then

$$f\,\bar{F}_L \phi\, U_R + f'\,\bar{F}_L (i\tau_2 \phi^*)\, D_R \tag{30'}$$

is invariant under the weak group (1). If $f = f'$, (30') can be written

$$f\,\bar{F}_L \Phi\, F_R \tag{31'}$$

where $F_R = \begin{pmatrix} U_R \\ D_R \end{pmatrix}$. Then we have invariance under (32) provided that

$$F_R \rightarrow V\,F_R \tag{32'}$$

(this explains the reason for the notation $SU(2)_R$) .

Thus we conclude that _if_ $\theta = 0$ and _if_ fermion doublets are degenerate, then there is invariance of the whole Lagrangian under global $SU(2)_{L+R}$.

We will now show that this invariance implies $\rho = 1$. To do this we return to (27). In the limit $\theta = 0$, we require to show that the W-mesons are degenerate. But this is implied by the global invariance under (29').

The $SU(2)_{L+R}$ group will prove to be important later, when we consider radiative corrections to the equation $\rho = 1$.

2. Generations, GIM and Kobayashi-Maskawa

2.1 Generations

At present 3 generations are required:

$$\left.\begin{matrix} \nu_e & e & u_\alpha & d'_\alpha \\ \nu_\mu & \mu & c_\alpha & s'_\alpha \\ \nu_\tau & \tau & (t_\alpha) & b'_\alpha \end{matrix}\right\} \quad \alpha = 1,2,3 \tag{34}$$

each employing exactly the same pattern of representations as in table 1. The average charge in each generation is zero. The top quarks, t_α , are in brackets because they have not been discovered.

2.2 Kobayashi-Maskawa mixing

The interaction of the fermions with the gauge bosons is completely fixed by the group and representation structure. The interaction of the Higgs fields with the fermions, on the other hand, is completely free, within the constraints of global $SU(2) \times U(1)$ invariance.

We seek all possible Yukawa type couplings between the fermion fields in (34) and the Higgs field (28), (29). There can be no cross-terms between quarks and leptons, since such terms cannot conserve charge (the Higgs field has integral charge). So the quark and lepton sectors are separate. Consider the quark sector. Let us call the left-handed doublets L^A $(A = 1,2,3)$ with

$$L^1 = \begin{pmatrix} u \\ d' \end{pmatrix}_L , \quad L^2 = \begin{pmatrix} c \\ s' \end{pmatrix}_L , \quad L^3 = \begin{pmatrix} t \\ b' \end{pmatrix}_L \tag{36}$$

and similarly call the right-handed doublets R^A , as in (31). We suppress colour indices. The required interaction is

$$\sum_{A,B} \bar{L}_A \Phi f^{AB} R^B + \text{c.c} , \tag{37}$$

where each of the coefficients f^{AB} $(A,B = 1,2,3)$ is a 2×2 matrix. (37) is automatically invariant under $SU(2)_L$. In order to be invariant under $U(1)_y$ (or to conserve charge) we require that each f^{AB} is a diagonal matrix.

The fermion mass matrix is obtained by replacing Φ by $\langle\Phi\rangle = FI$ in (36) where F is a number and I is the unit 2×2 matrix.

(37) contains 2×3^2 complex coefficients. We are at liberty to perform an arbitrary unitary transformation on the R^A hermitean (since any matrix can be written as a product of an hermitean matrix and a unitary matrix). Thus

$$f^{AB} = \begin{pmatrix} f_I^{AB} & 0 \\ 0 & f_{II}^{AB} \end{pmatrix}$$

where f_I^{AB} and f_{II}^{AB} are each hermitean. We can perform a unitary transformation in L^A and R^A (since their kinetic and vector boson terms are proportional to the unit matrix δ_{AB}) to make f_I^{AB} diagonal. In the notation (36) we suppose we have already done this, so that u, c, t are mass eigenstates.

We are not at liberty to perform an independent unitary transformation on the upper and lower components of L^A, so f_{II}^{AB} will in general <u>not</u> be diagonal. Let K^{AB} be the unitary matrix which does diagonalise f_{II}^{AB}. Then K connects the mass eigenstates d, s, b with d', s', b' (which are defined to be the weak partners of u, c, t).

We have one remaining freedom, which is to adjust the phases of d', s', b' and d, s, relative to b. This gives us 5 phases which we can remove from K.

Let us see how the numbers work out for n generations. K, being unitary, has n^2 real elements, but 2n-1 phases can be absorbed, leaving $(n-1)^2$ elements. An orthogonal matrix has $\frac{1}{2}n(n-1)$ elements. For n = 2 these two numbers are equal. For n = 3, we get 4 and 3 respectively. Thus for n = 2 we can reduce K to the Cabibbo form

$$\begin{pmatrix} d' \\ s' \end{pmatrix} = K \begin{pmatrix} d \\ s \end{pmatrix} = \begin{pmatrix} \cos\theta_c & \sin\theta_c \\ -\sin\theta_c & \cos\theta_c \end{pmatrix} \begin{pmatrix} d \\ s \end{pmatrix} . \tag{38}$$

For n = 3, we have a rotation in 3-dimensions plus an extra phase. One way of parameterizing this is

$$\begin{pmatrix} 1 & 0 & 0 \\ 0 & c_2 & s_2 \\ 0 & -s_2 & c_2 \end{pmatrix} \begin{pmatrix} 1 & 0 & 0 \\ 0 & 1 & 0 \\ 0 & 0 & e^{i\delta} \end{pmatrix} \begin{pmatrix} c_1 & s_1 & 0 \\ -s_1 & c_1 & 0 \\ 0 & 0 & 0 \end{pmatrix} \begin{pmatrix} 1 & 0 & 0 \\ 0 & c_3 & s_3 \\ 0 & -s_3 & c_3 \end{pmatrix} \tag{39}$$

where $c_i = \cos\theta_i$, $s_i = \sin\theta_i$. When $\delta = 0$, this reduces to the standard Euler-angle parameterization of a rotation. (39) comes out to be

$$K = \begin{pmatrix} c_1 & s_1c_3 & s_1s_3 \\ -c_2s_1 & c_1c_2c_3 - s_2s_3e^{i\delta} & c_1s_2s_3 + s_2c_3e^{i\delta} \\ +s_1s_2 & -c_1s_2c_3 - c_2s_3e^{i\delta} & -c_1s_2s_3 + c_2s_3e^{i\delta} \end{pmatrix} . \tag{40}$$

What is known about these parameters? c_1 determines the $u \leftrightarrow d$ transition, as measured in neutron decay, which gives

$$c_1 \simeq 0.97 \tag{41}$$

Radiative corrections are involved in the determination of (41). Strangeness-changing decays $s \to d$ give

$$|s_1c_3| = 0.22 \pm 0.01 , \tag{42}$$

where the error is remarkably small considering that approximate SU(3) invariance goes into the determination of (42). Comparing (41) and (42), we can say that s_3 is not too big. Charm production measures $|c_2s_1|$, and it seems that $|s_2|$ is also not too big. (See the data tables [6]).

Let us remind ourselves that the top left-hand corner of (40), which is presumably approximately the same as (38), contains the GIM mechanism, which predicted charm. Charm-changing decays behave at least qualitatively right. For example, the predominance of strangeness in charm decays, and the $\Delta S = \Delta C$ rule which forbids $D^+ \to K^+\pi^+\pi^-$, in agreement with observation.

Now consider the Higgs-lepton section. If there are no right-handed neutrinos, so that the neutrinos are exactly massless, the analogues of f_I^{AB} do not appear at all. Then we can diagonalise f_{II}^{AB}, and no mixing arises. On the other hand, if there are right-handed neutrinos, the situation is exactly the same as for the quarks in general.

We thus have 10 parameters for the quarks (6 masses and θ_1 θ_2 θ_3 δ) and either 3 or 10 parameters for the leptons, depending on whether or not the right-handed neutrinos exist.

2.3 CP Violation

The phase $e^{i\delta}$ is CP and T violating. It arises because we have not imposed CP violation on anything. The vector boson interactions happen to be CP conserving because interactions generated by gauge invariance are automatically CP conserving. If you try to smuggle an i in to violate CP, you need gauge transformations e^{θ} or $e^{\gamma_5\theta}$ instead of $e^{i\theta}$ or $e^{i\gamma_5\theta}$, and these are obviously no good. The Yukawa Higgs interactions, however, are much more arbitrary, and so are CP violating in general. Even then, we have shown that the phases can be transformed away unless there are at least 3 generations (or more Higgs's).

CP violation in $K^o_{L,S}$ decay can be explained entirely by having a complex $K^o \leftrightarrow \bar{K}^o$ transition amplitude. In the K-M theory, this is generated by diagrams like

which gives an imaginary part proportional to

$$c_1c_2c_3s_1s_2s_3 \sin\delta(m_t - m_c) \tag{43}$$

Note that the effect must vanish when $m_t = m_c$, since then the phase can be transformed away.

The observed CP violation requires that (43) be of order 10^{-3} (though the kaon matrix elements of the operator cannot be calculated reliably). This probably requires a small value of $\sin\delta$.

CP violation is in general predicted in decay amplitudes. The reason it is not supposed to be important in $K^o_L \to 2\pi$ is that there are only two channels, isospin $T = 0$ and 2, and both the normal and CP violating operators have $T = \frac{1}{2}$ and only feeds the $T = 0$ channel. But interference between two channels with different phases is required for CP violation.

The multitude of parameters, with widely different magnitudes, including the smallness of the phase δ, are completely unexplained in the standard model.

3. Anomalies

In a gauge-theory, it is crucial that the radiative corrections preserve the gauge invariance; for it is only gauge invariance that allows us to have a theory with spin 1 particles which is both unitary and renormalizable.

Gauge-invariance is assured if we can regularize in a manifestly gauge-invariant way. Mostly, dimensional regularization has this property: it preserves gauge-invariance, Lorentz invariance, translational invariance, and even unitarity. It doesn't work with dilation invariance or γ_5-invariance. For γ_5-invariance, this is because of the impossibility of building a γ_5-invariant theory in a general number of dimensions.

There are two obvious ways of generalizing γ_5 to d dimensions. One is to use

$$\gamma_{[\mu_1} \gamma_{\mu_2} \ldots . \gamma_{\mu_d]}$$

(totally anti-symmetrical). This is a pseudo-scalar in d dimensions. The difficulty is that quantities like

$$\mathrm{Tr}(\gamma_5 \gamma_{\mu_1} \gamma_{\mu_2} \gamma_{\mu_3} \gamma_{\mu_4}) ,$$

which naturally arises in Feynman diagrams, involve a Kronecker δ_{d4}, which cannot be analytically continued to general d.

The other possibility is to use

$$\gamma_{[\mu_1} \gamma_{\mu_2} \ldots . \gamma_{\mu_4}] .$$

In d-dimensions, this is not a pseudo-scalar, but a 4th rank totally anti-symmetric tensor. This cannot be used to define an invariance group.

Other regularization methods are no better. For instance, Pauli-Villars regularization involves artificial masses, which obviously break γ_5-invariance. A method invented by Slavnov [7] works, but only for diagrams with more than one closed loop.

It is easy to prove that <u>no</u> method of regularizing γ_5 can exist. If it did, it would guarantee the correctness of the formal manipulations which embody the Ward identities.

Consider the famous triangle

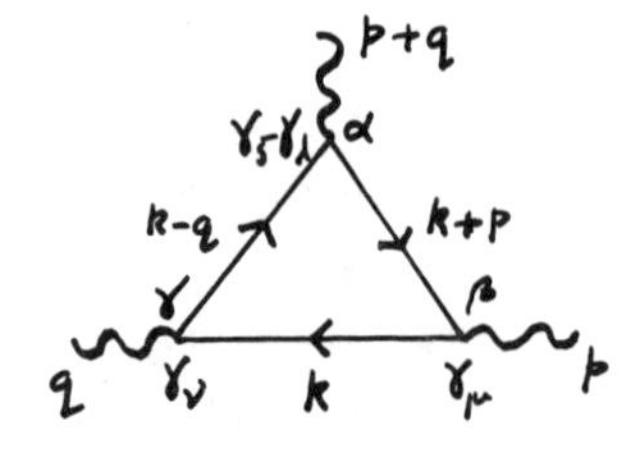

(α,β,γ are group indices)

which gives

$$\int d^4k \; \mathrm{Tr}\,[\gamma_5\,\gamma_\lambda\,(\not p+\not k+m)\,\gamma_\mu\,(\not k+m)\,\gamma_\nu\,(\not p-\not q+m)\,]$$

$$\times\,[\,(p+k)^2-m^2]^{-1}\,[k^2-m^2]^{-1}\,[(q-k)^2-m^2]^{-1}$$

$$\equiv\; R_{\lambda\mu\nu} \tag{43}$$

One Ward identity involves $(p+q)^\lambda R_{\lambda\mu\nu}$. Using

$$\gamma_5(\not p+\not q) = \gamma_5\,(\not p+\not k-m) - \gamma_5\,(\not p-\not q-m)$$

$$= \gamma_5(\not p+\not k-m) + (\not k-\not q-m)\,\gamma_5 + 2m\gamma_5\;, \tag{44}$$

the contributions from the first two terms here are

$$\int d^4k\;\mathrm{Tr}[\gamma_5\gamma_\mu(\not k+m)\;\gamma_\nu(\not k-\not q+m)\;\;[k^2-m^2]^{-1}\;[(q-k)^2-m^2]^{-1}$$

$$+\int d^4k\;\mathrm{Tr}[\gamma_5\;(\not p+\not k+m)\;\gamma_\mu\;(\not k+m)\,\gamma_\nu\,]\;[(p+k)^2-m^2]^{-1}\;[k^2-m^2]^{-1}\;.$$

Each of these is a pseudo-second-rank-tensor function of a <u>single</u> vector (q in the first case, p in the second), and so must be zero.

Suppose now we try to prove a Ward identity involving $p^\mu R_{\lambda\mu\nu}$ or $q^\nu R_{\lambda\mu\nu}$. If we leave the integral in the form (43), the last step of the argument will fail, because we will get functions of p <u>and</u> q . To make the argument we must first shift the origin of the k-integration in (43). If the integral were convergent, or if a good regularization existed, this would be harmless. As it is, it is easy to compute the effect of such a shift of origin. If we shift by a vector a_μ, we get a change

$$a_\rho \int d^4k \, \frac{\partial}{\partial k_\rho} \, [\text{integrand}]$$

$$= a_\rho \int_S dS^\rho \, [\text{integrand}] \tag{45}$$

where S is a "sphere" at infinity.

To calculate (45), we can replace the integrand by

$$(k^2)^{-3} \, \mathrm{Tr}[\gamma_5 \gamma_\lambda \not{k} \gamma_\mu \not{k} \gamma_\lambda \not{k}] \tag{46}$$

independent of masses and momenta. This must be proportional to

$$(k^2)^{-2} \, \varepsilon_{\lambda\mu\nu\sigma} \, k^\sigma$$

and the constant of proportionality is easily fixed by choosing $\lambda = 1$, $\mu = 2$, $\nu = 3$, $\sigma = 4$, and is equal to 1. Thus (45) gives simply

$$a^\sigma \, \varepsilon_{\lambda\mu\nu\sigma} \, 2\pi^2 i \tag{47}$$

where the i comes from a Wick rotation and $2\pi^2$ is the area of a 3-dimensional sphere.

Armed with (47), we see that if we insist on some Ward identities, then others must fail. In fact, we can impose each of the vector-Ward identities, and then

$$(p+q)^\lambda \, R_{\lambda\mu\nu} = 2\pi^2 \, \varepsilon_{\lambda\mu\nu\sigma} \, p^\lambda q^\sigma + (\text{expected terms}) \, . \tag{48}$$

The first term here is the anomaly. Note that it is independent of masses.

Since the anomaly is symmetric under

$$p \leftrightarrow q \, , \quad \mu \leftrightarrow \nu \, ,$$

it must be symmetric under the interchange of the group indices $\beta \leftrightarrow \gamma$. In orthogonal groups, there are no such symmetric numerical tensors. For SU(3) there is $d_{\alpha\beta\gamma}$, and for SU(2) $\times$ U(1) there is

$$\mathrm{Tr}\,[(t_\beta t_\gamma + t_\gamma t_\beta)y] \propto \delta_{\alpha\beta}\,\mathrm{Tr}\,y = \tfrac{1}{2}\delta_{\alpha\beta}\,\mathrm{Tr}\,Q\,.$$

In the standard electroweak theory, anomalies are cancelled by arranging to have $\mathrm{Tr}Q = 0$. This is true for each generation separately, as can be seen by adding up the charges in Table 1 (remembering that there are three colours).

With the anomalies cancelled, the gauge-invariance is presumably saved. However, it is still true that there is no systematic regularization of γ_5 — no simple set of rules like dimension regularization provides for vector-like theories. For some discussion of this, see [7,8]. In my opinion, this is another scandal of modern quantum field theory.

Note that, since anomalies are mass-independent, there could be arbitrarily large mass-differences within a generation while one still has anomaly cancellation. This is rather surprising. If the anomalies had not cancelled, we could have invoked 100 TeV fermions to cancel them! We return to this point in section 5.3.

4. The Effective Potential and the Higgs' Mass

For the remainder of this article, we will be concerned with one-loop corrections to the tree diagrams (which are what we have dealt with hitherto). To tree-diagram approximation, $\langle\phi\rangle$ is determined by minimizing the last two terms in (36), i.e. the potential terms in the Higgs Lagrangian. The one-loop corrections to this are not entirely straightforward, since the spontaneous symmetry breaking is in a sense non-perturbative $(\langle\Phi\rangle \propto \lambda^{-1})$. The simplest way to proceed is via the one-loop "effective potential" — i.e. we calculate one-loop diagrams with external Φ lines carrying zero-momentum, and we do it "before" spontaneous symmetry breaking, i.e. with massless vector bosons, leptons and quarks.

Take first loops made of vector bosons. We have diagrams like

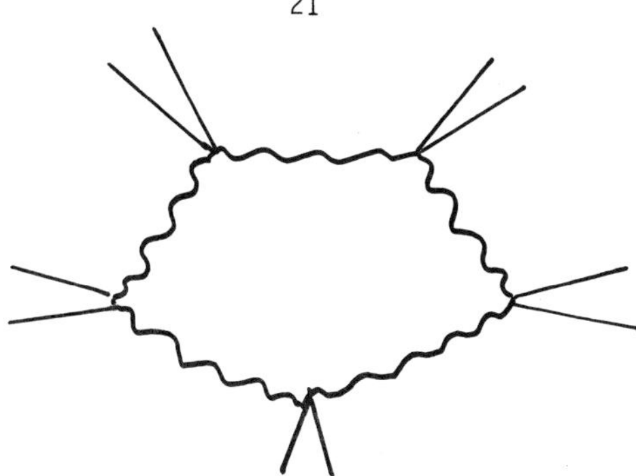

We use the Landau gauge, with a projection or $-g_{\mu\nu} + \frac{k_\mu k_\nu}{k^2}$, since then diagrams like

are zero (since the vertex gives a factor k_μ). In this way we obtain

$$\frac{i}{(2\pi)^4} \sum_n \frac{3}{4} \int d^4k \, \frac{1}{n} \left(\tfrac{1}{2}g^2\phi^2\right)^n (k^2)^{-n}$$

$$= \frac{2\pi^2}{(2\pi)^4} \frac{3}{4} \int k^2 \, dk^2 \, \ell n \left(\tfrac{1}{2}g^2\phi^2 + k^2\right)$$

where the $\frac{3}{4}$ has come from leaving out the projector $\left(-g_{\mu\nu} + \frac{k_\mu k_\nu}{k^2}\right)$ from the trace and we have done a Wick rotation. Note that the infrared divergences have gone, but it is highly ultraviolet divergent. The UV divergences can be absorbed into the original Lagrangian (26), leaving

$$3 \frac{1}{64\pi^2} \left(\tfrac{1}{2}g^2\phi^2\right)^2 \, \ell n(\phi^2 / \mu'^2)$$

where μ'^2 is a renormalization subtraction point. Really there are W_1 , W_2 and Z eigenstates of the mass-matrix, so g^4 is replaced by

$$2g^4 + (g^2 + g'^2)^2 \; .$$

Fermion loops contribute with an opposite sign and with strength reduced by $\frac{m_f}{m_W}$. We shall neglect these contributions. Contributions from Higgs loops are order λ^2 , and we shall neglect these also.

Thus the total potential is

$$V(\phi) = -\tfrac{1}{2}\mu^2 \phi^2 + A\,|\phi|^4\,\ell n(|\phi|^2/\mu'^2) .$$

We can leave out the λ term by absorbing it into μ'^2 . This function can take the shapes

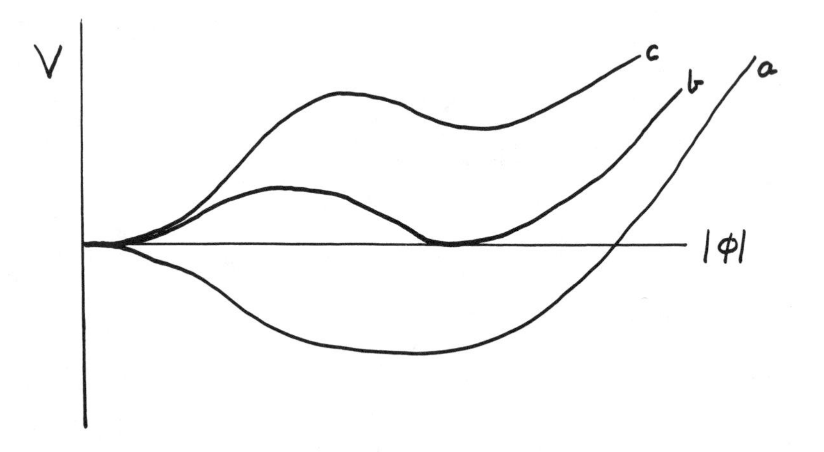

a is the normal Goldstone curve. In case c the symmetry breaking vacuum is unstable. b is the transitional case. We want to find the condition for the symmetry breaking vacuum to be stable.

Assuming that, at $|\phi| = F$,

$$\frac{\partial V}{\partial\phi} = 0 \qquad \frac{\partial^2 V}{\partial\phi^2} = M_H^2 > 0 ,$$

we can re-express V in terms of m_H and F (eliminating μ and μ') . We get

$$V = -(\frac{1}{4}m_H^2 - 2AF^2)\phi^2 + (\frac{m_H^2}{8F^2} - \frac{3}{2}A)\,\phi^4$$

$$+ A\phi^4\,\ell n(\phi^2/F^2) .$$

Then the condition for $V(F) < 0$ is

$$m_H^2 > 4AF^2 \ .$$

From (23) we know that

$$M_W^2 = \tfrac{1}{2}g\langle\phi\rangle^2$$

and hence from (11)

$$\langle\phi\rangle^2 = \frac{1}{2\sqrt{2}} \ \frac{1}{G}$$

so that

$$F = 174 \text{ GeV} \ . \tag{49}$$

Remembering that

$$A = \frac{3}{16} \left(\frac{e^2}{4\pi} \right)^2 \left[2 \frac{1}{\sin^4\theta} + \frac{1}{\sin^4\theta\cos^4\theta} \right] ,$$

one obtains the Linde-Weinberg [9] bound

$$m_H > 6 \text{ GeV} \ . \tag{50}$$

Heavy fermions, since they contribute to A with opposite sign, weaken this bound [10] .

It may be that the vacuum, while not stable, is metastable. This possibility also relaxes the bound (50) somewhat [11].

5. Heavy Fermions and Heavy Higgs

5.1 The decoupling theorem

What would be the effect of very heavy fermions (or bosons) in conventional quantum field theory? For example, would a lepton of mass 1 TeV alter the anomalous magnetic moment of the electron? It would be rather disastrous if this were so, and in fact we are protected from such trouble by the "decoupling theorem" [12]. The proof of this runs very roughly as follows.

Consider a convergent fermion loop. If the fermion mass tends to zero the integrand tends to zero. Since the integral is convergent, the integral tends to zero. Now take a divergent fermion loop. In this case, it is sensitive to the large mass, but the mass-dependence can be absorbed into the renormalization constant, which is arbitrary, and so the mass-dependence has no effect.

5.2 Decoupling in gauge theories

The decoupling theorem raises very interesting questions in gauge theories:-

(a) Can we let the mass of the Higgs tend to infinity, so that it has essentially no observable consequences?

(b) Is the decoupling theorem valid, in view of the fact that gauge theories have many more divergent integrals than they have arbitrary constants? For example, the $W^{\pm}$ and Z each have divergent self-masses, but there is only one mass parameter (ignoring fermions) $\langle\phi\rangle$; so that M_W and M_Z are in tree approximation constrained by $M_Z^2 \cos^2\theta = M_W^2$. (Of course θ is also renormalized, but it is an independently observable parameter.)

It will turn out that the answer to (a) is: No, not if perturbation theory remains valid. And the answer to (b) is: Yes, provided perturbation theory remains valid.

We begin with a simple physical discussion about the high-energy behaviour of tree graphs and its dependence on Higgs and fermion masses.

The polarization vector for a vector boson with helicity 0 (in a given frame) is

$$\frac{1}{M}\left(|\underset{\sim}{k}|\ ,\ k_o\hat{\underset{\sim}{k}}\right) = \frac{1}{M}\left(k_o,\underset{\sim}{k}\right) + \frac{M}{k_o + |\underset{\sim}{k}|}\left(-1,\hat{\underset{\sim}{k}}\right) . \tag{51}$$

Since this introduces a factor $\frac{1}{M}$, we expect, on dimensional grounds, bad high-energy behaviour. Gauge theories are so constructed that these $\frac{1}{M}$ terms cancel.

Consider the process $f^+f^- \rightarrow W^+W^-$, where f is a hypothetical heavy fermion. We have three tree diagrams

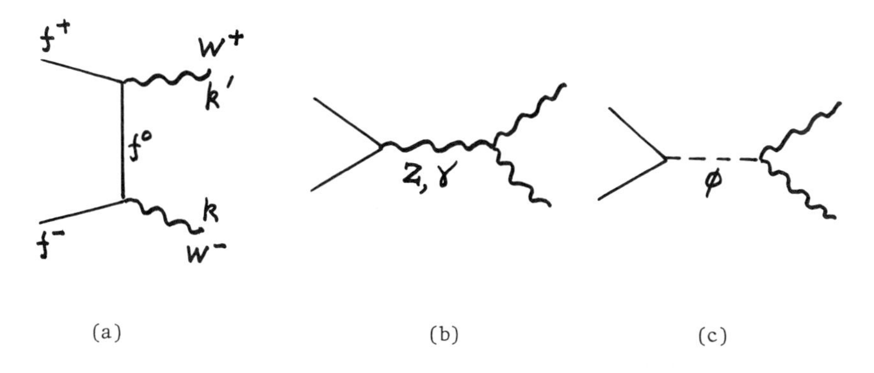

(a) (b) (c)

Suppose that both W^+ and W^- have zero helicity (in the centre of mass frame, say). Take the most dangerous contribution, from the $M^{-1}k_\lambda$ term in (51) for each W. If $m_f = 0$, we can use elementary Ward identities to prove that these contributions to (a) and (b) cancel one-another. For $m_f \neq 0$, there is a contribution from (a) proportional to $\frac{1}{2}g^2 m_f/M_W^2\ (\bar{v}u)$ and also one from (c) proportional to

$$\frac{f}{\sqrt{2}}\ \frac{1}{(k+k')^2 - m_H^2}\ g^2 \langle\phi\rangle\, k.k'/M_W^2\ (\bar{v}u) \tag{52}$$

where f here is the Higgs-fermion coupling strength. At high-energies these two contributions cancel in virtue of the relation $m_f = 2^{-\frac{1}{2}} f\langle\phi\rangle$ but it is clear that the cancellation only takes place for $(k+k')^2 \gg m_H^2$. $\bar{v}$ and u are Dirac spinors.

If we make m_H large enough, the contribution from (a) and (b) will exceed the unitarity limit before it is cancelled by (c). Just on dimensional grounds, we can see that this will happen when

$$g^2 m_f\ E/M_W^2 \sim 1 \quad \text{for} \quad E \sim m_H$$

or

$$m_f m_H \sim \langle\phi\rangle^2 \sim \frac{1}{\sqrt{2}G}\ . \tag{53}$$

Similar arguments applied [13] to WW → WW, with a more careful application of unitarity, lead to

$$m_H^2 \lesssim \frac{8\pi}{\sqrt{2}G} \sim (1\ \text{TeV})^2 \qquad (54)$$

which can be improved in favourable cases [14].

An even simpler application of the same ideas occurs in $f^+f^- \to f^+f^-$. If a W is exchanged, the $\frac{q_\mu q_\nu}{M_W^2}$ in the numerator of the propagator (in the unitary gauge) produces a term proportional to

$$g^2 \frac{m_f^2}{M_W^2} \sim \frac{m_f^2}{\langle\phi\rangle^2} , \qquad (55)$$

which will violate unitarity if m_f is too big. More careful applications of partial wave unitarity [15] insert multiples of π into (55) and give a limit to m_f of order 1 TeV.

Whenever unitarity is violated by a perturbation theory calculation, it means that the use of perturbation theory was unjustified — there is a strong coupling situation. In the present case, this can be seen particularly directly. If we express (54) in terms of the Higgs self-coupling strength λ, we find simply

$$\lambda = \sqrt{2}G\, m_H^2 \lesssim 8\pi ; \qquad (56)$$

so that at the limit the Higgs self-coupling becomes strong. Thus we do have a strong-coupling situation, <u>although</u> the Higgs self-coupling λ does <u>not</u> appear <u>explicitly</u> in the tree diagrams that lead to (54).

In a similar way, (54) is associated with the combination $\lambda^{\frac{1}{2}}f$ becoming of order 1, i.e. either the Higgs self-coupling or the Higgs-fermion coupling or both. (55) is associated with f^2 becoming large.

The appearance of the Higgs self-coupling λ in (56) is not really so surprising. In (51) the helicity 0 state of a W was represented by a polarization vector for the field W_μ. But, in a Higgs-Kibble gauge theory, we can make gauge transformations which mix $W\mu$ with the Goldstone components ϕ_1, ϕ_2 of ϕ. In this way, we can transfer the first term in (51) entirely to ϕ, if we wish. (This is essentially what happens in 't Hooft's gauge). Then all the previous results (53) to (56) are statements about the Higgs-fermion sector, and the gauge couplings strengths g, g' are irrelevant [16].

5.3 Radiative corrections to $\rho = 1$

Now we turn from tree-diagrams to one-loop diagram corrections to the formula $\rho \equiv M_Z^2 \cos^2\theta / M_W^2 = 1$, which has been much discussed, particularly by Veltman and his co-workers. We ask the question whether heavy fermion or heavy Higgs loops could give very large corrections to this formula, in violation of the decoupling theorem.

The parameter ρ measures the ratio of the neutral to charged current interactions. We consider the corrections arising from self-energy diagrams (evaluated at $q^2 = 0$) on the Z or W propagator. Of course, there are many other radiative corrections [17], but these are not so sensitive to heavy masses. [Warning! - Ref.(17) uses different renormalization conventions from ours. In their conventions, $\rho = 1$ by definition, and the corrections under discussion become corrections to the relation $M_W^2 = \pi\,\alpha\,/\,(\sqrt{2}G\,\sin\theta)$.]

We consider heavy fermions first, because they are simpler than heavy Higgs's. Let the masses be m_1 and m_2. We can immediately say that the correction must vanish when $m_1 = m_2$. This follows from section 1.7. To get $m_1 = m_2$, we put $f = f'$ in (30) and thus get the accidental global symmetry $SU(2)_{L+R}$, which we showed implied $\rho = 1$.

The self-energy diagram for W^+ gives (at zero q)

$$\frac{ig^2}{2(2\pi)^4}\int d^4k\;\frac{\mathrm{Tr}[\gamma_\mu \not{k}\gamma_\nu \not{k}\,\frac{1}{2}(1+\gamma_5)]}{(k^2-m_1^2)(k^2-m_2^2)}\,. \tag{57}$$

The masses have disappeared from the numerator because of the $(1+\gamma_5)$ projection operators. The trace in numerator gives simply $\frac{1}{2}k^2 g_{\mu\nu}$ (since $k_\alpha k_\beta$ can be replaced by $\frac{1}{4}k^2 g_{\alpha\beta}$). For simplicity we work in the approximation $\theta = 0$. The self-energy for W^o is similar, except that there are two contributions, one with m_1^2 for each mass and one with m_2^2 for each mass. The difference gives

$$\delta M^2_{W^o} - \delta M^2_{W^+} = \frac{ig^2}{8(2\pi)^4}\int d^4k\;k^2\left[\frac{1}{k^2-m_1^2} - \frac{1}{k^2-m_2^2}\right]^2. \tag{58}$$

We note that this does vanish for $m_1 = m_2$ as expected, and that it is convergent. According to (27), (58) is the same thing as $\delta M_Z^2 - \delta M_{W^+}^2$ and so we get (in the limit $\theta = 0$)

$$\delta M_Z^2 - \delta M_{W^+}^2 = \frac{\pi^2 g^2}{4(2\pi)^4} \int dk^2 \, (k^2)^2 \left[\frac{1}{k^2 - m_1^2} - \frac{1}{k^2 - m_2^2} \right]^2$$

$$= \frac{\pi^2 g^2}{4(2\pi)^4} \int dk^2 \left[\frac{m_1^2}{k^2 - m_1^2} - \frac{m_2^2}{k^2 - m_2^2} \right]^2$$

$$= \frac{\pi^2 g^2}{4(2\pi)^4} \left[m_1^2 + m_2^2 - \frac{2m_1^2 m_2^2}{m_1^2 - m_2^2} \ln(m_1^2/m_2^2) \right] . \tag{59}$$

Thus we do find contributions proportional to the masses (provided they are not equal), in apparent contradiction to the decoupling theorem. But (59) can only become comparable to M_W^2 if f or f_1 in (30′) is of order 1 . So the large effects are associated with large <u>coupling constants</u>, which is no surprise.

Once again, this can be understood by going back to the field theory before spontaneous symmetry breaking, i.e. before $\langle\phi\rangle$ is extracted from ϕ . Then the type of diagram contributing to (57) is

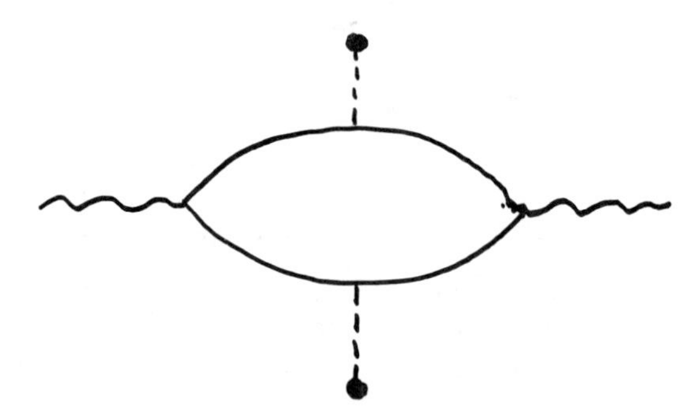

where ---● is a Higgs field disappearing into the vacuum. The dependence on the Higgs-fermion coupling f is clear.

What happens if we try to give the fermion a mass m directly <u>not</u> by spontaneous symmetry breaking. We can do this if there is a heavy fermion transforming as a vector (rather than a left-handed) representation of SU(2) . We can also produce extra mass-differences δm by spontaneous symmetry breaking. In a model studied in [18] it was found that

$$\delta M_Z^2 - \delta M_W^2 \propto (\delta m)^4/m^2 ,$$

which is in agreement with the decoupling theorem as $m \to \infty$, whereas large δm requires large coupling-constants as before.

Consider now Higgs loops. With the minimal Higgs field, large values of m_H do not produce large corrections to ρ. This is because the Higgs sector maintains the accidental $SU(2)_{L+R}$ symmetry which enforces $\rho = 1$. This shielding from m_H (to one-loop order), was called the "shielding theorem" by Veltman [19]. One can produce more complicated Higgs models [20] where there are large contributions to ρ. Again there is no exception to the rule that large corrections only occur when there are Higgs coupling constants.

From the examples treated in this section, together with the work of the last section, we conclude that: there are no exceptions to the decoupling theorem; apparent exceptions only occur when there is a large coupling constant rendering perturbation theory useless.

I don't know of any general proof of this result, but it seems to me very likely that it is true generally.

Although the principles seem clear, practice may be a little muddier. First of all, we should remember that in QED we live with a situation where strongly interacting quarks do appear in the photon self-energy function. Yet we manage to control the situation fairly well, and get accurate predictions for the anomalous magnetic moment. So perhaps we should not give up all hope of calculating if the Higgs couplings turned out to be strong.

Second, although the corrections to ρ may not be of order 1 without introducing a strongly-interacting sector, nevertheless the corrections may be sizeable. For example, since it is known that $\rho = 1$ to within about 3 we can deduce [5,15] from (59) that, for a heavy lepton with accompany light neutrinos,

$$m^2 \lesssim \left(\frac{64\pi^2}{g^2} \times .03 \right) M_W^2 \sim (400\ \mathrm{GeV})^2 .$$

Finally we return to the point raised at the end of section 3. Anomaly cancellation appears to be independent of fermion mass. What happens if we make a fermion mass arbitrarily large? In Higgs-Kibble theories with axial currents, the Ward identities are maintained by couplings of the Higgs fields to the fermions. If the mass became arbitrarily large, so would the coupling. Thus again we would be in a situation where perturbation theory would be useless. In this connection, see also [21].

References

[1] W. Bacino et al. Phys. Rev. Lett. 45, 329 (1980).
[2] J. Bernstein & T.D. Lee, Phys. Rev. Lett. 11, 512 (1963).
[3] P. Dittmann & V. Hepp, Tests of QED and Electroweak Theories at PETRA, DESY 81/030.
[4] J.J. Sakurai, Electroweak Physics of the '80s, UCLA/80/TEP/27 (Lectures at Erice 1980).
[5] J.D. Bjorken, Electroweak Interactions, FERMILAB-Conf-80/86-THY.
[6] C. Bricman et al. Rev. Mod. Phys. 52, No.2 (1980).
[7] P.H. Frampton, Phys. Rev. D20, 3372 (1979).
[8] M.S. Chanowitz, M.A. Furman, I. Hinchliffe, Nucl. Phys. B159, 225.
[9] A.D. Linde, Zh. Eksp. Teor. Fiz. Pis'ma. 23, 73 (1976);
S. Weinberg, Phys. Rev. Lett. 36, 294 (1976).
[10] H.D. Politzer and S. Wolfram, Phys. Letts. 82B, 242 (1979).
[11] P.H. Frampton, Phys. Rev. D15, 2922 (1977).
[12] T. Appelquist & J. Carazzone, Phys. Rev. D11, 2856 (1975).
[13] B.W. Lee, C. Quigg & H.B. Thacker, Phys. Rev. D16, 1519 (1977).
[14] M.S. Chanowitz, M.A. Ferman, I. Hinchliffe, Nucl. Phys. B.
[15] M.S. Chanowitz, M.A. Furman & I. Hinchliffe, Phys. Letters. 78B, 285.
[16] T. Appelquist & C. Bernard, Phys. Rev. D22, 200 (1980).
[17] W.J. Marciano & A. Sirlin, Phys. Rev. D22, 2695 (1980).
[18] M.B. Einhorn, D.R.T. Jones, M. Veltman, "Heavy particles and the parameter in the standard model", NSF-ITP-81-43, UM HE 81-25.
[19] M. Veltman, Acta Physica Polonica, B8, 475 (1978).
[20] D. Toussaint, Phys. Rev. D18, 1626 (1978).
[21] T. Sterling & M. Veltman, "Decoupling in Theories with Anomalies" UM HE 81-2.

The structure of self dual monopoles

David Olive : Blackett Laboratory, Imperial College.

INTRODUCTION

Self dual monopoles are the stable classical field configurations, carrying magnetic charge, which arise in gauge theories [1] which are spontaneously broken by a Higgs field [2] which lies in the adjoint representation of the gauge group, and which has vanishing self-interaction. Such theories are of interest both for the resemblance to the currently fashionable grand unified theories[3] (accounting for all fundamental forces other than gravity) and for the remarkable theoretical properties they enjoy.

The study of such theories is a rapidly expanding subject and of necessity I can talk only about one aspect in any depth. This I choose, arbitrarily, to be the aspect that has occupied me recently, the study of spherically symmetric solutions, for arbitrary simple gauge group G. As will be seen this involves Lie algebra theory in a fundamental way. It seems to be a completely soluble problem, though this can only be an interim report. The progress so far is of interest in mathematical physics because it relates to the theory of integrable dynamical systems and the Toda molecule in particular.

My first lecture will review the salient features of the theory of self dual monopoles paying particular attention to a conjectured duality property which may be the most important feature of all, "explaining" all the verified properties. Unfortunately this duality is a quantum property and so impossible to verify at the present level of understanding. Nevertheless, it does have some consequences which can be checked at present. Some have been checked and more will be accessible when the programme of constructing self dual solutions is completed and analysed.

The reason this duality is important is that similar or closely related concepts arise in many areas of theoretical physics; the Kramers Wannier duality transformation in spin or gauge lattice theories[4]; the Sine-Gordon Thirring model equivalence[5]; the theories of quark confinement[6]; and in supergravity theories[7]. Thus the proper understanding of "duality" appears to be a central problem in theoretical physics, if not the central problem. It is likely that it will be based on fundamental advances in pure mathematics. The point about the duality in the theories we discuss is that the four dimensional nature of spacetime is fundamental (since only then are magnetic monopoles point objects like electric charges) and that the theory we consider is perhaps the simplest with a dual property in four dimensions and that there is a real prospect of a precise mathematical formulation.

The encouragement comes from the rich mathematical structures which unexpectedly emerge in the work described here and in other lectures at this conference[8].

In Lecture II, I explain how the requirements of spherical symmetry and self duality lead to Lax pair equations which in turn imply simplifications in the form of conserved quantities.

The types of spherical symmetry can be classified by Lie algebra theory as explained in Lecture III and one natural choice leads to the generalised Toda molecule equations which are discussed in Lecture IV. In these equations and in others, generalising the Sinh-Gordon equation, also discussed, the Cartan matrix plays an important role.

In Lecture V, I present the general solution to the two dimensional Toda molecule equations and explain how it can be used to construct explicit monopole solutions for general gauge groups. An interesting extra symmetry of this solution occurs when the condition of regularity at the centre of the monopole is applied. This lecture ends at the frontier of research with more awaiting to be discovered. It seems clear that at the very least the different approaches to self-dual magnetic monopole theory are working towards a synthesis with many apparently diverse branches of mathematics, mathematical physics and physics.

For background to these Lectures, there exists review articles explaining the earlier work not discussed here [9,10] and giving further references.

LECTURE 1. Review of self dual monopoles

The basic Lagrangian density to be discussed will be :

$$= -\frac{1}{4}(F_{\mu\nu})^2 + \frac{1}{2}(D_\mu\Phi)^2 \qquad (1)$$

The gauge field strength $F_{\mu\nu}$ refers to any simple Lie group G but the "Higgs" field Φ lies in the adjoint representation of G and has vanishing self interactions.

This apparently innocuous Lagrangian seems to describe a mysterious kind of symmetry limit in which there occurs a wonderland of novel features to be listed below. Maybe (1)(or modifications described below) will turn out to be a "soluble field theory" in D=3+1 in some sense, yet to be formulated, somewhat analogous to known examples in D=1+1 space time. When (and if) these insights come to pass, they are likely to shed light on the outstanding problems of theoretical particle physics - "confinement"[6], "unification" and "superunification".

We now consider, in turn, some of the special properties consequent on (1):

1. The energy following from (1) can assume its minimum value, zero, for a field configuration in which Φ does not vanish, but has a constant length, a. We can thus regard Φ as a Higgs field. Since the self interaction of Φ vanishes the lagrangian itself does not specify a nor the orbits of Φ in vacuo. This information is to be regarded as coming from some sort of boundary condition imposed from outside.

The exact gauge symmetry of the theory has a special physical importance, being the symmetry of both the action and the vacuum. It consists of the subgroup of G which leaves Φ invariant in vacuo. Let us define the generator of G in the direction of Φ.

$$Q = e\hbar\, \Phi.T/a. \tag{2}$$

Then from the definition the Lie algebra of H, $\mathcal{L}(H)$ is the subalgebra of $\mathcal{L}(G)$ commuting with Q. It follows that $\mathcal{L}(H)$ can be expressed as the sum of two mutually commuting parts Q and $\mathcal{L}(K)$[11]. This structure can be exponentiated to give:

$$H = \text{"}U(1)_Q \times K\text{"} \tag{3}$$

The inverted commas indicate that the direct product structure need not hold globally; there can be points of intersection between the two factors[12].

Thus H automatically possesses an invariant U(1) subgroup generated by Q. This is the situation in nature with the Maxwell gauge group when the eigenvalues q of Q are electric charges. Notice that Q is a K singlet. We shall be interested in choices of Φ for which K is semisimple (as in nature).

2. When Φ does not vanish in vacuo the second term in (1) provides a mass term for those gauge potentials not corresponding to H. This is the Higgs-Kibble-Brout-Englert mechanism [2]. Specifically the mass squared matrix is:

$$M^2_{\alpha\beta} = e^2\hbar^2\, \Phi\, T_\alpha T_\beta \Phi = a^2 Q^2_{\alpha\beta}$$

using (2) and the fact that in the adjoint representation the generators T can be represented by totally antisymmetric structure constants.

Hence diagonalizing Q and hence M^2 the masses of the gauge particles are given by:

$$M = a \left| q \right| \tag{4}$$

where q is the U(1) charge born by them [13]. The scalar particles which survive correspond to H gauge particles and so have q=0. Since there is no mass term in (1) they likewise satisfy (4). Thus this mass formula applies to all the quantum excitations of the elementary fields occuring in the Lagrangian (1).

3. These are not the only particle like states in the theory. There are classical field configurations whose stability is guaranteed by the topological quantum number $\pi_1(H)$ [14]. Since H has the form (3).

$$\pi_1(H) \supset \pi_1(U(1)) \equiv \mathbb{Z} \tag{5}$$

where $\mathbb{Z}$ is the group of integers. As t'Hooft and Polyakov showed the static solutions are magnetic monopole solitons whose U(1) magnetic charge g is proportional to the integer in (5) [15].

4. The configuration with given magnetic charge g and least possible energy were shown by Bogomolny [16], Coleman, Parke, Neveu and Sommerfield [17] to be static in the sense.

$$F_{io} = E_i = 0 \quad , \qquad D_o \Phi = 0. \tag{6}$$

and to satisfy "self duality" equations

$$\tfrac{1}{2}\varepsilon_{ijk} F_{ijk} = B_i = \pm D_i \Phi \tag{7}$$

Further the mass of these configurations can be expressed as:

$$M = a \left| g \right| \tag{8}$$

where a is again the length of the Higgs field Φ in vacuo.

The significance of these solutions is that they are stable in the sense that no deformation small or large can lower their energy. These are the self dual magnetic monopoles of the title of this Lecture.

In addition if one allows electric charge q which is also held fixed there are "dyon" solutions [18] minimizing the energy with mass.

$$M = a\sqrt{q^2 + g^2} \tag{9}$$

Notice that in these results the vanishing of the self interaction of Φ in equation (1) is crucial, unlike the previous properties (1-3).

5. Now one sees that the masses of all the particle states are given by a single universal formula (9) in terms of their electric and magnetic U(1) charges [19]. This is irrespective of whether the particles are quantum excitations of the original fields of (1) i.e. gauge particle and scalar particles, or whether they are soliton solutions, i.e. stable magnetic monopoles and dyons. That the derivation of the result is so different in the different cases makes the result so remarkable [19].

6. The mass formula (9) is invariant with respect to rotations in the (q,g) plane. This appears to realize in the presence of matter the duality invariance of the Maxwell theory, $\underline{E} + i\,\underline{B} \to e^{i\theta}(\underline{E} + i\,\underline{B})$ which has fascinated physicists for nearly 100 years [20].

We have seen that magnetic charge, being proportional to the topological integer in expression (5) is quantized. The question arises as to whether electric charge is also quantized and, if so, how the units of electric and magnetic charge are related. It has been shown by Corrigan and Olive [12] that:

$$e^{igQ/\hbar} = k \in K \tag{10}$$

where Q is as in equation (2).

Now, since Q is a K singlet, so is the element K on the right hand side of equation (10). It thus must lie in the centre of K. This is a finite subgroup if K is semisimple as we already wanted to assume. In fact the solution to (10) must lie in the cyclic subgroup of the centre of K. If this has order Z we have

$$e^{igQ|Z|/\hbar} = 1$$

which tells us that Q is quantised and that if q_0 is the smallest unit, then g must be an integer multiple of

$$(2\pi\hbar)/(|Z|q_0)$$

In the case that K = 1 this reduced to Dirac's classical result [21]. However when K is nontrivial fractional charges are allowed. In fact there is more information in equation (10); one can deduce a relation between the fractional charges and K transformation properties [12]. The simplest example concerns quarks. The most detailed analysis of this has recently been given by Goddard and Olive [22].

7. The universal mass formula equation (9) is very suggestive of a hidden role played by two extra dimensions of space and time. If one introduces a six momentum

$$\mathbb{P}^{\mu} = (p^{\mu}, aq, ag)$$

composed or ordinary four momentum and the electric and magnetic charges, then equation (9) can be written as the condition that $\mathbb{P}^{\mu}$ be lightlike

$$\mathbb{P}^2 = 0 .$$

In fact it can be proven that the identification of the fifth component with electric charge is a consequence of the identification of the Higgs field Φ with the fifth component of the gauge potential [13].

This is only possible because Φ lies in the adjoint representation like the gauge potential, and because the self interaction of Φ vanishes in equation (1) which can be written as a single gauge kinetic energy in five dimensions when it is understood that the fields are independent of x^5, the fifth coordinate. This situation is quite similar in concept to the old Kaluza Klein theory[23].

8. The identification of magnetic charge with the sixth component of momentum is more difficult and has been understood only in the context of a radical new principle, supersymmetry [24]. The Lagrangian (1) can be made supersymmetric by adding terms containing extra spinor and scalar fields. When these vanish we have the original equations so that their solutions are still valid.

There are two candidate supersymmetries; N=2 and N=4 extended supersymmetry. These lagrangians can be written in terms of a single gauge and spinor field in six and ten dimensions respectively.

When this is done and the theory interpreted in four dimensions the following virtues emerge: (i) at least if N=2 the two extra components of momentum are indeed the electric and magnetic charges. So the duality rotation is a rotation in the 56 plane [13]; (ii) the quantum corrections to the universal mass formula appear to vanish owing to supersymmetric cancellations [24]. This is necessary if the duality is to survive at all in the quantum theory. However, there is still some controversy over this cancellation even in analogous two dimensional models [25]. (iii) The monopole states must occur in supermultiplets. Since the supercharges carry spin $\frac{1}{2}$ some of the monopole states must carry spin.

9. Actually the work on supersymmetry was motivated by a conjecture which preceded it and concerned the possible local field theory of the monopole states. For Sine Gordon solitons in D = 1+1 such a field theory exists and its equations of motion are that of the massive Thirring model [5]. Could a similar thing happen here? Since the monopole antimonopoles have Coulomb interactions we expect a U(1) magnetic gauge group to play a role. Montonen and Olive [19], suggested that if G = SU(2) the underlying Lagrangian was similar to (1)

but with the magnetic charge replacing the electric charge in the gauge coupling constant. The "magnetically"charged gauge fields create the monopoles and the original, electrically charged gauge particles arise as solitons of this second, dual Lagrangian. Thus the heavy gauge particles are "composite" but in a bootstrap sense.

This conjecture cannot be proved but circumstantial evidence in its favour can be given. The mass formulae (4) and (8) agree with it because of the similarity with g replacing q. So does the fact that for G=SU(2) there are precisely two self dual monopoles with equal and opposite signs.

The most striking prediction is that the self dual monopoles have spin 1. This has to be a quantum effect. If the theory is quantized supersymmetrically then the monopole states must represent the algebra and hence, as remarked earlier, different components carry different spins.

It is not quite clear whether the relevant multiplet carries a spin one component or not. If the supersymmetry is N=4 there is only one multiplet with spins less than or equal to one, i.e: which could satisfy renormalizable equations of motion in flat space time. This necessarily contains spin one and is necessarily isomorphic to the one in which the original heavy gauge particles lie. Thus this case, N=4 supersymmetry, is particularly favourable to the conjecture as Osborn has pointed out[26].

The evidence concerning the spins and the mass formulae were independent of the choice of G but the argument concerning the quantum numbers applied only to G = SU(2). Goddard and Olive[27] recently tried to full this gap by developing earlier work of Bais[28] and trying to show that the self dual monopoles carried magnetic charges consistent with their being gauge particles for the group G^V dual to G. (G^V has roots α/α^2 where α is a root of G.)[29,30]. In the argument it was essential to discard states which could be interpreted as overlapping multimonopole states. However E. Weinberg has found an error in the argument and it is not clear at present whether the desired result is true or not.

10. If the duality conjecture is true one might expect q and g to renormalize similarly. Yet their product must presumably remain constant in order to respect the quantization condition(10). Apparently the only resolution of this is that the Callan Symanzik beta function which describes their renormalization should vanish. This practically never ever happens except that by an amazing "coincidence" it does in N=4 supersymmetric gauge theory, at least to the first three loop orders which have been calculated[32]. Here is more evidence for the duality conjecture as has been pointed out by Nahm[33] and Rossi [34].

Actually, the N=4 supersymmetry gauge theory is a limiting case of the fermionic dual string model and as such could be expected to have such miraculous properties[35].

It seems that beta functions also vanish in certain supergravity theories (again related to the string model) [36]. The vanishing seems to be fortuituous but if the ideas of this lecture are correct, this circumstance is related to the quantum validity of a duality symmetry which cannot be realized directly on the fields of the original Lagrangian (1).

11. Closed expressions for spherically symmetric solutions to the self duality equation (7) were discovered by Prasad and Sommerfield for G = SU(2), essentially by trial and error [37]. It would be interesting to understand the structure of the solution and how it generalises to any gauge group G.

It is now known from the work of Leznov and Saveliev[38,39,40] that in all cases the demands of self duality (7) and spherical symmetry lead to equations which can be written as a Lax pair.

$$dA/dr = \left[A,B\right] \tag{11}$$

where A and B are Lie algebra valued. This immediately yields radius independent quantities $TR(A^2)$, $TR(A^3)\cdots$.

For a special choice of spherical symmetry, equation (11) turns out to be the integrability condition for the equations of motion associated with a type of Toda lattice or molecule :

$$d^2\rho_\alpha \;/dr^2 \;= \sum_\beta K_{\alpha\beta}\, e^{\rho_\beta} \;.$$

K is the Cartan matrix for G. If G = SU(N) it has entries equal to 2 on the diagonal, - 1 just above and below and zero elsewhere. These results will be explained in the subsequent lectures as well as methods of solution of such equations[38], which have been used to find explicit solutions satisfying the boundary condition at the origin appropriate to a monopole [41,42], thereby generalizing the Prasad Sommerfield solution. Much interesting structure emerges as a result.

12. Generalizations of the Prasad Sommerfield solutions have recently been made in another direction; keeping G=SU(2) but enlarging the number of monopoles from one to any number arranged in any way. This uses techniques borrowed from instanton theory and has been carried out by R. Ward[43] the Hungarian School,[44], Corrigan and Goddard,[45] W. Nahm [46] and N. Hitchin [46]. These solutions satisfy the self duality equations (6) and (7) and so are static. This means that the like monopoles exist in equilibrium without moving whatever their separations and relative angles. Hence they exert no forces on each other at all. This was known approximately from earlier work of Manton[47] and Nahm [47].

On the other hand monopoles and antimonopoles attract each other with twice the expected Coulomb force and so their solutions would have to be time dependent and cannot be self dual and have not been found.

These results can also be regarded as circumstantial evidence for the duality conjecture. The reason is that it would imply an analogous property for forces between heavy gauge particles in the G = SU(2) thory. Since the Lagrangian for these particles is given by equation (1) the forces between them can be evaluated approximately, in the non-relativistic limit by Feynman rules [19]. Inverse square forces are due to neutral zero mass particle exchange. Lagrangian (1) describes two neutral zero mass particles, the photon and the scalar particle. The scalar exchange (like graviton exchange) is always attractive whereas photon exchange is attractive for unlike particles and repulsive for like charges. In fact an exact cancellation occurs for like charges and this can be seen directly from the six dimensional formalism where there is a single six vector exchange[13]. The numerator of the exchange graph is in the nonrelativistic limit

$$\mathbb{P}_1 . \mathbb{P}_2 = (m_1,0,0,0,aq_1,0) \quad (m_2,0,0,0,aq_2,0)$$

$$= (m_1 m_2 - a^2 q_1 q_2)$$

$$= a^2 (|q_1 q_2| - q_1 q_2)$$

by equation 4 which is $\mathbb{P}_1^2 = \mathbb{P}_2^2 = 0$. This vanishes when q_1 and q_2 have like sign. The numbers one and two refer to the two charged particles concerned.

Finally let us remind the reader that duality arguments play an important role in confinement and supergravity theories. However the models there are complicated and difficult to handle exactly. The model presented here seems to have shown a remarkable degree of solubility and could therefore shed light on the more complicated but possibly more realistic models.

LECTURE II : Self duality, spherical symmetry and Lax pairs.

There have been interesting developments in the construction of spherically symmetric monopole solutions for theories as mentioned above in which the Higgs field lies in the adjoint representation and has vanishing self interaction. When the gauge symmetry of the Lagrangian was SU(2), Prasad and Sommerfield(37) found the exact solution, essentially by trial and error:

$$\phi_a = r_a(\xi \coth \xi - 1)/(er^2) \quad \text{where} \quad \xi = aer$$

$$W_a{}^i = -\varepsilon_{aij}\, r_j \left[1 - \xi/(\sinh\xi) \right] /(er^2) \tag{1}$$

Owing to work of Wilkinson and Goldhaber(48), Bais(49,50), Weldon(49) and Leznov and Saveliev(38) we now know that this is the simplest application of a systematic method which can be applied to any semisimple gauge symmetry G.

The method is extremely interesting in itself since it relates to two dimensional field theories which are zero curvature conditions and in particular to generalized Toda molecule equations.

The Lie algebra structure of G plays a very important role as we shall see in subsequent lectures.

Our first step is to formulate a concept of spherical symmetry in a gauge theory. Ordinarily the three rotation generators about the origin are $-i r \wedge \nabla$ and the demand that a scalar function f(r) be spherically symmetric is simply that $\left[-i r \wedge \nabla , f(r)\right]$ vanishes. This implies that f is a function of radius $r = |r|$ only. In a gauge theory a function is counted as spherically symmetric if, after a space rotation, the original form can be restored by a gauge rotation.

In effect we demand that the rotation generators take the form:

$$\underset{\sim}{J}_o = -i\,\underline{r}\wedge\nabla + \underline{t} \tag{2}$$

where t_1, t_2, t_3, are three generators selected from the Lie algebra of G such that they satisfy an angular momentum algebra

$$\left[t_i, t_j\right] = i\,\varepsilon_{ijk}\, t_k \tag{3}$$

Thus J_o is constructed out of two mutually commuting angular momenta and thus itself satisfies angular momentum commutation relations (11, 9).

The Higgs field Φ (r), (and the time component of the gauge potential) are rotational scalars and hence are assumed to satisfy

$$\left[J_i^o\,,\,\Phi\right] = 0 \tag{4}$$

The space components of the gauge potential form a rotational vector and hence are assumed to satisfy

$$\left[J_i^o, W_j\right] = i\varepsilon_{ijk}W_k \tag{5}$$

(Here W_j and Φ , lying in the adjoint representation, are taken to be Lie algebra valued i.e:

$$\Phi = \sum_{\alpha=1}^{\dim G} \Phi^\alpha T^\alpha \text{ ,(where } T^\alpha \text{ are the generators of G)}$$

The transverse components of the equations in the index i determine the angular dependence of Φ and W_j, leaving the radial dependence to be determined by the (Bogomolny) equation of motion. Ordinarily the radial component plays no role since $-ir\wedge\nabla$ is purely transverse, but J_o given by equation (2) has a radial component r.t. This implies an extra consistency condition on Φ from equation (4).

$$\left[r.t,\Phi\right] = 0$$

and similarly for W. As we shall see these consistency conditions will play a crucial role in what follows.

Since the angular dependence is determined it is convenient to work on the z axis. Then A_3 is the radial component of any vector A and A_1 and A_2 the two transverse components. Let us define:

$$A^{\pm} = A_1 + iA_2$$

Then it is convenient to express the two transverse components of this gauge potential in terms of new functions $M^{\pm}$ by

$$eW^{\pm} = \mp i\,(M^{\pm} - t^{\pm})\,/\,r$$

The radial component of W is assumed to vanish as a result of a gauge transformation. Then the consistency condition on the Higgs and gauge fields read:

$$\left[t_3, \Phi\right] = 0$$
$$\left[t_3, M^{\pm}\right] = \pm\, M^{\pm}. \qquad (6)$$

Now it is possible to evaluate the radial and transverse components of $\underline{B}$, the non Abelian magnetic field and the covariant derivative of the Higgs field. The result, originally due to Wilkinson and Goldhaber is [48]:

$$D_3\Phi = d\Phi\,/dr, \qquad D^{\pm}\Phi = \mp \left[M^{\pm}, \Phi\right] /r ,$$

$$eB_3 = (\tfrac{1}{2}\left[M^{+}, M^{-}\right] - t_3)/r^2, \quad erB^{\pm} = dM^{\pm}\,/dr.$$

Notice that the expressions involving transverse derivatives, namely $D^{\pm}\Phi$ and B_3 are determined by the spherical symmetry assumption.

The Bogomolny equations, the simplified version of the field equations to be satisfied by the monopole solutions in the absence of self interaction read (see equation (7) in lecture I).

$$\underline{B} = \eta\, \underline{D\ \Phi} \qquad (8)$$

where $\eta = \pm 1$ is the sign of the topological charge. Hence inserting the result of the spherical symmetry ansatz, the radial and transverse components of equation (8) read, respectively:

$$er^2 d\Phi/dr = \eta(\tfrac{1}{2}\left[M^+,M^-\right] - t_3)$$

$$dM^\pm/dr = \pm\eta\left[\Phi,M^\pm\right]$$

The first point is that these equations simplify if we make the further substitutions:

$$\left.\begin{aligned} e\Phi &= \eta(\psi + t_3/r) \\ M &= \eta r\, N^\pm \end{aligned}\right\} \qquad (9)$$

to

$$\left.\begin{aligned} d\psi/dr &= \tfrac{1}{2}\left[N^+,N^-\right] \\ dN^\pm/dr &= \pm\left[\psi,\, N^\pm\right] \end{aligned}\right\} \qquad (10)$$

So far the "consistency conditions" (6) have not come into play. If a given quantity X satisfies

$$\left[t_3,X\right] = nX \qquad (11)$$

we say X has "grade n". Since "n" is an eigenvalue of an angular momentum component it must be an integer or half integer so we have what is called a "Z" (integer) grading (rather than the Z_2 grading occurring in supersymmetry algebras).

Now comes the second and unexpected step; define

$$A = \psi + \tfrac{1}{2}\,(N^+ - N^-) \qquad (12)$$

i.e. as a linear combination of Higgs and gauge potential components. Then

$$\left[t_3,A\right] = \tfrac{1}{2}\,(N^+ + N^-) = B$$

as N^+, ψ, N^- have grade +1, 0 and -1 respectively. Further

$$\left[A,B\right] = \frac{dA}{dr} \qquad (13)$$

using the Bogomolny equations in the form (10). Now this equation has three parts with distinct grades 0, 1 and -1 which must be linearly independent. Hence because of the grading the single equation (13) is precisely equivalent to the three equations (10).

This equation (13) has a very special structure which underlies its solubility. In any finite dimensional representation of the Lie algebra of G, A is an ordinary finite dimensional matrix and so Tr (A^n) is perfectly well defined. But by (13)

$$\frac{d}{dr} \operatorname{Tr}(A^n) = \operatorname{Tr}([A^n, B]) \equiv 0$$

since the trace of a commutator of finite matrices vanishes. Hence $\operatorname{Tr}(A^n)$ is r independent for any integer n, and we have found a series of integrals of the equations (10).

A system of equations such as (13) is almost what is known as a Lax pair[51]. We say almost because A and iB are not hermitian in the usual sense. This is because r is a spatial rather than a time variable.

If A were hermitian we could say that its eigenvalues were real and time independent. This is the basis of the "inverse scattering method" when A is a Schrödinger operator. Actually if we define a modified conjugation operation

$$X^x = (-1)^{t_3} X^\dagger (-1)^{-t_3} \tag{14}$$

we see that this is indeed an involution on the integer grade subalgebra of the Lie algebra of G. This is the subalgebra in which A and B lie. Further A and iB are hermitian with respect to this operation[41] and this is sufficient to guarantee the reality of $\operatorname{Tr}(A^n)$.

The precise system of equations resulting depends on the choice of t_3, or more precisely the SU(2) subalgebra t_1, t_2, t_3 of G. The choices of this 'embedding" of SU(2) in G have been completely classified by Dynkin[52]. This is where the Lie algebra structure enters as we now explain. For one special choice of t_3 the resultant equations will relate to the generalized Toda molecules, as explained below.

LECTURE III : SU(2) Embedding and Lie Algebra Theory.

Let us illustrate some embeddings of SU(2) into low dimensional groups G.

To embed SU(2) in SU(2) is trivial: $(t_1,t_2,t_3) = (\sigma_1,\sigma_2,\sigma_3)/2$ where the σ_i are Pauli matrices. Any other choice is related by conjugation and so the given choice is unique up to conjugation.

Possible embeddings of SU(2) in SU(3) are familiar as the "isospin embedding"

$$(t_1,t_2,t_3) = (\lambda_1,\lambda_2,\lambda_3)/2 \tag{1}$$

or the "nuclear physics embedding"

$$(t_1,t_2,t_3) = (\lambda_2,\lambda_5,\lambda_7) \tag{2}$$

where the λ_i are Gellman lambda matrices. These are clearly inequivalent since the eigenvalues are different, $(\pm\frac{1}{2},0)$ and $(\pm 1,0)$ respectively. Up to conjugation these are the only possibilities.

Since the SU(2) is a subgroup of G, the adjoint representation can be decomposed into SU(2) irreducible representations. This is the same thing as forming the generators of G into SU(2) multiplets. This is important because it will tell us the t_3 grades of the G generators. For example in the two SU(3) possibilities (1) and (2), we have, respectively,

$$8 = 3 \oplus 2 \oplus 2 \oplus 1$$

$$8 = 3 \oplus 5$$

Notice 3 must always occur since this is furnished by t_1, t_2, t_3 themselves. Sometimes there are half integer spin multiplets (the twice 2) and sometimes not.

The "t_3 grades" of the generators occurring in a (2J+1) dimensional SU(2) multiplet are simply -J, -J+1.... . J. It follows that the number of SU(2) multiplets $\geq$ the number of integer spin SU(2) multiplets = the number of grade 0 generators. But this latter number must exceed the rank of G, r(G) (the maximal number of mutually commuting generators). Hence the number of multiplets is bounded below by r(G). SU(3) has rank 2 so that the "nuclear physics" embedding (2) with the two multiplets 3 and 5 achieves this lower bound. Such an embedding is called a "maximal embedding". It can be constructed for any G as we shall see, and will lead to a particularly nice system of equations.

To describe it we shall need more Lie algebraic concepts.

The choice of basis for a Lie algebra, familiar to physicists, is the Cartan-Weyl basis: H_i i=1....r(G) are the mutually commuting Cartan subalgebra generators and $E_{\pm\underline{\alpha}}$ the "step operators corresponding to a root $\underline{\alpha}$. So

$$\left.\begin{aligned} \left[H_i, H_j\right] &= 0, \quad \left[H_i, E_{\pm\underline{\alpha}}\right] = \pm\, \alpha_i\, E_{\pm\underline{\alpha}} \\ \left[E_\alpha, E_{-\alpha}\right] &= \alpha . H. \end{aligned}\right\} \qquad (3)$$

We need not specify the other commutators. The root system completely characterizes the Lie algebra and the mathematical classification of Lie algebras proceeds by classifying the possible root systems[53,54].

The roots lie in a space of dimension r(G) and it turns out that there is a particularly convenient basis of "simple roots", $\alpha_1 \ldots \alpha_{r(G)}$ with the magic property that any root can be expressed as

$$\alpha = \sum_{i=1}^{r(G)} n^i \alpha^i \qquad (4)$$

with the coefficients n^i integers which are either all positive (≥ 0) or all negative (≤ 0). For SU(3) a pair of roots subtending an angle of $(2\pi/3)$ constitutes a system of simple roots as the reader can check.

In fact the root system (and hence the Lie algebra) can be reconstructed from the "Cartan matrix" K_{ij} which roughly speaking specifies the angles between the simple roots. It is the $r(G) \times r(G)$ matrix

$$K_{ij} = 2\alpha^i.\alpha^j/(\alpha^j)^2 \tag{5}$$

For example, for SU(2); $K=2$
for SU(3) $K=\begin{pmatrix} 2 & -1 \\ -1 & 2 \end{pmatrix}$

while for SU(N) it is an $(N-1)\times(N-1)$ matrix with 2 on the diagonal, -1 just above and below the diagonal and zero elsewhere. For the exceptional group G_2, $K=\begin{bmatrix} 2 & -1 \\ -3 & 2 \end{bmatrix}$. The entries in the Cartan matrix are always integers.

In physical applications we study not just the structure of the group but its irreducible matrix representations. The possible eigenvalues of the Cartan subalgebra generators are called "weights"

$$H_i|\lambda> = \lambda_i|\lambda> \qquad i = 1 \ldots . r(G)$$

It is a theorem that any such λ_i must satisfy

$$2\lambda.\alpha/\alpha^2 = \text{integer, for any root } \alpha. \tag{6}$$

Conversely any solution λ to these equations is a weight of some irreducible representation. The "fundamental weights" are essentially reciprocal to the simple roots.

$$2\lambda^j.\alpha^i/(\alpha^i)^2 = \delta^{ij} \qquad i,j = 1 \ldots . r(G) \tag{7}$$

Any quantity of the form $\Sigma n^i\lambda^i$, (n^i integers) is a weight and vice versa.

A relatively recent mathematical development is the realization that there is another basis to the Lie algebra, called the "Chevalley basis" (54). This provides a rather convenient alternative to the Cartan Weyl basis defined earlier. Define

$$H_\alpha = 2H.\alpha/\alpha^2$$

so that $H_\alpha|\lambda\rangle = (2\lambda.\alpha/\alpha^2)|\lambda\rangle$ = integer $|\lambda\rangle$, by eqn (6) .

Thus H_α always has an integer spectrum and is a good candidate for a quantum number in a quantum theory application. Then rescale the step operators E_α so

$$\left[H_\alpha, H_\beta\right] = 0, \quad \left[H_\alpha, E_{\pm\beta}\right] = \pm E_{\pm\beta} K_{\beta\alpha} \tag{8}$$

$$\left[E_\alpha, E_{-\beta}\right] = H_\alpha \delta_{\alpha\beta} \quad , \quad \alpha,\beta \text{ simple roots.}$$

Notice that the structure constants displayed here are always integers, (since the Cartan matrix $K_{\beta\alpha}$ has integer entries). This also applies to the other commutators which we have not written and shall not need. The last commutator needs some explanation. The Jacobi identities would tell us that $\left[E_\alpha, E_{-\beta}\right]$ must be proportional to the step operator for the root $\alpha-\beta$. Since α and β are simple there is no such root and so the commutator must vanish unless $\alpha=\beta$ when the result follows from equations (3). Thus the $E_{\pm\alpha}$'s behave a bit like annihilation creation operators for independent harmonic oscillators.

Now we have assembled enough machinery to return to the problem of specifying t_3. Without loss of generality it can always be gauge rotated to lie in the chosen Cartan subalgebra so that

$$t_3 = \frac{1}{2} f^i H^i$$

The gauge freedom is exhausted by requiring

$d_\alpha = (f.\alpha) \geq 0$, α a simple root.

Since $\left[t_3, E_\alpha\right] = \frac{1}{2}(f.\alpha)E_\alpha$

we see that $(f.\alpha)$ is twice the grade of the step operator associated with the simple root α, and hence an integer. The integers d_α are called the "Dynkin characteristic" (52) and characterise the embedding up to a gauge transformation.

Consider $t_3 = 2\delta^V.H$ when $\delta^V = \sum_i \lambda^i/(\alpha^i)^2$ so that $2\alpha.\delta^V=1$ for any simple root α, and the Dynkin characteristic is (2,2,....2). We shall show that this is the maximal embedding alluded to earlier. Any positive root β can be written (equation (4)),

$$\beta = \sum_i^r m_i\alpha_i \qquad m_i \text{ an integer} \geq 0.$$

so

$$\left[t_3, E_\beta\right] = 2\delta^V.\beta\, E_\beta = \left(\sum_i^r m_i\right) E_\beta$$

This shows that each step operator has an integer ($\neq 0$) grade, never a half integer grade. The only zero grade generators are the r H_i. Further the only grade 1 generators are the r step operators for the simple roots. Hence there are precisely r integer spin SO(3) multiplets, each with spin greater than or equal to one. This then saturates the lower bound on the number of multiplets characteristic of the maximal embedding.

Now t^+ has grade 1 and so satisfies $\left[t_3, t_+\right] = t_+$ providing $t^+ = \Sigma_{i=1}^r C_{\alpha_i} E_{\alpha_i}$. So it is easy to choose the coefficients C to ensure that $\left[t_+, t_-\right] = 2t_3$, and the SO(3) algebra is indeed realised.

Let us now return to our spherically symmetric monopole problem, choosing this embedding, and working out the radial equations of motion.

The most general solution to the consistency requirement eqn. (6) in lecture II stating that N^+, N^- and ψ have grade 1, -1 and 0 respectively is

$$N^+ = \Sigma f^\alpha(r)E_\alpha \; , \quad N^- = (N^+)^* = \Sigma f^{\alpha *}(r)E_{-\alpha} \tag{9}$$

$$\psi = \Sigma \psi^\alpha(r)H_\alpha \quad \text{(all sums over simple roots } \alpha)$$

The expansion coefficients are functions of radius only. Inserting into the radial equations (10) of lecture II, and using the commutators (8) appropriate to the Chevalley basis we find

$$\begin{aligned} d\psi_\alpha/dr &= \tfrac{1}{2} f^*_\alpha f_\alpha \\ df_\alpha/dr &= \Sigma_\beta \; f_\alpha K_{\alpha\beta}\psi_\beta \\ df^*_\alpha/dr &= \Sigma_\beta \; f^*_\alpha \; K_{\alpha\beta} \; \psi_\beta \end{aligned} \tag{10}$$

so in particular $d/dr\,(\ell n\, f_\alpha/f^*_\alpha) = 0$, ie the phase of f_α is r independent, and can be chosen zero by a suitable gauge choice. The structure we have described is preserved if we apply gauge transformations generated by grade 0 generators, the H_α . The quantity

$$\rho_\alpha = \ell n\, |f_\alpha|^2 \tag{11}$$

is invariant with respect to these transformations and turns out to be a good variable since it satisfies the second order differential equations

$$\begin{aligned} d^2\rho_\alpha/dr^2 &= \frac{d}{dr}\left(\frac{d}{dr}(\ell n f_\alpha) + \frac{d}{dr}(\ell n f^*_\alpha)\right) \\ &= \frac{d}{dr}(K_{\alpha\beta}\psi_\beta + K_{\alpha\beta}\psi_\beta) \\ &= 2K_{\alpha\beta}d\psi_\beta/dr \;\; = K_{\alpha\beta}e^{\rho_\beta} \end{aligned}$$

i.e

$$\rho''_\alpha = K_{\alpha\beta}e^{\rho_\beta} \tag{12}$$

i.e. a set of r coupled non linear equations involving the Cartan matrix. For the case of G = SU(N) these equations closely resemble some equations which have been intensively studied in theoretical physics: the Toda lattice equations[55]. Without understanding the underlying group theory it was discovered that these equations have remarkable properties, integrability and solubility. Now we know that these properties apply whenever K is the Cartan matrix for any group, not just SU(N).

LECTURE IV : Toda lattices, Toda molecules and related equations.

The subject of the title originates with the famous computer experiment of Fermi, Pasta and Ulam [56], who considered a finite number of mass points arranged in line interacting with their nearest neighbours via anharmonic forces. The system was started by displacing the end point with the others at rest. Soon all were moving but after a finite time, the initial situation recurred with all but one pendulum stopping instantaneously. This surprise meant the system was not ergodic.

Toda[55] suggested a variant of the above system susceptible to analytic solution. The tensions in the springs connecting adjacent mass points were exponential functions of their extension. In dimensionless units, Newton's equations for the ith and i+1 th points $(1<i<i+1<N)$ read

$$\ddot{x}_i = e^{x_{i+1}-x_i} - e^{x_i - x_{i-1}}$$

$$\ddot{x}_{i+1} = e^{x_{i+2}-x_{i+1}} - e^{x_{i+1}-x_i}$$

If ρ_i is the separation $x_{i+1} - x_i$ between two adjacent points:

$$\ddot{\rho}_i = e^{\rho_{i+1}} - 2e^{\rho_i} + e^{\rho_{i-1}} \qquad 1<i<N-1$$

while only one spring acts on mass point 1 so

$$\ddot{\rho}_1 = e^{\rho_2} - 2e^{\rho_1}$$

and similarly for point N.

In matrix notation

$$\ddot{\rho}_i = -\sum_{j=1}^{N-1} K_{ij}\, e^{\rho_j} \qquad i = 1 \ldots N-1 \tag{1}$$

where K is the (N-1)x(N-1) matrix with 2 on the diagonal -1 just above and below the diagonal and zero elsewhere. But this is precisely the Cartan matrix for SU(N) described earlier in Lecture III after equation (5).

Equation (1) is the Toda molecule equation[57]. Apart from a sign equation (1) coincides with equation (12), derived above in Lecture III (and subsequently referred to as equation (III-12)), as describing the radial structure of a self dual monopole in a G= SU(N) gauge theory. The sign difference is accounted for by the fact that the independent variable in equation (1) is time while in equation (III-12) it is a spatial variable, r. These equations relate by the Wick rotation, $r \to it$, and there is a correspondence between the integrals of the two systems (the quantities TrA^n considered in lecture II). Historically these constants of the motion for equation (1) were discovered in a laborious way. Enlightenment dawned when Flaschka [58] and others [59,60] discovered the Lax pair (A,B) which works for (1) just as (A,B) in equations (II-12), (II-13) (III-9) work for equation (III-12). In Flashka's Lax pair, A and iB are hermitian owing to the Wick rotation.

Notice that the monopole analysis lead to the spacelike Toda molecule equation via the Lax pair. It also leads to other equations, corresponding to other choices of SU(2) embedding, not yet explored.

The monopole also teaches us something else. Equation (III-12) has been formulated when G (the original guage group) is any simple Lie group, and K its corresponding Cartan matrix. The work in the previous lectures showed that there were always integrals, $Tr(A^n)$. Correspondingly there must be a generalised Toda molecule equation (1) where K is the Cartan matrix for any simple Lie group, which has a Lax pair and integrals of the motion (conserved quantities). This has only been realised quite recently[38,61].

The question arises as to whether there are other choices of matrix K in equation (1) for which there are integrals of the motion and, if so, whether there is a Lie algebraic interpretation of the matrix. The answers to this question are very interesting (and incomplete).

The original Toda lattice equation(55) was the infinite system of equations obtained by taking N to infinity in equation 1 i.e. the $N=\infty$ limit of the SU(N) Toda molecule equation. These equations have soliton solutions and constitute a lattice version of the KdV equation(55). One can seek solutions which are periodic in the sense that the Kth mass point executes the same oscillations as the N+Kth one (for all K).

This is the same as identifying mass points 1 and N in the SU(N) Toda molecule equation above. In the resultant equation the matrix K is modified by the addition of two entries -1 in the corners furthest from the main diagonal

$$K = \begin{bmatrix} 2 & -1 & 0 & 0 & \cdot & 0 & -1 \\ -1 & 2 & -1 & 0 & \cdot & 0 & 0 \\ 0 & -1 & 2 & -1 & \cdot & 0 & 0 \\ \cdot & \cdot & \cdot & \cdot & \cdot & \cdot & \cdot \\ 0 & 0 & 0 & 0 & \cdot & 2 & -1 \\ -1 & 0 & 0 & 0 & & -1 & 2 \end{bmatrix} \qquad (2)$$

with (N-1) rows and columns.

Unlike the previous matrices K this is singular, reflecting the fact that the sum of the spring extensions is now fixed. This matrix also has an interpretation in terms of root systems.

Let $\alpha_0 = -\psi$ be adjoined to the r = rank G simple roots of G where ψ is the highest root, so that any other root β differs from ψ by a sum of positive roots. Then

$$\left[E_\alpha, E_{-\beta}\right] = 0 \qquad \alpha \neq \beta \qquad (3)$$

for any distinct roots α and β selected from $\alpha_0, \alpha_1, \ldots\ \alpha_r$, thereby generalising equation (III-8). Now define the extended Cartan matrix $K_{i,j}$ as the (r+1) x (r+1) matrix $2\alpha_i.\alpha_j/(\alpha_j)^2$. This is automatically singular as the $\alpha_0,\alpha_1.....\alpha_r$ are linearly dependent. For G = SU(N-1) the extended system of simple roots sums to zero and it is not difficult to verify that the Cartan matrix is the matrix (2) above.

Because of the property (3) the Lax pair can be set up, very much as before. Flashka[58] actually considered the SU(N-1) periodic Toda lattice, but the procedure works for any Lie group and again the equations support unexpected conserved quantities.

The extended Cartan matrix defined above can be regarded as the ordinary Cartan matrix of the infinite dimensional algebra called a Kac-Moody algebra[62], obtained by considering the light cone components of the current algebra[63] corresponding to G when space is a circle and the Schwinger term is included. The algebra without the Schwinger term is called a "loop algebra". There are also Kac-Moody algebras (with Schwinger terms) corresponding to "twisted loop algebras". These also have singular Cartan matrices for which equation (1)yields Lax pairs and constants of the motion[64].

Let us list the known 2x2 Cartan matrices, ordinary and generalized: the non-singular ones, corresponding respectively to the Lie groups SO(4), SU(3), SO(5) = Sp(4) and G_2 are

$$K = \begin{bmatrix} 2 & 0 \\ 0 & 2 \end{bmatrix}, \begin{bmatrix} 2 & -1 \\ -1 & 2 \end{bmatrix}, \begin{bmatrix} 2 & -1 \\ -2 & 2 \end{bmatrix}, \begin{bmatrix} 2 & -1 \\ -3 & 2 \end{bmatrix},$$

whereas the two singular ones are

$$\begin{bmatrix} 2 & -2 \\ -2 & 2 \end{bmatrix} \begin{bmatrix} 2 & -1 \\ -4 & 2 \end{bmatrix}, \tag{4}$$

the first being the SU(2) extended Cartan matrix.

Let us generalise equation (1) by supposing the ρ_α depends on two independent variables, t and x so

$$\partial^2 \rho_\alpha = \frac{\partial^2 \rho_\alpha}{\partial t^2} - \frac{\partial^2 \rho_\alpha}{\partial x^2} = \sum_\beta K_{\alpha\beta} e^{\rho_\beta} \tag{5}$$

Time independent solutions satisfy equations (III-12) and x independent ones equation (1). Equation (5) is the Toda field equation . To illustrate the singular case let K be the first matrix (4) so

$$\partial^2 \rho_1 = -2\, e^{\rho_1} + 2\, e^{\rho_2}$$
$$\partial^2 \rho_2 = 2\, e^{\rho_1} - 2\, e^{\rho_2}$$

Evidently $\partial^2 (\rho_1+\rho_2) = 0$. Let us consider $\rho_1 = -\rho_2 = \rho$ say. Then

$$\partial^2 \rho = -2e^{\rho} + 2\, e^{-\rho} = -4 \sinh \rho$$

This is the sinh-Gordon equation and is known to have many interesting properties. By a similar calculation the second matrix (4) yields

$$\partial^2 \rho = 2e^{\rho} - e^{-2\rho}$$

which is the Dodd-Bullough equation[65], also known to have interesting properties.

LECTURE V : Toda molecule field theory solutions and monopoles.

Leznov and Saveliev[38] discovered that it is easier to solve the Toda molecule field theory equation (eq. 5 in Lecture IV) with two independent variables rather than the single variable equations (IV-1) or (III-12). In light cone coordinates

$$U = \frac{r+t}{2} \qquad V = \frac{r-t}{2} \tag{1}$$

equation (IV-5) reads

$$\partial^2 \rho_\alpha / \partial U \partial V = \sum_{\beta=1}^{r} K_{\alpha\beta} \exp \rho_\beta \quad . \tag{2}$$

Now introduce an auxiliary gauge potential a_μ (in two dimensions, and not to be confused with the original four potential) for the gauge group G, with light cone components

$$\begin{aligned} a_U &= \sum_{\alpha=1}^{r} u_\alpha \mathbf{H}_\alpha + f_\alpha E_\alpha \\ a_V &= \sum_{\alpha=1}^{r} v_\alpha H_\alpha + g_\alpha E_{-\alpha} \quad . \end{aligned} \tag{3}$$

Then the requirement of zero curvature

$$\left[\partial_\mu + a_\mu \, , \, \partial_\nu + a_\nu \right] = 0 \tag{4}$$

is equivalent to the variable

$$\rho_\alpha = \ln f_\alpha g_\alpha \tag{5}$$

satisfying equation (2) above. This is a slight generalisation of the work at the end of Lecture III. The simplification compared to that is that the light cone components a_U, a_V in (3) are composed of only two grades (0,1) and (0,-1) respectively.

Thus we have the remarkable results that there is a correspondence between solutions of the Toda molecule field theory equation (2) and gauge potentials which simultaneously possess the grading structure in equation (3) and which, by virtue of condition (4) can be expressed as a pure gauge.

$$a_\mu = g^{-1}\partial_\mu g \quad . \tag{6}$$

Notice that the structure given by equation (3) is unaffected by a grade 0 gauge transformation. Even though f_α and g_α individually transform ρ_α(equation (5)) is invariant.

Let us evaluate the consequences for g in equation (6) following from the grade structure (3) using the fact that it can be written as a Dyson ordered exponential :-

$$g\ (U,V) = T \exp \int_0^{(U,V)} a_\mu \, dx^\mu \tag{7}$$

with factors ordered according to their position along the integration contour, the upper limit being on the right. Any path of integration between the origin and (U,V) yields the same g (because of (4)), but, in view of the grading structure (3) it would be best to work along characteristic surfaces of the differential equation (2), namely surfaces on which either U or V are constant. The two simplest routes of this kind are labelled (i) and (ii) in the diagram:

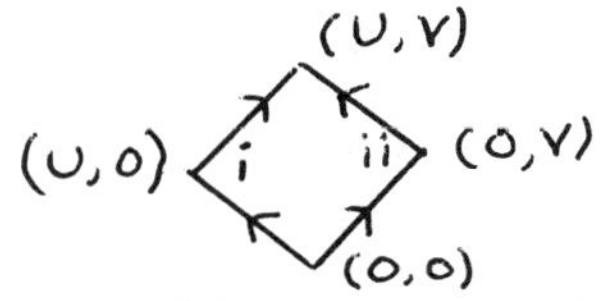

According to route (i) and equation (7)

$$g(U,V) = T \exp\int_0^{(U,0)} a_U \, dU \quad T \quad \exp \int_{(U,0)}^{(U,V)} a_V dV.$$

Because of the grading property (3) of a_U and a_V this can be written

$$g(U,V) \quad = \quad M^+(U,0)\ N^-(U,V)\ e^{\Sigma H_\alpha t_\alpha(U,V)} \tag{8}$$

where M^+ and N^- are elements of the groups G^+ and G^- obtained by exponentiating the step operators for positive and negative roots respectively. Notice that $M^+(U,0)$ depends only on U and not V. Similarly route (ii) implies

$$g(U,V) = M^-(0,V)N^+(U,V)e^{\Sigma H_\alpha s_\alpha(U,V)} \tag{9}$$

with M^- and N^+ in G^- and G^+ respectively and $M^-(0,V)$ depending only on V and not U. Equating the two expressions (8) and (9) for g yields :

$$M^+(U,0)^{-1}\, M^-(0,V) = N^-(U,V)\, e^{\Sigma H_\alpha(t_\alpha - s_\alpha)}\left[N^+(U,V)\right]^{-1} \tag{10}$$

Now the right hand side can be thought of as a "normal ordered" version of the left hand side with positive grade generators moved to the right. The three factors involving grades negative, grades 0 and grades positive should therefore be uniquely determined by the left hand side. This means that given $M^+(U,0)$ and $M^-(0,V)$, $g(U,V)$ is determined up to multiplication on the right by an exponential of grade 0 generators. This is precisely the grade 0 gauge transformation mentioned above which preserves the grade structure (3) and leaves ρ_α(equation (5)) invariant. It follows that ρ_α is uniquely determined by $M^+(U,0)$ and $M^-(0,V)$ which can therefore be regarded as parameterising the solution of the Toda molecule field equation.

Let us explicitly solve equation (10) for $t_\alpha - s_\alpha$ in terms of the left hand side. This will be useful in obtaining the expression for ρ . Let $|\lambda_\beta\rangle$denote the normalised state of weight $\lambda_\beta$in the irreducible representation of G defined by having λ_β as its highest weight. Then

$$\begin{aligned} E_\alpha\,|\lambda_\beta\rangle &= 0 \quad , \quad \alpha > 0 \\ H_\alpha\,|\lambda_\beta\rangle &= \delta_{\alpha\beta}|\lambda_\beta\rangle \end{aligned} \tag{11}$$

using the properties (III-7) of fundamental weights. Then taking the expectation value of equation (10) with respect to this state yields

$$e^{t_\beta - s_\beta} = \langle\lambda_\beta|M^+(U,0)^{-1}\; M^-(0,V)|\lambda_\beta\rangle \tag{12}$$

We still have to impose the conditions on M^+ and M^- that a_U (a_V) has positive (negative) grades +1 (-1) only. By equations (3) and (8) the positive grade part of a_U is

$$\Sigma\, f_\alpha E_\alpha = e^{-\Sigma H_\alpha t_\alpha}(M^+)^{-1}\; \frac{\partial M^+}{\partial U}\, e^{\Sigma H_\alpha t_\alpha}$$

Hence the logarithmic derivative of $M^+(U,0)$ has grade 1 only and therefore can be written

$$M^+(U,0)^{-1}\; \frac{\partial}{\partial U}\; M^+(U,0) = \Sigma\, m_\alpha(U) E_\alpha \tag{13}$$

Further, using the commutators (II-8), and equating coefficients of E_α

$$f_\alpha(U,V) = e^{-\Sigma K_{\alpha\beta} t_\beta(U,V)}\; m_\alpha(U) \tag{14}$$

Similarly we have

$$M^-(0,V)^{-1}\; \frac{\partial}{\partial V}\, M^-(0,V) = \Sigma\; \bar{m}_\alpha(V)\; E_{-\alpha} \tag{15}$$

and

$$g_\alpha(U,V) = e^{\Sigma K_{\alpha\beta} s_\beta(U,V)}\; \bar{m}_\alpha(V) \tag{16}$$

So by equations (5), (14) and (16)

$$e^{\rho_\alpha} = f_\alpha g_\alpha = e^{\Sigma K_{\alpha\beta}(s_\beta - t_\beta)}\; m_\alpha(U)\; \bar{m}_\alpha(V)$$

$$= m_\alpha(U)\; \bar{m}_\alpha\; (V)\; \prod_\beta \left[\langle\lambda_\beta|M^+(U,0)^{-1}\; M^-(0,V)|\lambda_\beta\rangle\right]^{-K_{\alpha\beta}}$$

This is the general solution[38] when one remembers that M^+ and M^- are easily found from $m_\alpha(U)$ and $\bar{m}_\alpha(V)$ by equations (13) and (15). It can be simplified by introducing a new variable θ_α by :

$$\rho_\alpha = K_{\alpha\beta}\,\theta_\beta$$

This satisfies

$$\partial^2\,\theta_\alpha = -\exp \Sigma K_{\alpha\beta}\theta_\beta$$

and can be expressed as

$$e^{-\theta_\alpha(U,V)} = \prod_\beta \left[m_\beta(U)\bar{m}_\beta(V)\right]^{-(K^{-1})_{\alpha\beta}} \langle\lambda_\alpha|M^+(U,0)^{-1}M^-(0,V)|\lambda_\alpha\rangle \tag{17}$$

If G = SU(2) so that equation (1) reduces to the Liouville equation we can write $M^+(U,0) = 1+\mu(U)\,\sigma^+$, $M^-(0,V) = 1+\bar{\mu}(V)\sigma^-$. Then solution (17) reduces to the familiar form

$$\theta(U,V) = \tfrac{1}{2}\;\ell n\left[\mu(U)'\;\bar{\mu}(V)'/(1-\mu(U)\bar{\mu}(V))^2\right] \tag{18}$$

In order to solve the monopole equations (III-12) we need to choose functions $m_\alpha(U)$, $\bar{m}_\alpha(V)$ in such a way that the time dependence cancels out leading to ρ_α or θ_α depending only on radius r. For the G = SU(2) solution (18) the choice $\mu(U) = \exp(c+aU)$ $\bar{\mu}(V) = \exp(c+aV)$ yields, by equation (1), $\mu(U)\bar{\mu}(V) = \exp(2c+ar)$ so that t has indeed cancelled. In fact

$$\exp\;-\theta_\alpha(u,v) = \left[\exp(-(c+ar/2))-\exp\,(c+ar/2)\right]/a \tag{19}$$

The same trick works in general[41]: choose

$$m_\alpha(U) = \exp -\;\alpha.(qU+c)$$

$$\bar{m}_\alpha(V) = \exp -\;\alpha.(qV+c)$$

where q and c constitute the 2 rank(G) constants of integration, to be fixed by boundary conditions. After some simplification it is found that θ_α depends only on r and not t (see ref.[41] for details):

$$\exp(-\theta_\alpha) = \langle\lambda_\alpha|X^X(0)\;\exp\left[(2c+rq).H\right]X(o)|\lambda_\alpha\rangle \tag{20}$$

where

$$X^{-1}(z)\ dX(z)/dz = \sum_{1}^{r} \exp(-q.\alpha)\ E_{-\alpha} \tag{21}$$

and $X(\infty) = 1$

providing $q.\alpha$ is strictly positive for each simple root α. See equation (II-14) for the definition of X^{X}. Inserting a complete set of orthonormal states $|\mu,r\rangle$ for the irreducible representation with highest weight λ_α yields

$$\exp(-\theta_\alpha) = \sum_{\mu,r} (-1)^{\ell(\lambda_\alpha-\mu)} e^{\mu.(qr+2c)} |\langle\mu,r|\ X(0)|\lambda_\alpha\rangle|^2 \tag{22}$$

where μ labels the weight of the state and r its degeneracy. Thus $\exp(-\theta_\alpha)$ is a sum of exponentials of radius.

At large radius the Higgs field must approach a prescribed limiting value $\phi(\infty)$ which determines the exact gauge symmetry as mentioned in Lecture I. At small radius the monopole must be regular so that the energy integral converges. These two boundary conditions are satisfied if [41]

$$q = -2\eta\ e\phi(\infty) \tag{23}$$

$$e^{2c.\mu} = \prod_{\alpha>0} (q.\alpha)^{-2\mu.\alpha/\alpha^2} \tag{24}$$

This fixes the constant of integration. The derivation of these results exhibits much interesting mathematical structure but is too complicated to give here.

The remaining problem, not yet completely solved, is to evaluate $|\langle\mu|X(0)|\lambda_\alpha\rangle|^2$, a purely Lie algebraic combinatorial problem.

It has been shown that [42]

$$|\langle\mu,r|X(0)|\lambda_\alpha\rangle|^2 = Q_\mu(q)/\prod_{\alpha>0} (\alpha.q)^{2p(\mu,\alpha)} \tag{25}$$

where $Q_\mu(q)$ is a polynomial in q and $p(\mu,\alpha)$ is defined in terms of the root string through μ, namely

$$\mu-m\alpha \;, \;\ldots\ldots\; \mu-\alpha, \mu, \mu+\alpha \;,\ldots\ldots\; \mu+p\alpha$$

It is easy to show that[54]

$$m-p = 2\mu.\alpha/\alpha^2 \tag{26}$$

Hence putting together (22), (23), (24), (25) and (26)

$$\exp(-\theta_\alpha) = \sum_{\mu,r} (-1)^{\ell(\lambda_\alpha-\mu)} e^{r\mu.q} \; Q_\mu(q) / \prod_{\alpha>0} (\alpha q)^{m+p}$$

It seems that if μ is a non degenerate weight, then the polynomial $Q_\mu(q)$ has degree zero, in other words, is constant and further takes the same value for all points μ on a given Weyl orbit. Now if λ_α is "minimal" the weights μ all lie on a single Weyl orbit[27], which must, of course, contain λ_α itself for which $Q_{\lambda_\alpha}(q) = 1$. Then we can write

$$\exp(-\theta_\alpha) = \sum_\mu e^{r\mu.q} / \prod_\alpha (\alpha.q)^{m(\mu,\alpha)} \tag{27}$$

where the product is now over all roots α (whether positive or negative) and the minus signs have been absorbed.

This generalizes and clarifies the result of Wilkinson and Bais[49] who studied G=SU(N), all of whose fundamental weights are minimal[27]. In particular when G=SU(2) there are two roots ± 1 and the fundamental (minimal) representation is the spinor one with two weights $\pm\frac{1}{2}$. So expression (27) reduces to

$$\exp(-\theta) = (2/q) \sinh (rq/2)$$

which yields the Prasad Sommerfield solution (II-1) when the substitutions relating θ_α to the fields are traced back.

It is not difficult to show that expression (27) has an extra symmetry with respect to

$$q \to \sigma(q) \qquad \sigma \varepsilon W(G)$$

where W(G) is the Weyl group of G. This symmetry is a consequence of the special choice of integration constants (24), ensuring regularity at the origin. (For G=SU(2) this is $q \to \pm q$)

Equation (27) somewhat resembles the Weyl character formula (53,54) which is also W(G) invariant and expressed as a quotient. In fact one can show(42) that

$$\exp(-\theta_\alpha) = \lim_{n \to \infty} \left[r/(d+n) \right]^{4\delta^V . \lambda^\alpha} X^{n\lambda_\alpha}(rq/(n+d)) \qquad (28)$$

where X^λ is the character of the irreducible representation with highest weight λ, and d a specific constant. Formula (28) makes manifest the power behaviour in r for r small needed for regularity at the origin(41), as well as the Weyl symmetry.

We mentioned earlier that $\alpha.q$ and hence by equation (23), $-\eta\phi(\infty).\alpha$,had to be strictly positive. This would be unfortunate in that it would force the exact symmetry group H to be a product of rank(G) U(1) factors. Physically we wish H to be U(1)xK with K semisimple, as we mentioned in Lecture I. Now it is possible to achieve this situation by a limiting procedure. The problem is that the denominators in equation (25) vanish and one has to establish compensating zeros in the numerator. This corresponds to the property of the Weyl character formula that at zero argument it must yield the dimensionality of the representation. Equation (28) shows that the desired limit exists for $\exp(-\theta_\alpha)$, since it does for the character.

This brings me to the limit of my present knowledge. Clearly much remains to be done, but I have tried to convey that physical problems can involve interesting and relatively modern mathematics.

I am grateful for discussions with P. Goddard and N. Ganoulis.

REFERENCES

1. C.N. Yang and R.L. Mills, Phys. Rev. 96, 191 (1954),
R. Shaw: Ph.D thesis, Cambridge University (1955).
M. Gell-Mann and S. Glashow: Ann. Phys (N.Y) 15, 437 (1961).

2. P.W. Higgs: Phys. Rev. Lett. 12, 132 (1964); Phys. Rev. Lett. 13, 508 (1964); Phys. Rev. 145, 1156 (1966)
T.W.B. Kibble: Phys. Rev. 155, 1557 (1967)
F. Englert and R. Brout: Phys. Rev. Lett. 13, 321 (1964)

3. D. Olive: in "Unification of the fundamental particle interactions", edited by S. Ferrara, J. Ellis and P. van Nieuwenhuizen, Plenum, New York, 1980, p.451 and Imperial College preprint ICTP/81/82-6.

4. H.A. Kramers and G.H. Wannier, Phys. Rev. 60, (1941) 252.
G.F. Newell and E.W. Montroll, Rev.Mod.Phys. 25 (1953) 353.
F.S. Wegener, J. Math. Phys. 12 (1971) 2259.
R. Balian, J.M. Drouffe and C. Itzykson, Phys. Rev. D11 (1975) 2098.

5. T.H.R. Skyrme: Proc. Roy. Soc. A262, 237 (1961).
R. Streater and I.F. Wilde: Nucl. Phys. B24, 561 (1970).
S. Coleman: Phys. Rev. D11, 2088 (1975).
S. Mandelstam: Phys. Rev. D11, 3026 (1975).

6. S. Mandelstam, Phys. Rep. 23C, 245, (1976).
S. Mandelstam, Lecture given at Les Houches Winter Meeting February 1980.
G. 'tHooft, Nuclear Physics B138, 1 (1978).
Nuclear Physics B153, 141 (1979).

7. S. Ferrara, J. Scherk and B. Zumino: Nucl. Phys - B121, 393 (1977).
E. Cremmer, J. Scherk and S. Ferrara; Phys. Lett. 74B,61 (1978)

8. M.F. Atiyah: Lectures at this meeting.

9. P. Goddard and D. Olive, Rep. Prog. Phys. 41, 1357 (1978).

10. S. Coleman "New Phenomena in Subnuclear Physics", (Proc. 1975 Int. School of Physics "Ettore Majorana"). ed. A. Zichichi (New York: Plenum) p297.

11. E. Corrigan, D. Olive, D.B. Fairlie and J. Nuyts: Nucl. Phys. B106, 475 (1976).

12. E. Corrigan and D. Olive: Nucl. Phys. B110, 237 (1976).

13. D. Olive: Nucl. Phys. B153, 1 (1979).

14. M.I. Monastyrsky and A.M. Perelomov: JETP Letts. 21, 43 (1975).
Yu. S. Tyupkin, V.A. Fateev and A.S. Shvarts: JETP Letts. 21, 42 (1975).
T.T. Wu and C.N. Yang: Phys. Rev: D12, 3845 (1975); Nuclear Physics B107, 365 (1976).

15. G. 't Hooft: Nucl. Phys. B79, 276 (1974).
A.M. Polyakov: JETP Lett. 20, 194 (1974).

16. E.B. Bogomolny: Sov. J. Nucl. Phys. 24, 449 (1976).

17. S. Coleman, S. Parke, A. Neveu and C.M. Sommerfield: Phys. Rev. D15, 554 (1977).

18. B. Julia and A. Zee: Phys. Rev. D11, 2227 (1975).

19. C. Montonen and D. Olive: Phys. Lett. 72B, 117 (1977).

20. O. Heaviside: Collected Works.
E. Schrödinger: Proc. Roy. Soc. A150, 465 (1935)

21. P.A.M. Dirac: Proc. Roy. Soc. A133, 60 (1931).

22. P. Goddard and D. Olive: Nucl. Phys. B191, 511 (1981).

23. T. Kaluza: Sitzungber, Preuss. Akad.Wiss.Berlin, Math.Phys. KA966 (1921).
O. Klein: Z. Physik: 37, 895 (1926).

24. A. d'Adda, R. Horsley and P. Di Vecchia:
Phys. Lett. 76B, 298 (1978).
E. Witten and D. Olive: Phys. Lett. 78B, 97 (1978).

25. P. Di Vecchia and S. Ferrara: Nucl. Physics B130, 93 (1977).
J. Hrubý: Nucl. Physics B131, 275 (1977).
A. D'Adda and P. Di Vecchia: Physics Lett. 73B, 162 (1978)
J. Shonfeld: Nucl. Physics B161, 125 (1979).
S. Rouhani: Nucl. Physics B182, 462 (1981).

26. H. Osborn: 83B, 321 (1979).

27. P. Goddard and D. Olive: Nucl. Physics B191, 528 (1981).

28. F.A. Bais: Phys. Rev. D18, 1206 (1978).

29. F. Englert and P. Windey: Phys. Rev. D14, 2728 (1976).

30. P. Goddard, J. Nuyts and D. Olive: Nucl. Physics B125, 1 (1977)

31. E. Weinberg: private communication.

32. S. Ferrara and B. Zumino: Nucl. Phys. B79, 413 (1974).
D.R.T. Jones: Phys. Lett. 72B, 199 (1977).
H.N. Pendleton and E.C. Poggio: Phys. Lett. 72B, 200 (1977)
L.V. Avdeev, O.V. Tarasov and A.A. Vladimirov:
Phys. Lett. 96B, 94 (1980).
M.T. Grisaru, M. Roček and W. Siegel:
Phys. Rev. Lett. 45, 1063 (1980).
W.E. Caswell and D. Zanon: Phys. Lett. 100B, 152 (1980).
M. Sohnius and P. West. Phys. Lett. 100B, 245 (1981).

33. W. Nahm: Seminar at Imperial College, Spring 1979.

34. P. Rossi, Phys. Lett. 99B, 229 (1981).

35. F. Gliozzi, J. Scherk and D. Olive. Nucl. Phys. B122, 253 (1977).

36. S.M. Christensen, M.J. Duff, G.W. Gibbons and M. Roček
Phys. Rev. Lett. 45, 161 (1980).

37. M.K. Prasad and C.M. Sommerfeld: Phys. Rev. Lett. 35, 760 (1975).

38. A.N. Leznov and M.V. Saveliev: Lett in Math.Physics 3,489 (1979).

39. A.M. Leznov and M.V. Saveliev: Comm. Math. Phys. 74, 111 (1980).

40. D. Olive: Imperial College preprint ICTP/80/81-1, to be published in Proceedings of International Summer Institute on Theoretical Physics organized by Wuppertal University at Bad Honnef - Plenum Press. New York and London.

41. N. Ganoulis, P. Goddard and D. Olive: "Self dual monopoles and Toda Molecules". Imperial College preprint ICTP/81/82-4.

42. N. Ganoulis, P. Goddard and D. Olive: Work in progress.

43. R. Ward: Comm. Math. Phys. 79, 317 (1981).
Phys. Lett. 102B, 136 (1981).

44. P. Forgács, Z. Horváth and L. Palla. Phys.Rev.Lett.45,505 (1980)
Phys. Lett. 99B, 232 (1981).
Phys. Lett. 102B, 131 (1981).
Nucl. Phys. 192B, 141 (1981).
Phys. Rev. D23, 1876 (1981).

See also -
M.K. Prasad and P. Rossi: Phys. Rev.Lett. 46, 806 (1981)
M.K. Prasad, A.Sinha and Ling-Lie Chau Wang.
Phys. Rev. D23, 2321 (1981).

45. E. Corrigan and P. Goddard: Comm.Math.Phys.80, 575 (1981).

46. W. Nahm: CERN preprint: TH3172 (CERN) 1981.
N. Hitchin: "Monopoles and Geodesics".
Oxford preprint (1981).

47. N. Manton: Nucl. Phys. B126, 525 (1977).
W. Nahm: Phys. Lett. 79B, 426 (1978).

48. D. Wilkinson and A.S. Goldhaber, Phys.Rev. D16, 1221 (1977).

49. F.A. Bais and H.A. Weldon, Phys. Rev. Lett. 41, 601 (1978).
D. Wilkinson and F.A. Bais. Phys. Rev. D19, 2410 (1979).

50. F.A. Bais, Talk presented at International Workshop on High Energy Physics and Field Theory at Serpukhov 1979.
Leuven preprint: KUL-TF-79/021.

51. P.D. Lax, Comm. Pure. Appl. Math. 21, 467 (1968).

52. E.B. Dynkin, Amer.Math.Soc. Transl.Series. 2, 6, 111 (1965).

53. G. Racah: "Lectures on Lie groups" CERN yellow report 61-8 and in "Group Theoretical concepts and methods in Elementary Particle Physics" p1-36, Gordon and Breach New York (1964).

54. J.E. Humphreys, Introduction to Lie algebras and representation theory (Springer, Berlin 1972).

55. M. Toda: Progress of theoretical Physics. (Suppl) 45, 174 (1970); Physics Reports, 18, 1 (1975); "Theory of Non-Linear Lattices", Springer series in Solid State Physics, 20 (Berlin) (1981).

56. E. Fermi, J. Pasta and S. Ulam, Los Alamos Report LA-1940 (1955); Collected papers of Enrico Fermi Vol: II (University of Chicago Press, 1965) p.978.

57. J. Moser: Lecture Notes in Physics 38, Springer p.467, Adv. Math. 16, 197 (1975).

58. H. Flaschka: Phys. Rev. B9, 1924 (1974).

59. M. Henon: Phys. Rev. B9, 1921 (1974).

60. O.I. Bogoyavlensky: Comm. Math. Phys. 51, 201 (1976).

61. B. Kostant, Advances in Mathematics, 34, 195 (1979).

62. V.G. Kac: Math. USSR Isvestija V2, 1271 (1968).
R. Moody: J. Algebra 10, 211 (1968).

63. S.L. Adler and R.F. Dashen: "Current Algebras"
W.A. Benjamin, New York 1968.
H. Sugarawa: Phys. Rev. 170, 1659 (1968).
C.M. Sommerfield: Phys. Rev. 176, 2019 (1968).

64. G. Wilson: "The modified Lax and two dimensional Toda lattice equations associated with simple Lie algebras" to appear in Ergodic Theory and dynamical systems.

65. R.K. Dodd and R.K. Bullough: Proc. Roy. Soc. A352, 481 (1977).

SUPERSYMMETRY, GAUGE SYMMETRY AND UNIFIED FIELD THEORIES

S. Ferrara
CERN, Geneva, Switzerland

January 1982

1. INTRODUCTION

The phenomenological success of the Glashow-Weinberg-Salam [1] model in describing electromagnetic and weak interactions in a unified fashion and in embedding the Fermi theory of weak interactions in a renormalizable field theory has dramatically reproposed unified field theories as the correct theoretical framework for describing all elementary particle interactions.

Fermions and bosons, the building blocks of matter, are the essential ingredients of any unified field theory based on renormalizable gauge interactions. Matter fields are usually described by spin $\frac{1}{2}$ fermion fields while fields related to the geometry of the gauge group, such as the gauge vector bosons (connections) and Higgs fields, are described by spin 1 and spin 0 bosons.

In gravity, where the gauge principle is enlarged to the space-time symmetry, an additional bosonic tensor field, the spin 2 graviton, is also needed. In relativistic quantum mechanics the connection between spin and statistics and the underlying geometry of a gauge Lie group lead unavoidably to a dichotomy between matter and geometrical fields. More specifically, in the context of a Lie algebra symmetry it is impossible to achieve a complete unification of particle interactions due to the different spin (and statistics) of the elementary quanta which describe the fundamental forces of Nature.

Supersymmetry [2] is a space-time symmetry which embeds bosons and fermions in irreducible multiplets. This symmetry, although not manifest at present energies, has some appealing properties when applied to local quantum field theories and it may achieve the ultimate goal of a complete unification of particle interactions including gravitation [3]. Supersymmetry is in fact at present the only known symmetry consistent with local quantum field theory which relates particles of different spin [4]. This property allows previous theorems on the impossibility of non-trivial mixing between space-time and intermal symmetries [5] to be overcome. Most surprisingly, supersymmetric field theories are the least divergent quantum field theories so far constructed. Supersymmetries are usually labelled by the number of fermionic invariances, i.e., by the number of spinorial generators Q^i_α. When i = 1,...,N we call N-extended supersymmetry the corresponding supersymmetry algebra. For reasons which will become transparent in the next section renormalizable local field theories admit up to N = 4 supersymmetries.

In N = 1 supersymmetry two kinds of multiplets and two kinds of interactions are possible. Chiral multiplets containing a left-handed (or Majorana) spin $\frac{1}{2}$ fermion and a complex scalar can interact through a super-Yukawa interaction (to be described later). Vector multiplets containing a spin 1 vector particle and a left-handed (or Majorana) fermion are associated with ordinary gauge interactions, such as QED on Yang-Mills interactions, and can interact among themselves and with the chiral multiplets if the latter transform non-trivially under the gauge group.

In N = 2 supersymmetry [6] two kinds of multiplets also exist, but only gauge interactions are possible. Chiral multiplets contain in this case a Dirac spinor and two complex scalars, while vector multiplets contain a vector particle, a Dirac spinor and a complex scalar.

In N = 3 and 4 supersymmetry [7] only one multiplet, the vector multiplet, exists with the only possible Yang-Mills gauge self-interaction. This multiplet describes a vector particle, four left-handed (or Majorana) spin $\frac{1}{2}$ fermions and six real scalars (or three complex ones).

Concerning the ultra-violet properties of these theories, particularly remarkable is the N = 4 Yang-Mills theory where the $\beta(g)$ function is known to vanish up to three loops by explicit calculations, and arguments have been given that it may in fact vanish to all orders of perturbation theory [8]. The softening of quantum divergences in supersymmetric Yang-Mills theories with respect to the ordinary renormalizable theories has recently proposed [9] these theories as candidates for the solution of the so-called hierarchy problem [10] that one encounters in grand unified theories of electroweak and strong interactions and even in the standard model of electroweak interactions [11]. In Section 4 we will describe attempts to use supersymmetric theories in the framework of GUTs and of electroweak interactions.

Some phenomenological problems which arise in this approach will also be considered, This survey is organized as follows:

In Section 2 we will describe the supersymmetry algebra and its representations on one-particle states. In Section 3 we will derive, using superspace, the most general $N = 1$ supersymmetry renormalizable Lagrangian and we will describe its component form. In Section 4 spontaneous supersymmetry breaking and attempts to build supersymmetric models of electroweak and strong interactions will be described.

2. SUPERSYMMETRY ALGEBRAS AND PARTICLE MULTIPLETS

Supersymmetry is a graded Lie algebra (GLA) whose multiplication rules contain both commutators and anticommutators. In general, for GLA's one has

$$[X_A, X_B\} = f_{AB}^{C} X_C \quad \text{with} \quad [X_A, X_B\} = X_A X_B - (-)^{AB} X_B X_A \tag{2.1}$$

where A,B = 0 if X is an even generator and A,B = 1 if X is an odd generator. The Jacobi identities are $[[X_A, X_B\}, X_C\} = 0$ and f_{AB}^{C} are the structure constants of the graded Lie algebra. The even part A of the X generators is an ordinary N-dimensional Lie algebra while the odd part S is the so-called grading representation of A. The GLA with generators X = (A,S) has dimension n + N where N is the dimension of the representation of A under which the S transforms.

In the case of supersymmetry the Lie algebra part has the structure $T \otimes G$ where T is the space-time symmetry and G is an internal symmetry group. The odd part S consists of a set of N self-conjugate spin $\frac{1}{2}$ generators Q_α^i (i=1,...,N) ($CQ_\alpha^T = Q_\alpha$) which provide the grading of A.

The most interesting possibility most used in particle physics is the grading of the Poincaré algebra [4,12]. In this case the most general grading is given by

$$\{Q_\alpha^i, Q_\beta^j\} = -\frac{1}{2}(\gamma^\mu C)_{\alpha\beta}\delta^{ij}P_\mu + Z^{ij}C_{\alpha\beta} + \tilde{Z}^{ij}(\gamma_5 C)_{\alpha\beta}$$

$$[Q_\alpha^i, M_{\mu\nu}] = (\sigma_{\mu\nu})_\alpha^\beta \, Q_\beta^i \tag{2.2}$$

$$[Q_\alpha^i, P_\mu] = 0 \quad [Q_\alpha^i, Z^{\ell m}] = [Q_\alpha^i, \tilde{Z}^{\ell m}] = [Z^{\ell m}, Z^{hk}] = 0$$

The antisymmetric Poincaré invariant operators $Z^{ij}, \tilde{Z}^{ij}$ belong to the centre of the GLA. Obviously, they can only appear for $N \geq 2$. It is often convenient to rewrite (2.2) in terms of the chiral projections

$$Q^i_{\alpha,L} = \frac{1}{2}(1+\gamma_5)^{\beta}_{\alpha}Q^i_{\beta} \qquad Q^i_{\alpha,R} = \frac{1}{2}(1-\gamma_5)^{\beta}_{\alpha}Q^i_{\beta} \tag{2.3}$$

In terms of chiral spinors Eq. (2.2) becomes

$$\begin{aligned}
\{Q^i_{\alpha,L},Q^j_{\beta,L}\} &= \frac{1}{2}[(1+\gamma_5)C]_{\alpha\beta}(Z^{ij}+\tilde{Z}^{ij}) \\
\{Q^i_{\alpha,R},Q^j_{\beta,R}\} &= \frac{1}{2}[(1-\gamma_5)C]_{\alpha\beta}(Z^{ij}-\tilde{Z}^{ij}) \\
\{Q^i_{\alpha,L},Q^j_{\beta,R}\} &= -\frac{1}{2}[(1+\gamma_5)\gamma^{\mu}C]_{\alpha\beta}P_{\mu}\delta^{ij}
\end{aligned} \tag{2.4}$$

It is also useful to use Van der Vaerden notation for chiral (Weyl) spinors [13]. One writes a four-component Majorana spinor as $Q = (Q_{\alpha},\bar{Q}^{\dot{\alpha}})$ where $\bar{Q}^{\dot{\alpha}} = \varepsilon^{\dot{\alpha}\dot{\beta}}(Q_{\beta})^*$ and $\varepsilon^{\alpha\beta} = \varepsilon^{\dot{\alpha}\dot{\beta}} = -\varepsilon_{\alpha\beta} = -\varepsilon_{\dot{\alpha}\dot{\beta}} = -\varepsilon^{\beta\alpha} = -\varepsilon^{\dot{\beta}\dot{\alpha}}$ $(\alpha,\beta = 1,2 \ \dot{\alpha},\dot{\beta} = 1,2)$. Then Eqs. (2.2) simplify as follows

$$\begin{aligned}
\{Q^i_{\alpha},\bar{Q}^j_{\dot{\beta}}\} &= \frac{1}{2}\sigma^{\mu}{}_{\alpha\dot{\beta}}P_{\mu} \\
\{Q^i_{\alpha},Q^j_{\beta}\} &= i\varepsilon_{\alpha\beta}(Z^{ij}+\tilde{Z}^{ij})
\end{aligned} \tag{2.5}$$

Once the algebra (2.4) [or (2.5)] is given, one can study its unitary irreducible representations using an extension of the Wigner method of induced representations [14]. In particular, one can study the general structure of the unitary (infinite dimensional) irreducible representations of the extended supersymmetry algebras acting on one particle states. We will consider first the algebra (3.4) in the zero central charge sector ($Z^{ij} = \tilde{Z}^{ij} = 0$).

For massive representations we can choose a Lorentz frame such that $P^{\mu} = (M,0)$ where M is the common mass of the multiplet components. Then (2.2) reduces to

$$\{Q^i_{\alpha},Q^j_{\beta}\} = \delta_{\alpha\beta}\delta^{ij} \qquad \alpha,\beta = 1,\dots,4 \ ; \ i,j = 1,\dots,N \tag{2.6}$$

The even part is given by the angular momentum operator

$$J_k = \varepsilon_{k\ell m}M^{\ell m} \qquad (\ell,m = 1,2,3) \tag{2.7}$$

The anticommutators (2.6) define the Clifford algebra for the group SO(4N). Its unique irreducible representation has dimension 2^{2N}. This is the spinor representation of SO(4N). Under SO(4N) this representation splits into two irreducible representations of dimension 2^{2N-1}. We can use two-component Weyl spinors as given in (2.5) and rewrite (2.6) as follows:

$$\{Q^i_\alpha, \bar{Q}^j_{\dot\beta}\} = \delta_{\alpha\dot\beta}\delta^{ij}$$
$$\{Q^i_\alpha, Q^j_\beta\} = \{\bar{Q}^i_\alpha, \bar{Q}^j_\beta\} = 0 \tag{2.8}$$

The Q^i_α and $\bar{Q}^i_{\dot\alpha}$ operators satisfy the algebra of 2N fermionic creation and destruction operators. If we start from a state Ω such that $Q^i\Omega = 0$ (Clifford vacuum) then the previously alluded to 2^{2N} states are obtained as follows:

$$\Omega,\ \bar{Q}^i_{\dot\alpha}\Omega,\ \bar{Q}^i_{\dot\alpha_1}\bar{Q}^{i_2}_{\dot\alpha_{i_2}}\Omega, \ldots, \bar{Q}^i_{\dot\alpha_1}, \ldots, Q^{i_k}_{\dot\alpha_k}\Omega, \ldots \tag{2.9}$$

If we further define the 2N-component spinors

$$Q^a_\alpha = Q^i_\alpha \quad a = 1,\ldots,N, \quad Q^a_\alpha = \bar{Q}^{\dot\alpha i} = \varepsilon^{\dot\alpha\dot\beta}\bar{Q}^i_{\dot\beta} \quad (a = N+1,\ldots,2N) \tag{2.10}$$

(2.8) becomes

$$\{Q^a_\alpha, Q^b_\beta\} = \varepsilon_{\alpha\beta}\Omega^{ab} \quad \alpha,\beta = 1,2 \quad a,b = 1,\ldots,2N \tag{2.11}$$

where

$$\Omega^{ab} = -\Omega^{ba} \begin{pmatrix} 0 & I \\ -I & 0 \end{pmatrix}, \quad (Q^a_\alpha)^* = \Omega_{ab}\varepsilon^{\alpha\beta}Q^b_\beta$$

The algebra (2.11) has a manifest covariance under SU(2) × USp(2N) ⊂ SO(4N). The 2^{2N} states of the spinor representation of SO(4N) decompose under SU(2) × USp(2N) as follows

$$2^{2N} = (N+1,1) + (N,2N) + \ldots (N-k+1,[2N\times\ldots 2N]_k) + \ldots$$
$$\ldots (1,[2N\times\ldots 2N]_N) \tag{2.12}$$

The two irreducible representations of SO(4N) of dimension 2^{2n-1} correspond to integral and half-integral SU(2) spin, respectively. Then we have proven that the lowest dimensional massive irreducible representation of N extended supersymmetry contains spin states from $j = 0$ up to $J = N/2$. Higher dimensional irreducible representations are obtained by relaxing the condition that Ω is an over-all singlet. In particular, Ω can carry any value of the spin and transform in some non-trivial (real) representation of a subgroup $G \leq U(N)$ where U(N) is the maximal group of automorphisms of the N extended supersymmetry algebra. Massive representations can be smaller in the presence of central charges. We give an example: in $N = 2n$ extended supersymmetry, massive (lowest dimensional) multiplets with a non-vanishing central charge have real dimension 2^{2N+1} instead of 2^{4N}. The spin runs from $J = 0$ up to $J = n/2$ (instead of $J = n$). These 2^{2n+1} states consist of a doublet of massive representations of n extended supersymmetry without central charges. The maximal group of automorphisms of the algebra is USp(2n) instead of U(2n).

In the application to quantum field theories the only interesting elementary massive multiplets are the lowest dimensional multiplet for $N = 1$ ($j = 0,1/2$) and the lowest dimensional multiplets with central charge for $N = 2$. According to the previous discussion this last multiplet consists of a pair of $J = 0,1/2$ chiral multiplets of $N = 1$ supersymmetry.

We now consider the more interesting case of massless representations. In this case we can choose a Lorentz frame $P^\mu = (P^0,0,0,P^0)$ for the light-like momentum $P^\mu(P^\mu P_\mu = 0,\ P^0 > 0)$. The stability subalgebra, written in terms of Weyl spinors, becomes

$$\{Q^i_\alpha,\bar{Q}^j_{\dot\beta}\} = \left(\frac{1+\sigma_3}{2}\right)_{\alpha\dot\beta}\delta^{ij} \quad , \quad \{Q^i_\alpha,Q^j_\beta\} = 0 \tag{2.13}$$

Equation (2.13) implies $Q^i_2 = \bar{Q}^i_2 = 0$. If we set $Q^i_1 = Q^i$ we see that (2.13) becomes the Clifford algebra for SO(2N). Its unique irreducible representation has dimension 2^N which again splits into two irreducible representations of dimension 2^{N-1} under SO(2N). We are interested in this case in the decomposition of SO(2N) under the subgroup $U(1)_{\text{helicity}} \times SU(N)$. We have

$$2^N = (\lambda,1) + (\lambda - \tfrac{1}{2},\bar{N}) + \ldots (\lambda - \tfrac{k}{2},[\bar{N}]_k) + \ldots (\lambda - \tfrac{N}{2},1) \tag{2.14}$$

where λ is the helicity of the Clifford vacuum defined by $Q^i\Omega = 0$. It is obvious that this irreducible representation does not have PCT-conjugate states because for the state of helicity λ there is no corresponding (antiparticle) state with helicity $-\lambda$.

This can only happen if $\lambda - N/2 = -\lambda$, i.e., $\lambda = N/4$. If $\lambda \neq N/4$ we must double the states and add to the multiplet with Clifford vacuum of helicity λ a PCT-conjugate multiplet with Clifford vacuum of helicity $N/2 - \lambda$. If the Clifford vacuum transforms under a non-trivial representation R of a subgroup $G \subset SU(N)$ then the PCT-conjugate Clifford vacuum must transform according to the complex conjugate representation $\bar{R}$. From the previous construction it follows that the minimum helicity range for a massless multiplet of N extended supersymmetry is from $\lambda = 0$ up to $\lambda = N/4$ ($N+1/4$ for odd N). Then if we confine ourselves to renormalizable interactions we find that $N \leq 4$ as claimed in the introduction. In the case of supergravity, when we allow for the spin 2 graviton to be the maximal helicity states, we have the bound $N \leq 8$.

3. SUPERSPACE, MULTIPLETS OF FIELDS AND INTERACTIONS

In the application of supersymmetry to weak and electromagnetic interactions it is important to establish the internal symmetry properties of left-handed fermions.

If we confine ourselves to renormalizable Yang-Mills interactions we have seen that only three supersymmetry algebras with $N = 1,2$ and 4 are possible (the $\lambda_{MAX} = 1$ representation of $N = 3$ and $N = 4$ actually coincide). Furthermore, if we want left-handed fermions in complex representations of the gauge group G which commutes with the supersymmetry generators, then only one possibility is left, namely the $N = 1$ supersymmetry algebra. This is because for $N = 4$ fermions are partners of vector bosons and must therefore belong to the adjoint representation of G. For $N = 2$ the left-handed fermion belonging to the chiral multiplet has a right-handed partner, both belonging to the same (arbitrary) representation R of G.

For $N = 1$, the chiral multiplet describes just one left-hand spin $\frac{1}{2}$ fermion which can therefore belong to a complex ($R \neq \bar{R}$) representation of the gauge group G.

Let us consider the relevant field representations of $N = 1$ supersymmetry. The simplest one is given by the chiral multiplet [5]

$$S = (A(x),\ B(x),\ \chi(x),\ F(x),\ G(x)) \tag{3.1}$$

A, B, F, G are four real scalar fields and $\chi(x)$ is a Majorana spinor. The infinitesimal transformation laws under supersymmetry are

$$\begin{aligned} &\delta A = i\bar{\varepsilon}\chi,\ \delta B = -\bar{\varepsilon}\gamma_5\chi,\ \delta\chi = \partial\!\!\!/(A + i\gamma_5 B)\varepsilon + (F + i\gamma_5 G)\varepsilon, \\ &F = i\bar{\varepsilon}\partial\!\!\!/\chi,\quad \delta G = -\bar{\varepsilon}\gamma_5\partial\!\!\!/\chi \end{aligned} \tag{3.2}$$

ε_α is a constant anticommuting Majorana spinor. It is useful to introduce complex notations:

$$Z = \frac{A-iB}{2}\ ,\ \chi = \frac{1}{2}(1+\gamma_5)\chi,\quad H = \frac{F+iG}{2} \tag{3.3}$$

Then (3.2) becomes

$$\begin{aligned}
\delta Z &= \varepsilon^{\alpha}\chi_{\alpha} \\
\delta\chi_{\alpha} &= 2i\sigma^{\mu}_{\alpha\dot{\beta}}\bar{\varepsilon}^{\dot{\beta}}\partial_{\mu}Z + 2H\varepsilon_{\alpha} \\
\delta H &= -i\partial_{\mu}\chi^{\alpha}\sigma^{\mu}_{\alpha\dot{\beta}}\bar{\varepsilon}^{\dot{\beta}}
\end{aligned} \tag{3.4}$$

To include vector particles a larger multiplet must be introduced [15] (vector multiplets). Its real components are

$$V = (C,\zeta,M,N,v_{\mu},\lambda,D) \tag{3.5}$$

C,M,N,D are real scalars, ζ and λ are Majorana spinors and v_{μ} a real vector field. The infinitesimal supersymmetry transformations are

$$\begin{aligned}
\delta C &= -\bar{\varepsilon}\gamma_5\zeta \\
\delta\zeta &= \gamma^{\mu}v_{\mu}\varepsilon - i\partial_{\mu}C\gamma_5\gamma^{\mu}\varepsilon + (M + i\gamma_5 N)\varepsilon \\
\delta M &= i\bar{\varepsilon}\lambda + i\bar{\varepsilon}\not{\partial}\zeta \\
\delta N &= -\bar{\varepsilon}\gamma_5\lambda - \bar{\varepsilon}\gamma_5\not{\partial}\zeta \\
\delta v_{\mu} &= i\bar{\varepsilon}\gamma_{\mu}\lambda + i\bar{\varepsilon}\partial_{\mu}\zeta \\
\delta\lambda &= -\frac{1}{2}F_{\mu\nu}\gamma^{\mu}\gamma^{\nu}\varepsilon + iD\gamma_5\varepsilon \\
\delta D &= -\bar{\varepsilon}\gamma_5\not{\partial}\lambda
\end{aligned} \tag{3.6}$$

This multiplet can be subjected to a supersymmetric U(1) gauge transformation by observing that the following multiplet [16]

$$C = B,\ \zeta = \chi,\ M = F,\ N = G,\ v_{\mu} = \partial_{\mu}A,\ \lambda = D = 0 \tag{3.7}$$

transforms as in (3.6). Then we can consistently demand that

$$\begin{aligned} &\delta C = B && \delta v_\mu = \partial_\mu A \\ &\delta\zeta = \chi && \delta\lambda = 0 \\ &\delta M = F && \delta D = 0 \\ &\delta N = G && \end{aligned} \tag{3.8}$$

Since the transformation (3.8) acts as a translation on the first four components V we can set them equal to zero by fixing a supersymmetric gauge and we are left with

$$\delta v_\mu = \partial_\mu A, \quad \delta\lambda = \delta D = 0 \tag{3.9}$$

i.e., an ordinary gauge transformation on v_μ, while λ and D are gauge invariant. The gauge for which $C = \zeta = M = N = 0$ is called the Wess-Zumino gauge. In this gauge we have

$$\delta v_\mu = i\bar{\varepsilon}\gamma_\mu\lambda \ , \quad \delta\lambda = -F_{\mu\nu}\sigma^{\mu\nu}\varepsilon, \quad \delta D = -\bar{\varepsilon}\gamma_5 \not{\partial}\lambda \tag{3.10}$$

The supersymmetry algebra is modified by the inclusion of gauge transmations, i.e.,

$$[\delta_1, \delta_2] v_\mu = -2i\bar{\varepsilon}_1\gamma^\mu\varepsilon_2 F_{\nu\mu} \tag{3.11}$$

For non-Abelian gauge transformations in the Wess-Zumino gauge, we have

$$\begin{aligned} \delta v_\mu &= i\bar{\varepsilon}\gamma_\mu\lambda \\ \delta\lambda &= -G_{\mu\nu}\sigma^{\mu\nu}\varepsilon + i\gamma D\varepsilon \\ \delta D &= -\bar{\varepsilon}\gamma_5 \not{\mathcal{D}}\, \lambda \end{aligned} \tag{3.12}$$

where $G_{\mu\nu}$ is the non-Abelian field strength of v_μ, $\mathcal{D}_\mu$ is the covariant derivative and λ and D belong to the adjoint representation of the gauge group G.

Let us now suppose that the chiral multiplet S belongs to some irreducible representation R of G. Its supersymmetry transformation in the Wess-Zumino gauge becomes:

$$\delta Z^a = \varepsilon^\alpha \chi^a_\alpha$$

$$\delta \chi^a_\alpha = 2i\sigma^\mu_{\alpha\beta}\bar{\varepsilon}^\beta(\mathcal{D}_\mu Z)^a + 2H^a\varepsilon_\alpha \tag{3.13}$$

$$\delta H^a = -i(\mathcal{D}_\mu \chi^\alpha)^a \sigma^\mu_{\alpha\dot{\beta}}\bar{\varepsilon}^{\dot{\beta}} - 2ig\bar{\lambda}^A_{\dot{\alpha}}\bar{\varepsilon}^{\dot{\alpha}}(R^A Z)^a$$

In order to establish interactions among the two types of multiplets we have introduced, it is convenient to go to superspace [17] where the supersymmetric tensor calculus becomes transparent.

Superspace is an extension of ordinary space-time to a space-time with spin degrees of freedom. The basic manifold of superspace for N = 1 supersymmetry has points parametrized by co-ordinates

$$Z^A = (x^\mu, \theta^\alpha, \bar{\theta}^{\dot{\alpha}}) \qquad \mu = 1,\ldots,4 \qquad \alpha,\dot{\alpha} = 1,2 \tag{3.14}$$

the θ variables have to be considered as odd elements of a Grassmann algebra, i.e., $[Z^A, Z^B\} = 0$.

Superspace is the quotient space $G/\mathcal{H}$ where G is the 14-dimensional graded Poincaré algebra (N = 1 supersymmetry algebra) and $\mathcal{H}$ is the homogeneous Lorentz group.

Supersymmetry transformations are realized as motions in superspace. Under an infinitesimal supertranslation of parameters (a^μ, ε^α, $\bar{\varepsilon}^{\dot{\alpha}}$) we have

$$\begin{aligned} \delta x^\mu &= -i\varepsilon\sigma^\mu\bar{\theta} + i\theta\sigma^\mu\bar{\varepsilon} + a^\mu \\ \delta\theta^\alpha &= \varepsilon^\alpha \;, \quad \delta\bar{\theta}^{\dot{\alpha}} = \bar{\varepsilon}^{\dot{\alpha}} \end{aligned} \tag{3.15}$$

The composition rule of two supertranslations of parameters ($a^\mu_1, \varepsilon^\alpha_1, \bar{\varepsilon}^{\dot{\alpha}}_1$, $a^\mu_2, \varepsilon^\alpha_2, \bar{\varepsilon}^{\dot{\alpha}}_2$) gives

$$[\delta_2, \delta_1]Z^A = (-2i\varepsilon_1\sigma^\mu\bar{\varepsilon}_2 + 2i\varepsilon_2\sigma^\mu\bar{\varepsilon}_1, 0, 0) \tag{3.16}$$

Superspace provides a representation of the supersymmetry algebra in terms of differential operators. The motion (3.15) is obtained in a standard way as the left action of a group element on the coset space $G/\mathcal{H}$. The infinitesimal generators of this motion are

$$P_\mu = i\frac{\partial}{\partial x^\mu},\; Q_\alpha = \frac{\partial}{\partial\theta^\alpha} - i\sigma^\mu_{\alpha\dot{\beta}}\bar{\theta}^{\dot{\beta}}\frac{\partial}{\partial x^\mu},\; \bar{Q}_{\dot{\alpha}} = -\frac{\partial}{\partial\bar{\theta}^{\dot{\alpha}}} + i\theta^\beta\sigma^\mu_{\beta\dot{\alpha}}\frac{\partial}{\partial x^\mu} \tag{3.17}$$

A scalar superfield is a scalar function in superspace, i.e., $\phi'(x',\theta') = \phi(x,\theta)$. In the infinitesimal one has

$$\delta\phi = \left[\varepsilon\frac{\partial}{\partial\theta} + \bar{\varepsilon}\frac{\partial}{\partial\bar{\theta}} + i(\theta\sigma^{\mu}\bar{\varepsilon} - \varepsilon\sigma^{\mu}\bar{\sigma})\partial_{\mu}\right]\phi \tag{3.18}$$

For a superfield operator, $\delta\phi$ stands for $[\varepsilon^{\alpha}Q_{\alpha} + \bar{\varepsilon}_{\dot{\alpha}}\bar{Q}^{\alpha},\phi]$; $\phi(x,\theta)$ is a finite collection of ordinary local fields because $\theta_{\alpha}\theta_{\beta} = -\theta_{\beta}\theta_{\alpha}$ hence $\theta_{\alpha_1}\ldots\theta_{\alpha_n} = 0$ for $n > 4$. It follows that

$$\phi(x,\theta) = \sum_{i=1}^{4} \theta^{\alpha_1}\ldots\theta^{\alpha_i}\ \phi_{\alpha_1\ldots\alpha_i}(x) \tag{3.19}$$

From (3.18) one can compute $\delta\phi_{\alpha_1}\ldots_{\alpha_i}(x)$ and get symbolically

$$\delta_{\varepsilon}\phi_{\alpha_1\ldots\alpha_i}(x) \rightarrow \phi_{\alpha_1\ldots\alpha_{i+1}}(x) + \partial_x\phi_{\alpha_1\ldots\alpha_{i-1}}(x) \tag{3.20}$$

Because of (3.16) it is evident that superspace has non-vanishing (super) torsion. However, one can construct covariant differential operators [13] which commute with supersymmetry variations. They are

$$D_{\alpha} = \frac{\partial}{\partial\theta^{\alpha}} + i\sigma^{\mu}_{\alpha\dot{\beta}}\bar{\theta}^{\dot{\beta}}\frac{\partial}{\partial x^{\mu}}\ , \qquad \bar{D}_{\dot{\alpha}} = -\frac{\partial}{\partial\theta^{\dot{\alpha}}} - i\theta^{\beta}\sigma^{\mu}_{\beta\dot{\alpha}}\frac{\partial}{\partial x^{\mu}} \tag{3.21}$$

Comparing (3.17) with (3.21) it is obvious that

$$D_{\alpha}\delta\phi = \delta D_{\alpha}\phi \tag{3.22}$$

The covariant derivatives satisfy the algebra

$$\begin{gathered} \{D_{\alpha},\bar{D}_{\dot{\alpha}}\} = -2i\sigma^{\mu}_{\alpha\dot{\alpha}}\partial_{\mu},\quad \{D_{\alpha},D_{\beta}\} = \{\bar{D}_{\dot{\alpha}},\bar{D}_{\dot{\beta}}\} = 0 \\ [D_{\alpha},\partial_{\mu}] = [\bar{D}_{\dot{\alpha}},\partial_{\mu}] = 0 \end{gathered} \tag{3.23}$$

From (3.23) it also follows that $D_{\alpha}D_{\beta}D_{\gamma} = 0$. We can impose on a complex superfield the condition

$$\bar{D}_{\dot{\alpha}}S(x,\theta,\bar{\theta}) = 0 \tag{3.24}$$

which is solved by

$$S(x - i\sigma\bar{\theta},\theta,\bar{\theta}) = \psi(x,\theta) = Z(x) + \theta^{\alpha}\chi_{\alpha} + \theta^{\alpha}\theta_{\alpha}H(x) \tag{3.25}$$

Equation (3.25) gives the chiral superfield S as defined in (3.4). In order to obtain the vector multiplet V it is sufficient to take a general scalar superfield subjected to the reality condition $\phi = \phi^{*}$. This condition is invariant under supersymmetry because δ is a real transformation, i.e., $(\delta\phi)^{*} = \delta(\phi^{*})$. The previous analysis can be extended to an arbitrary superfield endowed with Lorentz and internal symmetry indices. A general superfield operator can be written as

$$\phi(x,\theta) = L(x,\theta)\phi L^{-1}(x,\theta) \tag{3.26}$$

when $L(x,\theta)$ is an element of the 10-dimensional coset space $G/\mathcal{H}$. By definition $\phi = \phi(0,0)$. ϕ is a representation $\mathcal{H}$. Note that $\mathcal{H}$ leaves the superorigin $x = \theta = 0$ invariant. If ϕ is a representation of $\mathcal{H}$ we can induce on $\phi(x,\theta)$ a representation of G. If X is a generator of $\mathcal{H}$ the action of X on $\phi(x,\theta)$ is computed as follows

$$[X,e^{-T}\phi e^{T}] = e^{-T}[Y,\phi]e^{T} + e^{-T}[X,\phi]e^{T} \tag{3.27}$$

where

$$Y = \int_0^1 d\lambda e^{\lambda T}[T,X]e^{-\lambda T}, \quad T = i(\theta^{\alpha}Q_{\alpha} + \bar{\theta}_{\dot{\alpha}}\bar{Q}^{\dot{\alpha}} - x^{\mu}P_{\mu})$$

The infinite chain of commutators which defines Y stops after a finite number of steps because of the O'Raifeartaigh theorem.

We now attack the problem of constructing invariant interactions in a superfield formalism. A scalar field, in order to be a candidate for a supersymmetric action, must transform as a total divergence under supersymmetry variations. Because of (3.20) this is achieved by the last component of a superfield. If we use the integration over anti-commuting variables, defined by the Berezin recipe

$$\int d\theta_i\theta_j = \delta_{ij} \tag{3.28}$$

We can rewrite the last component of a superfield as

$$\int d^4\theta\ \phi(x,\theta) \quad \text{or} \quad \int d^2\theta\ \phi(x,\theta) \quad \text{if} \quad \bar{D}_{\dot{\alpha}}\phi = 0 \tag{3.29}$$

Given a chiral superfield S ($\bar{D}_{\dot{\alpha}}S = 0$) we can write two possible invariant bilinears as follows

$$\int d^4x \int d^4\theta \; S\bar{S} \tag{3.30}$$

$$\int d^4x \int d^2\theta \; S^2 \tag{3.31}$$

Because of the property $\int d\theta = \partial/\partial\theta$, (3.30) may also be rewritten as

$$\int d^4x \int d^2\theta \; S\overline{DDS} \tag{3.32}$$

(the imaginary part of $S\overline{DDS}$ being a total divergence).

Because dim S = 1 (in energy units) we see that dim $S\bar{S} = 2$ and dim $\int d^4\theta \; S\bar{S} = 4$ (dim $d\theta$ = dim $\partial/\partial\theta = 1/2$). Analogously dim $\int d^2\theta \; S^2 = 3$.

The only interacting term consistent with renormalizability is

$$\int d^2\theta \; S^3 \tag{3.33}$$

which has dimension four. An additional term is also possible with dimension two, namely

$$\int d^2\theta \; S \tag{3.34}$$

Therefore we conclude that the most general Lagrangian of interacting chiral multiplets consistent with renormalizability (dim $\mathcal{L} \leq 4$) is given by

$$\int d^4\theta \; S^a\bar{S}^a + \text{Re} \int d^2\theta \; f(S^a) \tag{3.35}$$

where

$$f(S^a) = \eta^a S_a + m^{ab} S_a S_b + g^{abc} S_a S_b S_c$$

$f(S_a)$ is often called the superpotential function. If a group G is acting on the index "a" one then gets a further constraint on f from the requirement of G invariance, namely

$$f_{,a}(S) R_{\alpha}{}^{a}{}_{b} S^b = 0 \tag{3.36}$$

where $T_{\alpha\ b}^{\ a}$ are the generators of the representation R of G under which S^a transforms.

In component form we have for the two terms in (3.33) respectively

$$\mathcal{L}_{KIN} = -2\partial_\mu Z^a \partial^\mu Z^{\star a}_a - i\chi^{\alpha a}\sigma_{\mu\alpha\dot{\beta}}\partial_\mu \bar{\chi}^{\dot{\beta}}_a + 2H^a H_a$$
$$\mathcal{L}_{POT} = H^a f_{,a} - \frac{1}{4}\chi^{\alpha a}\chi^b f_{,ab} + h.c. \tag{3.37}$$

when Z, χ and H have been defined in (3.3).

Elimination of H^a yields to the final component Lagrangian

$$\mathcal{L} = -2\partial_\mu Z^a \partial^\mu Z^\star_a - i\chi^{\alpha a}\sigma_{\mu\alpha\dot{\beta}}\partial_\mu \bar{\chi}^{\dot{\beta}}_a -$$
$$- \frac{1}{2}f_{,a}f^{\star,a} - \frac{1}{4}\chi^{\alpha a}\chi^b_\alpha f_{,ab} - \frac{1}{4}\bar{\chi}_{\dot{\alpha}a}\bar{\chi}^{\dot{\alpha}}_b f^{\star,ab} \tag{3.38}$$

The scalar potential is given by

$$V = \frac{1}{2}\ |f_{,a}|^2 \tag{3.39}$$

and it is semi-positive definite.

Let us now consider gauge theories [18], with a vector multiplet V. Let us endow V with an index A when $A = 1,\ldots,D$ and D is the dimension of the Lie algebra of a compact Lie group G. [We take SU(N) as an illustrative example.] More specifically we can assume that V belongs to the Lie algebra G, i.e., is a Lie algebra valued function. If we introduce a chiral superfield Λ ($\bar{D}_{\dot{\alpha}}\Lambda = 0$), which is also Lie algebra valued, then a finite Yang-Mills transformation in superspace becomes

$$e^{2gV} \rightarrow e^{-i\Lambda^+}e^{2gV}e^{i\Lambda} \tag{3.40}$$

V, Λ and Λ^+ are $N \times N$ Hermitean traceless matrices. It is easy to see that the chiral superfield

$$W_\alpha = \frac{1}{g^2}\bar{D}_{\dot{\alpha}}\bar{D}^{\dot{\alpha}}(e^{-2gV}D_\alpha e^{2gV}) \tag{3.41}$$

transforms as

$$W_\alpha \to e^{-i\Lambda} W_\alpha e^{i\Lambda} \tag{3.42}$$

under a Yang-Mills transformation. W_α is the super-Yang-Mills strength in the sense that it contains only the Yang-Mills covariant quantities $\lambda_\alpha, \sigma^\mu_{\alpha\dot{\alpha}} \mathcal{D}_\mu \bar{\lambda}^{\dot{\alpha}}$ and $G_{\mu\nu} = \partial_\mu v_\nu - \partial_\nu v_\mu + ig\,[v_\mu, v_\nu]$. Here $\mathcal{D}_\mu$ is the Yang-Mills covariant derivative $\mathcal{D}_\mu \lambda = \partial_\mu \lambda + ig\,[v_\mu, \lambda]$.

If V is a singlet superfield (3.40) and (3.42) simplify considerably and we obtain an Abelian U(1) gauge transformation and gauge field strength respectively

$$\begin{aligned} V &\to V + \frac{1}{2g} i(\Lambda - \bar{\Lambda}) \\ W_\alpha &= \bar{D}_{\dot{\alpha}} \bar{D}^{\dot{\alpha}} D_\alpha V \end{aligned} \tag{3.43}$$

In terms of W_α the pure Yang-Mills supersymmetric action is given by

$$\int d^4x \int d^2\theta \;\mathrm{Tr}\; W^\alpha W_\alpha \tag{3.44}$$

Let us now consider a (matter) chiral superfield S^a ($\bar{D}_{\dot{\alpha}} S^a = 0$) belonging to some irreducible representations R of G. Then under a Yang-Mills transformation it transforms as

$$\begin{aligned} S &\to e^{-i\Lambda} S \\ \bar{S} &\to \bar{S} e^{i\Lambda^+} \end{aligned} \tag{3.45}$$

when the Lie algebra valued parameter Λ now takes values in the representation space of S. One sees immediately that the following quantity

$$\bar{S}\, e^{2gV} S = \bar{S}_a (e^{2gV})^a{}_b S^b \tag{3.46}$$

is a Yang-Mills invariant and defines a vector multiplet whose last component gives rise to the supersymmetric minimal coupling. The most general superfield Lagrangian is therefore given by

$$\int d^4x \big[\int d^4\theta\; \bar{S}\, e^{2gV} S + \mathrm{Re} \int d^2\theta \big[\mathrm{Tr}\; W^\alpha W_\alpha + f(S)\big]\big] \tag{3.47}$$

where S denotes a collection of irreducible representations R^i of G and f(S) is a G invariant superpotential function as defined by (3.35) and (3.36) If the gauge group G contains p Abelian factors $U(1)^p$ p additional terms are also possible, namely

$$\int d^4x \int d^4\theta \sum_{i=1}^{p} \xi^i V_i \tag{3.48}$$

where V_i are the vector superfields associated with the p U(1) factors of G.

In component form (3.44) becomes

$$Tr(-\frac{1}{4}G^2_{\mu\nu} - \frac{i}{2}\bar{\lambda}\gamma^\mu \mathcal{D}_\mu \lambda + \frac{1}{2}D^2) \tag{3.49}$$

The component form of (3.46) gives

$$\begin{aligned} &-2|\mathcal{D}_\mu Z|^2 - i\chi^{\alpha^a}\sigma^\mu_{\alpha\dot{\beta}}(\mathcal{D}_\mu \bar{\chi}^{\dot{\beta}})_a + 2|H_a|^2 \\ &+2gi\lambda^{\alpha^A}\chi^a_\alpha R^A{}_a{}^b Z^*_b + 2gZ^a R^A{}_a{}^b Z^*_b D^A + h.c. \end{aligned} \tag{3.50}$$

while R^{Ab}_a are the Hermitean generators of the representation R of S^a. If more irreducible representations are present in (3.50) a sum over all inequivalent irreducible representations is understood. $\mathcal{D}_\mu$ is the ordinary Yang-Mills covariant derivative, i.e.,

$$\mathcal{D}_\mu Z^a = \partial_\mu Z^a + igv^A_\mu R^A{}_b{}^a Z^b$$

4. SPONTANEOUS SUPERSYMMETRY BREAKING AND APPLICATION TO UNIFIED THEORIES

In the present section we describe some attempts to use $N = 1$ supersymmetric gauge theories to describe models of electroweak and strong interactions. To have a realistic model we have first to discuss spontaneous supersymmetry breaking. In a quantum field theory the condition for symmetry breaking for a given generator X of a continuous symmetry is that $X|0\rangle \neq 0$. In the case of supersymmetry this implies that $Q_\alpha|0\rangle \neq 0$. From the transformation laws (3.4) and (3.6) we see that this demands that H^a or (and) D^A must acquire a non-vanishing expectation value on the classical solutions of the field equations. In a gauge theory defined by the Lagrangian (3.47) [with the possible modification given by (3.48)] this requires that one of these equations:

$$f_{,a} = 0 \tag{4.1}$$

$$D^A = -2g_A Z^a R^A{}_a{}^b Z^*_b - \xi_A \delta_{Ai} = 0 \qquad (i = 1,\ldots p) \tag{4.2}$$

does not admit a solution.

It is evident that if all chiral multiplets transform under G [$\eta^a = 0$ in (3.35)] and if $\xi_i = 0$, Equations (4.1) have always the G invariant solution $Z^a = 0$. The only way to allow for an inconsistency of (4.1) and (or) (4.2) is to have some singlet fields X^a under G ($\eta^a \neq 0$) or (and) some U(1) factor in G with $\xi_i \neq 0$. Therefore, a necessary condition to have spontaneous supersymmetry breaking at the tree-level in a supersymmetric gauge theory is to have one of the following two conditions satisfied:

i) The group G should contain at least a neutral field X with linear term in the superpotential f(S).

ii) The group G should contain at least an Abelian factor U(1) with a non-vanishing ξ in (3.48).

One can easily see that i) and ii) are necessary but not sufficient conditions to have spontaneous supersymmetry breaking at the tree level.

Let us consider an example which fulfils condition i). Consider a group G, a singlet X and two chiral superfields S_1, S_2 transforming according to a real representation R of G.

Then the following superpotential term [19]

$$f(S_i,X) = gXS_1^2 + \mu^2 X + mS_1S_2$$

undergoes spontaneous supersymmetry breaking. In fact we get

$$f_{,X} = gS_1^2 + \mu^2$$

$$f_{,S_1} = mS_2 + 2gXS_1 \tag{4.3}$$

$$f_{,S_2} = mS_1$$

and we see that the two equations $f_{,X} = 0$ and $f_{,S_2} = 0$ are incompatible. An obvious property of this model is that det $f_{,ab} = 0$ and that the tree level potential does not fix the three vacuum expectation values $\langle x \rangle$ $\langle s_1 \rangle$ and $\langle s_2 \rangle$. Let us consider the full gauge theory with the over-all potential

$$\frac{1}{2} f_{,a} f^*_{,a} + \frac{1}{2} D^A D^A \tag{4.4}$$

where

$$D_a^A = -2g_A Z^a R_a^{A\ b} Z_b^* - \xi_A \delta_{Ai} \qquad (i = 1,\ldots p)$$

Differentiating (4.4) we get the extremum condition

$$f_{,ab} f^{*a} + 2D^A D_b^A = 0 \tag{4.5}$$

Equation (4.5) can be written as

$$\mathcal{M}_{AB} \mathcal{F}^B = 0 \tag{4.6}$$

where

$$\mathcal{F}^B = (f^{*a}, D^A)$$

and M_{AB} is the matrix

$$\mathcal{M}_{AB} = \begin{pmatrix} f_{,ab} & 2D^A{}_b \\ 2D^A{}_a & 0 \end{pmatrix}$$

The condition that (4.6) has a solution with a non-vanishing eigenvector $\mathcal{F}^B$ implies [20] that det $\mathcal{M}_{AB} = 0$, i.e., the fermion mass matrix must have a vanishing eigenvalue, the Goldstone fermion. The Goldstone fermion is a combination of $f_{,a}\chi^a$ and $D^A\lambda^A$ with strength fixed by (4.5) and (4.6). Therefore, the Goldstone fermion is the mixture of those left-handed fermions χ_L^a, λ^A for which the corresponding auxiliary fields $f_{,a}$ and D^A are different from 0 at the absolute minimum of the potential determined by (4.5)

The main application of supersymmetric Yang-Mills theories to low energy physics is in the context of grand unified theories and of the conventional Weinberg-Salam model.

In GUTs one is faced with the hierarchy problem [10,11] which is connected with the fact that there is no natural way of having a small expectation value (or Higgs mass) for the Higgs weak SU(2) doublet due to quadratically divergent self-energy graphs which mix the SU(2) × U(1) breaking scale $M_W \sim 100$ GeV with the grand unification scale M_X (Planck scale $M_{P\ell}$?). In supersymmetric gauge theories, because of non-renormalization theorems [21], it is known that it is "natural" to set a scalar mass equal to zero. If we stipulate that the supersymmetry breaking be related to the SU(2) × U(1) breaking of the electroweak interaction group then one can find a perhaps unique mechanism that could explain naturally the smallness of the Higgs boson mass in spite of the fact that the complete theory contains a big scale M_X [9] (or $M_{P\ell}$). The only supersymmetric theories which have quadratic divergences [22] are theories where the gauge group G contains a $\tilde{U}(1)$ group with a non-vanishing ξV term as defined by (3.48) and with an over-all $\tilde{U}(1)$ non-vanishing trace over the chiral fields S^a

$$\mathrm{Tr}_a \tilde{Y} \neq 0 \qquad (4.7)$$

It turns out that in order to have a realistic mass spectrum for quarks, leptons and their scalar superpartners the effective gauge theory must contain at least the group [23,24]

$$SU(3) \times SU(2) \times U(1) \times \tilde{U}(1) \tag{4.8}$$

with the value of the $\tilde{Y}$ charge which is positive (or negative) over all left-handed matter fields. This turns out to be a consequence of the general mass matrix for quarks and leptons as derived by (3.50) under the assumption that colour and electromagnetism are unbroken symmetries at low energies. If we want to avoid quadratic divergences then in addition we must have [20-22]

$$\mathrm{Tr}_a \, \tilde{Y} = 0 \tag{4.9}$$

which means that other fields must exist, carrying colour and SU(2) quantum numbers with opposite values of the $\tilde{Y}$ charge with respect to the matter fields. The fact that these extra fields carry colour and charge does not come from (4.9) but from the triangle anomaly equations of the type

$$\mathrm{Tr}\ SU(2)^2\tilde{U}(1) = \mathrm{Tr}\ SU(3)^2\tilde{U}(1) = \mathrm{Tr}\ \tilde{U}(1)^3 = 0 \tag{4.10}$$

For each generation of quarks and leptons the first equation can be satisfied by the Higgs doublets, the second equation by a colour triplet and a colour antitriplet and the third equation by two additional scalars which are neutral under $SU(3) \times SU(2) \times U(1)$. However, one can show that even neglecting matter the most general potential, involving the previously mentioned 27 fields which may be embedded in the 27-dimensional representation of E_6, is not able to stabilize the colour triplet fields then leading to an unacceptable situation where colour is broken.

Recently Weinberg has given some criteria for finding physically relevant local minima in a general supersymmetric gauge theory. Unfortunately these criteria do not exclude the existence of lower supersymmetric minima, for any range of the parameters of the theory, which have physically unwanted properties [25]. It is possible to show that by extending the 27-dimensional representation of E_6 whose reduction under $SU(3) \times SU(2) \times U(1) \times \tilde{U}(1)$ is [25]

$$Q(3,2,\tfrac{1}{6},1)\ ;\quad u^c(\bar{3},1,-\tfrac{2}{3},1)\ ;\quad d^c(\bar{3},1,\tfrac{1}{3},1)$$

$$L(1,2,-\tfrac{1}{2},1)\ ;\quad e^c(1,1,1,1)\ ;\quad P(1,1,0,1) \tag{4.11}$$

$$H(1,2,-\tfrac{1}{2},-2)\ ;\ H^c(1,2,\tfrac{1}{2},-2)\ ;\ T(3,1,-\tfrac{1}{3},-2)\ ;\ T^c(\bar{3},1,\tfrac{1}{3},-2)\ ;$$
$$S(1,1,0,4)$$

to a larger set of 30 fields by adding three chiral fields singlets under $SU(3) \times SU(2) \times U(1)$ with $\tilde{U}(1)$ charge given by

$$R(1,1,0,4),\ R^c(1,1,0,-4),\ N(1,1,0,0) \tag{4.12}$$

it is possible to find a local minimum with all desired properties, which stabilizes the colour triplets and therefore does not break colour and charge [25].

Unfortunately, this model has always a supersymmetric solution which breaks colour and charge. It is also possible to show that any modification of this set of 30 chiral multiplets by addition of $\tilde{U}(1)$ anomaly free and traceless sets of chiral superfields not transforming under $SU(3) \times SU(2) \times U(1)$ does not help. A possible way out is to enlarge the gauge group to $SU(3) \times SU(2) \times U(1) \times \tilde{U}(1) \times U'(1)$ and to play with the three parameters ξ_Y, ξ and ξ' to see whether supersymmetric solutions can be avoided and physically acceptable supersymmetric breaking solutions exist.

A possible way out of these difficulties has recently been proposed [25] in a class of models in which the group $\tilde{U}(1)$ is broken at a very high scale ξ, while the $SU(2) \times U(1)$ group and supersymmetry are controlled by a much smaller scale μ. Higgs and boson squared masses are found of order μ^2 in these models while the $U(1)$ vector boson squared mass is of order ξ and the gravitino squared mass is of order $\mu^2\xi/M_{P\ell}^2$. Since the $\tilde{U}(1)$ group is broken at a high scale ξ ($\lesssim M_P$) the effect of possible anomalies is not dramatic and in any case does not disturb the low energy theory controlled by the much smaller scale μ. Interestingly enough, if $\xi \sim M_{P\ell}^2$ the gravitino has a mass of order μ and it is almost decoupled from low energy physics.

To illustrate the model we just consider the relevant fields which are H, H^c, S, R and R^c. The superpotential is

$$\lambda H H^c S + \mu R R^c \qquad (4.13)$$

and supersymmetry is broken if $\mu \neq 0$. The complete model is obtained with the 30 scalar fields previously described with the exclusion of T and T^c. Of course, the model has $\tilde{U}(1)$ anomalies. However, since $\tilde{U}(1)$ can be broken at super high energies where gravitational interactions become strong, the presence of these anomalies is not dramatic. Finally, it should be mentioned that even in a completely anomaly-free theory with a small scale ξ, gravitational interactions would introduce triangular anomalies and therefore not renormalizable effects due to a modification of the $\tilde{U}(1)$charge which is proportional to $\xi/M_{P\ell}^2$.

As a final comment we should mention what the scenario would be if the group G were semi-simple. In this case the terms $\xi^k V_k$ in the superspace Lagrangian [see (3.48)] are absent and the only possible supersymmetry breaking at the three level is obtained by a mechanism as given by i). It is a general property of such a kind of potentials that the mass spectrum of the scalar partners of the quarks and leptons is unrealistic because some of them are lighter than the usual quarks and leptons [27]. However, the potential is unstable at the tree level because some vacuum expectation values remain undertermined. It is not impossible that the effective potential resolves the mass splitting problem of the classical Lagrangian by the introduction of some higher scale [26] which would give radiatively a fermion-boson mass splitting at least of order M_W.

Finally, we mention that apart from conjectures of non-perturbative effects responsible for supersymmetry breaking [9] another possibility relies on explicit but soft supersymmetry breaking [27], which, although inelegant, would not spoil the ultra-violet improvement of supersymmetric theories [28].

REFERENCES

[1] S. Glashow, Nucl. Phys. 22 (1961) 579;
J.C. Ward and A. Salam, Phys. Lett. 13 (1964) 168;
S. Weinberg, Phys. Rev. Lett. 19 (1967) 1264.
A. Salam, "Elementary Particle Theory", ed. N. Svartholm (Almquist and Wiksell, Stockholm, 1968), p. 367.

[2] For reviews on supersymmetry see for instance:
P. Fayet and S. Ferrara, Phys. Rep. 32C (1977) 251;
A. Salam and J. Strathdee, Fortsch. Phys. 26 (1976) 57.

[3] For a review see for instance:
P. van Nieuwenhuizen, Phys. Rep. 68 (1981) 184.

[4] R. Haag, J.T. Lopuszanski and M. Sohnius, Nucl. Phys. B88 (1975) 257.

[5] S. Coleman and J. Mandula, Phys. Rev. 159 (1967) 1251;
L.O'Raifeartaigh, Phys. Rev. Lett. 14 (1965) 575; Phys. Rev. B139 (1965) 1052.

[6] P. Fayet, Nucl. Phys. B113 (1976) 135.

[7] F. Gliozzi, J. Scherk and D. Olive, Nucl. Phys. B122 (1977) 253;
L. Brink, J.H. Schwarz and J. Scherk, Nucl. Phys. B121 (1977) 11.

[8] S. Ferrara and B. Zumino, unpublished;
M. Sohnius and P. West, Phys. Lett. 100B (1981) 245;
K.S. Stelle, LPTENS preprint 81/24 (1981).

[9] L. Maiani in Proceedings of the Summer School of Gif-sur-Yvette (1979), p. 3;
E. Witten, Nucl. Phys. B188 (1981) 513;
S. Dimopoulos and S. Raby, Nucl. Phys. B192 (1981) 353.

[10] E. Gildener and S. Weinberg, Phys. Rev. D13 (1976) 3333;
S. Weinberg, Phys. Rev. Lett. 82B (1979) 387.

[11] G. 't Hooft, Cargèse Lectures 1979, to be published;
M. Veltman, Acta Physica Polonica, to be published.

[12] Y.A. Gol'fand and E.P. Likhtam, JETP Lett. 13 (1971) 323;
D.V. Volkov and V.P. Akulov, Phys. Lett. 46B (1973) 109;
J. Wess and B. Zumino, Nucl. Phys. B70 (1974) 39.

[13] S. Ferrara, B. Zumino and J. Wess, Phys. Lett. 51B (1974) 239.

[14] A. Salam and J. Strathdee, Nucl. Phys. B80 (1974) 499; Nucl. Phys. B84 (1975) 127; see also
S. Ferrara, CERN preprint TH.2957 (1981), Plenary talk given at the 9th Int. Conference on General Relativity and Gravitation, Jena (1980), to be published.

[15] See the third Ref. in [12].

[16] J. Wess and B. Zumino, Nucl. Phys. B78 (1974) 1.

[17] A. Salam and J. Strathdee, Nucl. Phys. B76 (1974) 477.

[18] S. Ferrara and B. Zumino, Nucl. Phys. B79 (1974) 413;
A. Salam and J. Strathdee, Phys. Lett. 51B (1974) 353.

[19] L. O'Raifeartaigh, Nucl. Phys. 56B (1975) 413; Nucl. Phys. B89 (1975) 41 - B96 (1975) 331.

[20] S. Ferrara, L. Girardello and F. Palumbo, Phys. Rev. D20 (1979) 403.

[21] J. Wess and B. Zumino, Phys. Lett. 49B (1974) 52;
J. Iliopoulos and B. Zumino, Nucl. Phys. B76 (1974) 310;
S. Ferrara, J. Iliopoulos and B. Zumino, Nucl. Phys. B77 (1974) 413;
S. Ferrara and O. Piguet, Nucl. Phys. B93 (1975) 261.

[22] W. Fishler, H. Nilles, J. Polchinski, S. Raby and L. Susskind, Phys. Rev. Lett. 47 (1981) 757.

[23] P. Fayet in "Unification of the Fundamental Particle Interactions", ed. by S. Ferrara, J. Ellis and P. van Nieuwenhuizen (Plenum Press, N.Y., 1980), p. 587.

[24] S. Weinberg, Harvard preprint HUTP 81/A047 (1981).

[25] R. Barbieri, S. Ferrara and D.V. Nanopoulos, CERN preprint TH.3226 (1982).

[26] E. Witten, Phys. Lett. 105B (1981) 267.

[27] S. Dimopoulos and H. Georgi, Nucl. Phys. B193 (1981) 150.

[28] L. Girardello and M.T. Grisaru, Brandeis University preprint (1981).

INTERNAL SUPERSYMMETRY AND DIMENSIONAL REDUCTION

Y. Ne'eman and S. Sternberg

INTRODUCTION

In a recent series of papers we have suggested the supergroup SU(7/1) as an internal symmetry for the fundamental particles in a grand unification scheme. For this purpose, we have described a family of fundamental representations of the supergroup SU(m/n), cf. [33]. These representations have also proved useful in the interacting boson model of the nucleus for adjacent elements, see the work of Bars and Iachello [3]. The supergroup SU(m/n) is the centralizer in the orthosymplectic supergroup OSp(2m/2n) of a certain one dimensional subgroup and the family of fundamental representations arise as irreducible types in the decomposition of the basic spin metaplectic representation of orthosymplectic group OSp(2m/2n). This suggests an approach to grand unification in terms of dimensional reduction.

The method of dimensional reduction as developed by Manton [24] Harnad, Shnider and Tafel [16] , is roughly speaking as follows - one starts with a principal bundle with a large structure group over a (higher dimensional) base manifold and looks at Yang Mills fields which are invariant under an automorphism group of this bundle. Under suitable hypotheses, the set of such invariant Yang Mills fields coincides with a set of spontaneously-broken Yang Mills fields with their Higgs fields over ordinary space time. The principal bundle is also reduced and its structure group arises as the centralizer of a certain subgroup of the structure group of the larger bundle. This suggests that our internal symmetry group SU(7/1) arises via dimensional reduction from an extended orthosymplectic theory. This is of interest in view of the recent use of the orthosymplectic groups in extended supergravity. Also, there might be a possibility of a unified theory to include gravity using dimensional reduction.

The plan of this paper is as follows: we begin by recalling the results of Bars and Iachello on the one hand and our results on the other, involving the representations of SU(m/n). We then summarize the results of Howe and Kashiwara-Vergne on the metaplectic representation and its reduction in the presence of a pair of mutually centralizing supergroups. We then present the results of Manton and Harnad Shnider and Tafel first in a geometrical setting and then in a form which admits generalization to the supergeometric case and to describe our conjecture as to the form of a possible theory.

In the last section, we discuss the issues arising when the supersymmetry is applied in Particle Physics, in the highly restrictive context of relativistic quantum field theory. One possible interpretation then involves the system of quantum ghost fields, whose presence guarantees the Unitarity of Gauge Theories. The supergroup is assumed to relate physical and ghost fields. The emergence of a new (spin 1) ghost may be due to the realization of spontaneous symmetry breakdown in the presence of gravity.

1. Internal su(m/n)

One of the principal ways of applying group theory to quantum mechanics, as introduced by Wigner over fifty years ago, is as follows: We try to organize a complex set of observed eigenvalues by assuming that there is a chain of (compact) groups

$$G_0 \supset G_1 \supset G_2 \supset \quad \ldots\ldots$$

and that the Hamiltonian of our system is of the form

$$H = H_0 + H_1 + H_2 + \quad \ldots..$$

where the H_i are invariant under G_i and of decreasing size. All the H_i, and hence H act on some fixed representation space of G_0. For instance, if the representation of G_0 were irreducible, then, by Schur's lemma, H_0 has a single eigenvalue. If under G, this space breaks up into three irreducibles, then H_1 can have at most three distinct eigenvalues. If the irreducibles under G_1 further each break up into 1, 2 and 3 irreducibles under G_2, then the eigenvalues split further into 1, 2, and 3 eigenvalues and so (at this stage) we can expect the form of the observed energy levels to be

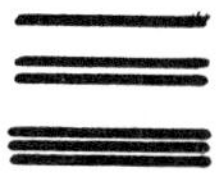

The representation theory of the chain of subgroups tells us the number of eigenvalues to expect and qualitatively how they should cluster. It does not predict quantitative information as to the actual eigenvalues.

One may make some restrictive group theoretical hypotheses about the form of H which then lead to more quantitative predictions. For example, in a scheme like the Gell-Mann-Okubo mass formula one assumes that H is an element (of relatively low order) in the universal enveloping algebra $U(g_0)$, and that the H_i must be invariants under G_i. This determines the H_i up to a small number of parameters. These parameters are fixed by a some of the empircally observed eigenvalues, which, in turn, determine the rest. In the Bars-Iachello model, one assumes that H is given in terms of "creation and annihilation operations" (we shall explain more precisely what this means later on). Then the same considerations apply to give quantitative predictions.

In the Bars-Iachello nuclear model and in our model for internal supersymmetry, the group G_0 is replaced by the supergroup SU(m/n) or, more precisely, by the superalgebra gl(m/n) which we now explain. Let F_0 and F_1 be two vector spaces. We combine them into one "super" vector space $F = F_0 + F_1$ and write

$$\text{End } F = \text{End}_0 \, F \oplus \text{End}_1 \, F$$

where

$$\text{End}_0 \, F = \{ \left(\begin{array}{c|c} A & 0 \\ \hline 0 & D \end{array}\right), \quad A \varepsilon \text{ End } F_0, \ D \varepsilon \text{ End } F_1 \} = \text{End } F_0 \oplus \text{End } F_1$$

and

$$\text{End}_1 \, F = \{ \left(\begin{array}{c|c} 0 & B \\ \hline C & 0 \end{array}\right) \} = \text{ Hom } (F_0, F_1) \oplus \text{Hom } (F_1, F_0).$$

We make End F into a super Lie Algebra via the super commutation.

$$[X_i, X_j] = X_i X_j - (-1)^{ij} X_j X_i \qquad X_i \in \mathrm{End}_i F$$
$$X_j \in \mathrm{End}_j F$$

(We refer the reader to [8] or [39] for the basic properties of Lie superalgebras.) This superalgebra is denoted by $gl(F_0/F_1)$. In case $\dim F_0 = n$ and $F_1 = m$ this superalgebra is also denoted by $gl(n/m)$. In [33] we introduced a family of representations of $gl(F_0/F_1)$ (depending on the integer k) on

$$\wedge^0(F_1) \otimes S^k(F_0) \oplus \wedge^1(F_1) \otimes S^{k-1}(F_0) \oplus \cdots\cdots$$

(the sum being finite and stopping either at $\wedge^m(F_1) \otimes S^{k-m}(F_0)$ or at $\wedge^k(F_1) \otimes S^0(F_0)$ if $k \leq m$.) We shall describe these representations from a slightly different point of view a little later on in this paper. For $\dim F_0 = 1$ we showed how we can replace the integer k by an arbitrary complex number, b with S^b replaced by the one dimensional space of homogeneous functions of degree b.)

In the Bars-Iachello model one takes $F_0 = \mathbb{C}^6$ and $F_1 = \mathbb{C}^4$ so the superalgebra in question is $gl(6/4)$. The k in question is chosen as follows: There are "magic numbers" corresponding to "closed shells" in nuclear theory - the first few of these are 28, 50, 82, 126, etc. For even nuclei ${}_pA_q$ with p and q close to these values, one assumes that the excess or missing protons and neutrons form pairs and that the k in question is given by

$$k = \tfrac{1}{2}(|m_1 - p| + |m_2 - q|)$$

where the m_i are the nearest magic numbers. So for example, for ${}^{190}_{76\ 114}\mathrm{Os}$ the $m_1=82$ and $m_2=126$ and $k=9$. The theory corresponds to the $j = 3/2$ protons (this is four dimensional which corresponds to the 4 in $gl(6/4)$). We add these protons one at a time getting the table

$^{190}_{76}\mathrm{Os}_{114}$ $^{191}_{77}\mathrm{Ir}_{114}$ $^{192}_{78}\mathrm{Pt}_{114}$ $^{193}_{79}\mathrm{Au}_{114}$ $^{194}_{80}\mathrm{Hg}_{114}$

$\Lambda^0 \otimes S^9$ $\Lambda^1 \otimes S^8$ $\Lambda^2 \otimes S^7$ $\Lambda^3 \otimes S^6$ $\Lambda^4 \otimes S^5$

where we have written $\Lambda^i \otimes S^j$ for $\Lambda^i(\mathbb{C}^4) \otimes S^j(\mathbb{C}^6)$.

The sequence of subalgebras

$$gl(6/4) \supset su(4) \times su(6) \supset so(6) \times su(4) \supset su(4)_{diag} \supset su(2)$$

describe very accurately hundreds of energy levels, where the Hamiltonian is given in terms of quartic expressions in creation and annihilation operators, the successive terms being invariant under successive subalgebras in the chain as described above.

In our particle classification scheme we start the observed charge values Q, of the leptons and quarks

	e^-	ν_e	u	u	u	d	d	d
Q	1	0	$\frac{2}{3}$	$\frac{2}{3}$	$\frac{2}{3}$	$-\frac{1}{3}$	$-\frac{1}{3}$	$-\frac{1}{3}$

,

eight in all. In the initial approximation one starts with $m = 0$ so the particles come in right and left handed versions, giving 16:

	e^+_R	$\bar{\nu}_{eR}$	u_R	u_R	u_R	d_R	d_R	d_R	e^+_L	$\bar{\nu}_{eL}$	u_L	u_L	u_L	d_L	d_L	d_L
Q	1	0	$\frac{2}{3}$	$\frac{2}{3}$	$\frac{2}{3}$	$-\frac{1}{3}$	$-\frac{1}{3}$	$-\frac{1}{3}$	1	0	$\frac{2}{3}$	$\frac{2}{3}$	$\frac{2}{3}$	$-\frac{1}{3}$	$-\frac{1}{3}$	$-\frac{1}{3}$
I_3	$\frac{1}{2}$	$-\frac{1}{2}$	0	0	0	0	0	0	0	0	$\frac{1}{2}$	$\frac{1}{2}$	$\frac{1}{2}$	$-\frac{1}{2}$	$-\frac{1}{2}$	$-\frac{1}{2}$

where I_3 is the third component of weak left handed isospin. There is a second "generation" of similar particles

μ ν_μ s s s c c c

which, in their right and left handed versions give another sixteen particles for a total of thirty-two. There is evidence for a third generation

$$\tau \quad \nu_\tau \quad t\ t\ t\ b\ b\ b$$

If we believe in four generations we would get a total of sixty-four. In our theory, we can allow for 2^{s+1} generations for any value of s where we take the superalgebra to be $gl(1/5+s)$ and the fundamental representation corresponding to the value of $b = \frac{1}{2}(s+4)$, on a space of dimension 2^{5+s}. There are various reasons for believing in eight generations: One has to do with the organization of right and left handedness within the individual Λ^k, cf. [38b]. Arguments from QCD suggest that asymptotic freedom no longer holds for more than eight generations, cf. [7] or [34]. Furthermore, it was shown in [51] one gets "critical QCD" which is the only version of QCD in which the S matrix displays the observed factorizable Pomeranchk Regge trajectories. Finally, it was shown in [35] that the $su(7) \times u(1)$ component of our theory is anomaly free.

In the theory, we take

$$Q = I_3 + \tfrac{1}{2} U$$

where I_3 and $U \in gl(7/1)$ are given by

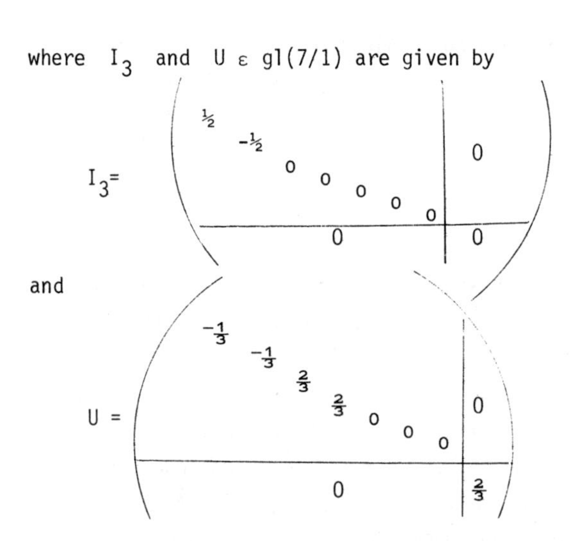

$$I_3 = \left(\begin{array}{ccccccc|c} \frac{1}{2} & & & & & & & \\ & -\frac{1}{2} & & & & & & 0 \\ & & 0 & & & & & \\ & & & 0 & & & & \\ & & & & 0 & & & \\ & & & & & 0 & & \\ & & & & & & 0 & \\ \hline & & & 0 & & & & 0 \end{array}\right)$$

and

$$U = \left(\begin{array}{ccccccc|c} -\frac{1}{3} & & & & & & & \\ & -\frac{1}{3} & & & & & & \\ & & \frac{2}{3} & & & & & \\ & & & \frac{2}{3} & & & & 0 \\ & & & & 0 & & & \\ & & & & & 0 & & \\ & & & & & & 0 & \\ \hline & & & 0 & & & & \frac{2}{3} \end{array}\right)$$

Then the image of Q has exactly the correct values of charge with the appropriate multiplicity.

The choices of I_3 and U are determined by the chain of subalgebras

$$gl(7/1) \supset sl(7/1) \supset su(7) x u(1) \supset su(2) x su(2) x su(3) x u(1)$$

where the first $su(2)$ corresponds to weak lefthanded isospin, the $su(3)$ corresponds to color, and the middle $su(2)$ corresponds to symmetry between generations. (For $\bar{s}$ generations, this $su(2)$ would be replaced by $su(s)$.) The element I_3 is specified by the requirement that it lie in the first $su(2)$ and take the standard form. The element U is specified by the requirement that it lie in $sl(7/1)$ and centralise $su(2) x su(2) x su(3) x u(1)$. These conditions force U to be of the form

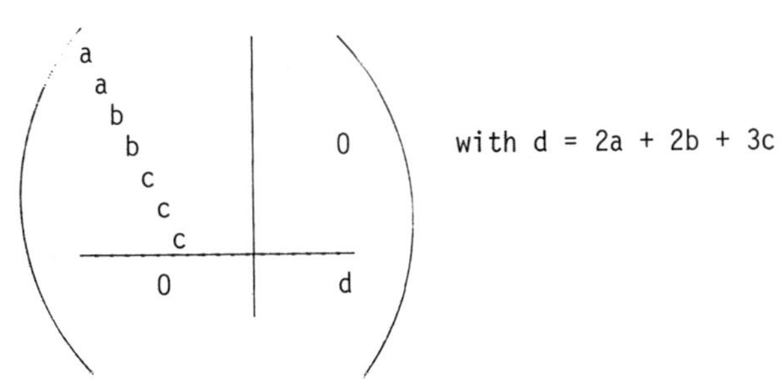

There are exactly two choices of the real parameters a,b and c which give all 128 correct charge values; and only one with $d \neq 0$, namely the values $a = -\frac{1}{3}$, $b = -\frac{2}{3}$ and $c = 0$. It can also be shown that if we postulate the above general form of the theory, but leave the number of colors arbitary, i.e. replace the su(3) by the su(m), then the only possibilities giving the observed charges are $m = 3$ or $m = 1$; i.e. the possible number of colors is specified by the theory. See [33c] for details.

We now describe our representations in detail, from the point of view of the orthosymplectic algebra, whose definition we also recall.

Let $V = V_0 + V_1$ be a graded vector space carrying a graded antisymmetric bilinear form $(\, , \,)$ of degree zero. Thus

$$(v_0,v_0') = -(v_0',v_0) \; , \; (v_0,v_1) = (v_1,v_0) = 0 \quad \text{and}$$

$$(v_1,v_1') = (v_1',v_1).$$

If we add a one dimensional vector space with generator z we obtain a super Lie algebra when we define the bracket relations by

$$[z,v_0] = [z,v_1] = 0 \quad ,$$

and

$$[u,w] = (u,w)z \quad \text{for } u \text{ and } w \text{ in } V_0 + V_1.$$

In case that $V_1 = \{0\}$ and $(\, , \,)$ is non-singular, this reduces to the standard Heisenberg algebra. We therefore denote this algebra by $h(V)$.

We can construct the universal enveloping algebra cf [8] of this super Lie algebra which we shall denote by $U(V)$. If we form the quotient algebra $W(V) = U(V)/I$ where I is the ideal generated by $z - 1$ we obtain an associative super algebra which we shall call the Weyl algebra. If $V_1 = 0$ and $(\, , \,)$ is non-singular, this reduces to the standard Weyl algebra. If $V_0 = 0$ then $W(V)$ is just the Clifford algebra of the space V_1 determined by the quadratic form associated to the (symmetric) form $(\, , \,)$. Thus $W(V)$ is an algebraic object which generalized both the standard Weyl algebra and the Clifford algebra.

The quadratic elements in the algebra $W(V)$ play a distinguished role. In view of the bracket relations on $W(V)$, the commutator of any two elements of V is a scalar; and hence we may identify the space of quadratic elements with

$$S^2(V) = (S^2(V_0) + \Lambda^2(V_1)) + V_0 \otimes V_1.$$

The bracket of a quadratic element with a linear element is again linear, so that $S^2(V)$ acts (via commutator bracket) as linear transformations on V which infinitesimally preserve the form (,). In fact $S^2(V)$ acts as a super Lie algebra of linear transformations, with $S^2(V_0) + \Lambda^2(V_1)$ being the even part and $V_0 \otimes V_1$ being the odd part. This algebra is one of the simple graded Lie algebras in the Kac classification [22], known as the orthosymplectic algebra. If $\dim V_0 = 2m$ and $\dim V_1 = k$ this superalgebra is denoted by $osp(2m/k)$.

Let U and W be vector spaces and let us set

$$V_0 = U + U^*$$

$$V_1 = W + W^*.$$

On V_0 we define the scalar product (,) by setting $(u_1,u_2) = (u_1^*,u_2^*) = 0$ for all u^i in U and u_i^* in U^* while $(u^*,u) = \langle u^*,u\rangle$ (natural pairing) with (,) taken to be antisymmetric. On V_1 we define (,) similarly except that it is taken to be symmetric.

We consider $U + W$ as a super vector space with U the even component and W the odd component. We consider the graded algebra $S(U+W)$ and recall that the definition of the super symmetric algebra is such that

$$S(U+W) = S(U) \otimes \Lambda(W).$$

Each element of $U^* + W^*$ determines a (super) linear function on $U + W$ which extends to a unique (super) derivation of $S(U+W)$. We denote the corresponding derivation by $D_\bullet$. Thus D_{u^*} is the super derivation corresponding to u^*. If we write the general element of $S(U+W)$ as a sum of terms of the form $f \otimes c$ $f \in S(U)$ and $c \in \Lambda(W)$ then D_{u^*} is determined by $D_{u^*}(f \otimes c) = \frac{\partial}{\partial u^*} f \otimes c$, where $\frac{\partial}{\partial u^*}$ is the usual directional derivative. Similarly D_{w^*} denotes the graded derivation determined by w^* and D_{w^*} acts only on the $\Lambda(W)$ component.

We let M_u and M_w denote multiplication by u in U and w in W respectively. It is practically immediate from the definitions that the mapping of $h(V) \to \text{End } S(U+W)$ defined by

$$z \rightsquigarrow \text{id}$$

$$u \rightsquigarrow M_u$$

$$u^* \rightsquigarrow D_{u^*}$$

$$w^* \rightsquigarrow M_w$$

$$w^* \rightsquigarrow D_{w^*}$$

is a representation of $h(V)$ as a super Lie algebra. It therefore extends by the universal property of the universal enveloping algebra to a representation of the algebra $U(V)$, and since z is sent into the identity transformation, this factors to give a representation of $W(V) = U(V)/(z-1)$. This is the <u>spin representation</u> of $W(V)$. For the case that $V_0 = 0$ this is the usual spin representation of the Clifford algebra of an even dimensional vector space on the exterior algebra of a space of half the dimension. In the case that $V_1 = 0$ this is the (infinitesimal) metaplectic representation in the Bargmann-Fock form with a slight change in notation, in that we are representing z by 1 and not by i/h. In the language of physicists, all we have done is construct the Fock space for a system of bosons and fermions, and realize the elements of our algebra $W(V)$ in terms of creation and annihilation operators.

It follows immediately from the Birkhoff-Witt theorem for super Lie algebras, cf. [8] that we have the vector space isomorphism

$$W(V) = S(U)\otimes\Lambda(W)\otimes S(U^*)\otimes\Lambda(W^*)$$

i.e. we can write the operators in $W(V)$ as "first apply the differentiations and then the multiplications". This has the following consequence: Suppose we choose a basis of U and W with corresponding dual bases of U^* and W^* with the induced basis of $S(U+W)$ and of $S(U+W+U^*+W^*)$. We assume that the constants have been chosen so that if e is a basis element of $S(U+W)$ then the corresponding dual basis element e^* has the property that $D_{e^*}e = 1$ and $D_{e^*}f = 0$ for all other basis elements. Thus if e_1 and e_2 are the basis elements

$$M_{e_2}D_{e_1^*}(e_1) = e_2$$

$$M_{e_2}D_{e_1^*}(f) = 0 \quad \text{if } f \text{ is any basis element other than } e_1.$$

Thus the representation of $W(V)$ on $S(U+V)$ is faithful. From now on we shall assume that we are in the situation that $V_0 = U + U^*$ and $V_1 = W + W^*$ and identify $W(V)$ as a subalgebra of End $S(U+V)$. It also follows from the preceeding remark that $W(V)$ contains all "finite matrices". Put in more invariant language: If X is any finite dimensional subspace of $S(U+W)$ then the restriction of $W(V)$ to $\mathrm{Hom}(X,S(U+V))$ is surjective.

Let G be a group with a given representation on U and on W. This determines a representation of G on $S(U+W)$ and on $W(V)$ preserving all the relevant algebraic structure. Let Γ' denote the space of G invariants in $S^2(V)$. Notice that Γ' is a super Lie subalgebra of $S^2(V)$. It is easy to use induction to prove the following:

Suppose that the group G *acts reductively on* $T(V)$ *and that the algebra of* G *invariants in* $T(V)$ *is generated by the space of* G *invariant tensors of degree* 2 *under multiplication and permutation. Then the associative subalgebra of* $W(V)$ *generated by* Γ' *is the full algebra of* G *invariants in* $W(V)$.

Let G be a group with a given representation on U and on W. This determines a representation of G on $S(U+W)$ and on $W(V)$ preserving all the relevant algebraic structures. Let Γ' denote the space of G invariants in $S^2(V)$. Notice that Γ' is a super Lie subalgebra of $S^2(V)$ when we regard $S^2(V)$ as the orthosymplectic algebra.

Recall that for a representation of a group G on a vector space, an isotypic component is the sum of all the subrepresentations equivalent to some fixed irreducible representation of G. Suppose that G acts reductively on $S(U+W)$ and let I_j denote the j-th isotypic component so that

$$S(U+W) = \bigoplus_{j=1}^{\infty} I_j .$$

The elements of Γ' commute with the action of G and hence must preserve each I_j. In [20] Howe proves that under suitable hypotheses the joint action of G and Γ' on each I_j is irreducible. Notice that since the representation of $W(V)$ on $S(U+W)$ is faithful, the G-invariants in $W(V)$ are precisely the operators in $W(V)$ which commute with G. We denote the algebra of these operators by $D(G)$, and observe that the underlined assertion, above, says that $D(G) = A(\Gamma')$ the algebra generated by Γ'. Howe assumes the following properties for G and its representations on $S(U+V)$ and $T(V)$

i) G acts reductively

ii) the invariants in $T(V)$ are generated by the invariants of degree two under multiplication and permutation

iii) there exists a linear projection operator $T \rightsquigarrow T^{\#}$ of $W(V)$ onto $D(G)$ such that

$$(ST^{\#})^{\#} = S^{\#}T^{\#}$$

and

iv) If $E \subset S(U+W)$ is a finite dimensional G invariant subspace and if $T \subset W(V)$ leaves E invariant, and commutes with G on E then $T_E = T^{\#}{}_E$.

For example, in the case that G is a reductive Lie group the existence of the above operator follows from the "unitary trick". Then Howe's theorem [20] is the following. <u>Under the above hypotheses the joint action of G and Γ' on I_j is irreducible, and $I_j = \sigma_j \otimes \tau_j$ where σ_j is an irreducible finite dimensional representation space of G and τ_j is an irreducible module for Γ'. Furthermore the σ_j and the τ_j determine each other so the correspondence $\sigma_j \leftrightarrow \tau_j$ is bijective</u>.

For the proof see [20] or [22] for the proof in the unitary case, and [23] for related theorems in symplectic geometry.

As we shall see, the superalgebra $gl(F_0/F_1)$ arises as the centralizer Γ' of $G = Gl(X)$ where X is a one dimensional vector space with $U = F_0 \otimes X$ and $W = F_1 \otimes X$. The fundamental representations are then exactly the representations τ_j(with $j = b$).

In fact, Howe [21] considers the following more general example: Let X be a finite dimensional vector space and $G = Gl(X)$.

$$U = X\otimes D_0 + X^*\otimes E_0 \quad \text{and} \quad W = X\otimes D_1 + X^*\otimes E_1$$

where D_0, E_0, D_1, and E_1 are auxiliary vector spaces. Since $V = U + U^* + W + W^*$ is a direct sum of copies of X and X^*, the tensor algebra $T(V)$ is isomorphic as a G module to sums of various tensor products of X and X^*. As is well known, the only invariants are those obtained by pairing various factors of X and X^* (so we must have as many factors of X as of X^*) and taking the invariant (a scalar multiple of the identity) in each $X\otimes X^*$. In other words all invariants occur only among tensors of even degree and are all obtained by applying a permutation to $I\otimes I\otimes I\otimes ..\otimes I$ in $X\otimes X^* \otimes X\otimes X^* .. \otimes X\otimes X^*$ and taking linear combinations. Thus the hypotheses about G are satisfied. We can write

$$V_0 = U+U^* = X\otimes F_0 + X^*\otimes F_0^* \quad \text{where} \quad F_0 = D_0+E_0^*$$

with

$$(x^*\otimes f^*,\ x\otimes f) = \langle x^*,x\rangle.\langle f^*,f\rangle$$

It is clear that G preserves (,) and that the centralizer of G in $End(V_0)$ is $End(F_1) + End(F_1^*)$. If $(T,T') \in End(F_1)+End(F_1^*)$ is to lie in $Sp(V_0)$ we must have $T \in Gl(F_1)$, and by symmetry, G is the centralizer of $Gl(F_1)$ in $Sp(V_0)$. At the infinitesimal level we see that the centralizer of G in $S^2(V_0)$ is $gl(F_1)$. Similarly, the centralizer of G in $O(V_1)$ is $gl(F_2)$ and the centralizer of G in $\Lambda^2(V_1)$ is $gl(F_2)$. Also

$$V_0\otimes V_1 = (X\otimes F_1+X^*\otimes F_1^*) \otimes (X\otimes F_2+X^*\otimes F_2^*)$$

and it is clear that the only G invariants can arise from pairing an X with an X^*, and hence the space of G invariants is $F_1^*\otimes F_2 + F_1\otimes F_2$.

Taking X one dimensional, $U = X\otimes F_0$ and $W = X\otimes F_1$ gives $\Gamma' = gl(F_0,F_1)$. The irreducible representations of G occuring in

T(X+X*) are all parametrized by an integer b (= # of X components - # of X* components) . The corresponding representations of Γ'(with the possible interchange of the roles of n and m in the notational convention) are our fundamental representations for integer b . For the case of sl(n/1) and non integer b (or for the reducible representations that we use in [33], the space S^b is taken to be the space of homogeneous functions of degree b. For example, for gl(7/1) the representation that we used was the one with b = 3 and was on the one hundred twenty eight dimensional space

$$\wedge^0 \otimes s^3 + \wedge^1 \otimes s^2 + \wedge^2 \otimes s^1 + \wedge^3 \otimes s^0 + \wedge^4 \otimes s^{-1} + \wedge^5 \otimes s^{-2} + \wedge^6 \otimes s^{-3} + \wedge^7 \otimes s^{-4}$$

and is reducible, the space

$$\wedge^0 \otimes s^3 + \wedge^1 \otimes s^2 + \wedge^2 \otimes s^1 + \wedge^3 \otimes s^0$$

being a sixty four dimensional invariant subspace without an invariant complement. We can realize the full space in an appropriate completion of the full spin metaplectic representation that we have been considering. Alternatively, in the particular case we are considering (and in all of the cases where s is even, with 2^{s+1} the number of generations) the invariant subspace is non-singularly paired with its quotient, and as we suggested in [33a], we might want to consider the sum of the subspace and its quotient as providing the correct representation, and the subspace does appear in our spin metaplectic representation as an irreducible component.

The hypothesis in the Bars-Iachello model is that the Hamiltonian lie in W(V) . This is the meaning of the requirement that it be built out of creation and annihilation operators. It is also a plausible requirement in our theory that a major component of the mass operator also be given in this way, and also by quartic elements.

Our representations of gl(7/1) arise from the identification of gl(1/7) as the centralizer of Gl(1,ℂ) in OSp(2/14) . Since we are using gl(n/m) as an internal symmetry, this suggest that we seek a geometrical setting in which a centralizer arises as an internal symmetry. This is precisely what happens in the method of dimensional reduction which we explain in the next section.

2. Dimensional reduction.

We begin this section under a geomtrical description of the method of dimensional reduction as developed by Manton [24] and Harnad, Shnider and Tafel [16]. We then show, following [40], how to give an algebraic version of this procedure which extends to supersymmetry.

Let $\pi : P \to X$ be a principal bundle with structure group K (not necessarily compact). We shall write $R_b : P \to P$ for right multiplication by $b^{-1} \varepsilon K$ so

$$R_{bp} = pb^{-1}$$

and

$$R_{b_1} o R_{b_2} = R_{b_1 b_2}.$$

(Thus R defines a left group action of K on P.)

By Aut P we mean the group of all diffeomorphisms ϕ, of P with itself which commute with the K action, i.e. satisfy

$$\phi(pb) = \phi(p)b \quad \text{for all } b \varepsilon K.$$

Each such ϕ defines a transformation, $\overline{\phi} : X \to X$ given by

$$\overline{\phi}(\pi p) = \pi\phi(p).$$

This makes sense because $\pi\phi(pb) = \pi(\phi(p)b) = \pi\phi(p)$ so $\pi\phi(p)$ is independent of the choice of p in $\pi^{-1}(\pi p)$. The map $\overline{\phi}$ is clearly a diffeomorphism of X and so we get a homomorphism

$$\text{Aut } P \to \text{Diff } X.$$

We shall be interested in the case where we are given a Lie subgroup G of Aut P.
This then gives an action of G on X.

Let G_x denote the isotropy group of a point $x\varepsilon X$. Each $a\varepsilon G\wedge$ maps the fiber $\pi^{-1}(x)$ into itself. In particular, each $p\varepsilon\pi^{-1}(x)$ determines

$$\wedge_p : G_x \to K$$

by

$$ap = p\,(\wedge_p(a)) \tag{2.1}$$

and $\wedge_p$ is clearly a homomorphism:

$$\begin{aligned}(a_1a_2)p&=a_1(a_2p)=a_1(p\wedge_p(a_2))\\ &=(a_1p)\wedge_p(a_2) \qquad\qquad \text{since } a_1\varepsilon\text{Aut } P\\ &=p\wedge_p(a_1)\wedge_p(a_2).\end{aligned}$$

Also

$$\begin{aligned}a(pb^{-1})&=p\wedge_p(a)b^{-1}\\ &=pb^{-1}b\wedge_p(a)b^{-1}\end{aligned}$$

or

$$\wedge_{pb^{-1}}(a)=b\wedge_p(a)b^{-1}$$

or

$$\wedge_{R_b(p)}=A_b\wedge_p \ , \tag{2.2}$$

where A_b denotes conjugation by b. In particular, <u>the set of all</u> p' <u>with</u> $\wedge_{p'}=\wedge_p$ <u>consist of all</u> p' <u>of the form</u> $p'b$ <u>where</u>

$b\Lambda_p(a)b^{-1}=\Lambda_p(a)$ <u>for all</u> $a\varepsilon G_x$, <u>i.e. of all</u> b <u>in the centralizer of</u> $\Lambda_p(G_x)$.

It is a theorem of Mostow [25] that for compact groups, G, there are, locally, only a finite number of orbit types on X. Suppose we make the drastic assumption that there is only a single orbit type - that there is a global cross-section for the action of G on X. More precisely let us assume that

$$X = M \times (G/B)$$

where G does not act at all on M and where B is some closed subgroup of G. Thus on each G orbit of X there is a unique point, corresponding to the coset of the identity in G/B, and the set of G orbits in X is thus parametized by points of M. The isotropy group at our distinguished points are all equal to B.

If the groups B and K are compact, we would know [6] that the set of conjugacy classes of homomorphisms of B into K would be discrete. Then, if M and K were connected, there would be only one conjugacy class of homomorphisms, Λ_p, of B into K. Let us adopt this also as an additional hypothesis. So we assume that there is a single conjugacy class of homomorphisms of B into K with all Λ_p, for $p\varepsilon\pi^{-1}(M \times B)$, belonging to this conjugacy class. But this means that we can fix one homomorphisms, $\Lambda : B \to K$, and consider the set of all p under $\Lambda_p=\Lambda$. Any two such p and p' lying over the same point $M \times B$ differ from one another by right multiplication by some $b\varepsilon\text{Cent}(\Lambda(B))$. Thus,

<u>Under the above hypotheses, we get a principal bundle,</u> $Q \to M$, <u>whose structure group is</u> $\text{Cent}(\Lambda(B))$.

The passage from $P \to X$ with structure group K to the smaller bundle $Q \to M$ with structure group $\text{Cent}(\Lambda(B))$ is one half of the geometrical part of the process of dimensional reduction. We now must look at G invariant connections, i.e. Yang-Mills fields, on P.

Since there are various conventions we first review the basic definitions in order to establish our notation: A connection on P is a choice of horizontal subspace at each point of P, the choice being invariant under the action of K. A choice of horizontal subspace at $p\varepsilon P$ is the same as projection V_p:

$TP_p \to TP_p$ where the image of V_p is the tangent space to the fiber of P at p. The condition of K invariance means that

$$dR_b o V_p = V_{pb^{-1}} o dR_b. \tag{2.3}$$

The action of K on P defines for each $p \varepsilon P$ a map

$$U_p : K \to P$$

$$U_p(c) = pc^{-1} \quad \text{for all } c\varepsilon K.$$

For any $b\varepsilon K$ we have

$$R_b o U_p(c) = pc^{-1}b^{-1} = U_p(bc)$$

Writing ℓ_b for the operation of left multiplication of b on K, we can write this last equation as

$$R_b o U_p = U_p o \ell_b.$$

We let $u_p = (dU_p)_e : k \to TP_p$ be the differential of the map U_p at e where $k = TK_e$, the Lie algebra of K. Thus

$$u_p(\xi) = \frac{d}{dt} U_p(\text{expt } \xi)\Big|_{t=o} \qquad \text{for } \xi\varepsilon k.$$

Now

$$R_b o U_p(\text{expt}\xi) = p(\text{expt}\xi)^{-1}b^{-1} = pb^{-1}b(\text{expt}\xi)^{-1}b^{-1}$$
$$= U_{R_b(p)}(b\text{expt}\xi b^{-1})$$

and, differentiating at $t=o$ gives

$$dR_b(u_p(\xi))=u_{R_b(p)}(Ad_b\xi). \tag{2.4}$$

Since u_p maps k onto the tangent space to the fiber we can form

$$\Theta_p=u_p^{-1}oV_p. \tag{2.5}$$

So $\Theta_p : TP_p \to k$ is a k valued linear differential form on P which is known as the connection form. By (2.4) and (2.5) it satisfies

$$\Theta_{R_bp}odR_b=Ad_b\Theta_p. \tag{2.6}$$

The action of G on P induces a map

$$\Psi_p : G \to P$$

for each $p\varepsilon P$ given by

$$\Psi_p(a) = ap.$$

Notice that

$$\Psi_{ap}(c)=cap=a(a^{-1}ca)p$$

$$= L_a o\Psi_p(A_{a^{-1}}c)$$

so

$$\Psi_{ap}=L_a o\Psi_p oA_{a^{-1}} \tag{2.7}$$

If $p\in\pi^{-1}(x)$ and $a\in G_x$ then

$$ap=p\Lambda_p(a)$$

so

$$\Psi_p(a)=u_p(\Lambda_p(a)^{-1}) \quad \text{for } a\in G_{\pi p}.$$

write

$$\psi_p=d\Psi_p \ : \ g \to TP_p.$$

It follows from (2.7) that

$$\psi_{ap}=dL_a o \psi_p o Ad_{a^{-1}} \tag{2.8}$$

We also write

$$\lambda_p \ : \ g_x \to k$$

$$\lambda_p=(d\Lambda_p)_e.$$

Then we can define

$$J_p : g \to k$$

$$J_p = \Theta_p \circ \psi_p . \qquad (2.9)$$

Then (2.5) and (2.9) imply that

$$J_p|g_{\pi p} = -\lambda_p$$

and

$$\Psi_{pb^{-1}}(a) = apb^{-1} = \Psi_p(a)b^{-1} = R_b \Psi_p(a)$$

so

$$\psi_{pb^{-1}} = dR_b \circ \psi_p$$

and hence

$$J_{pb^{-1}} = \Theta_{pb^{-1}} \circ \psi_{pb^{-1}} = \Theta_{R_b(p)} \circ dR_b \psi_p$$

$$= Ad_b \Theta_p \circ \psi_p$$

$$= Ad_b \, J_p .$$

Thus, if we define the function $J : P \to Hom\,(g,h)$ by $J(p)=J_p$ we see that J satisfies the two conditions

$$J \circ R_b = Ad_b J \qquad (2.10)$$

and

$$J(p)_{|g_{\pi p}} = -\lambda_p. \tag{2.11}$$

This result is due to Wang [49]. In the case that G acts transitively on X, the image of ψ_p, together with the tangents to the fiber, space all of TP_p and hence we can recover Θ_p from J_p. In this case giving a J satisfying (2.10) (2.11) is the same as giving a connection. In the general case there is more information contained in Θ than in J.

Let us now examine the condition that a connection be G invariant: let $L_a: P \to P$ denote the (left) action of $a \varepsilon G$, so

$$L_a(p) = ap.$$

Then

$$U_{ap}(c) = apc = L_a U_p(c)$$

so

$$u_{ap} = dL_a o u_p \tag{2.12}$$

as maps from $k \to TP_{ap}$. We say that the connection is invariant, under G if

$$V_{ap} o dL_a = dL_a o V_p \tag{2.13}$$

for all $a \varepsilon G$ and $p \varepsilon P$. By (2.5), (2.13), and (2.12)

$$\Theta_{ap} o dL_a = u_{ap}^{-1} V_{ap} o dL_a$$

$$= u_p^{-1} dL_a^{-1} dL_a V_p = \Theta_p$$

so G invariance is equivalent to

$$\Theta_{ap} o dL_a = \Theta_p \tag{2.14}$$

This implies, by (2.8) that

$$J_{ap}=\Theta_{ap}\Psi_{ap}$$

$$=\Theta_{ap}dL_a\Psi_p Ad_{a^{-1}}$$

$$= \Theta_p\Psi_p Ad_{a^{-1}}$$

so

$$J_{ap}=J_p o Ad_{a^{-1}} \tag{2.15}$$

In case $a\varepsilon G_{\pi p}$ we can combine (2.6) with (2.14) to get

$$\Theta_p o dR_{\Lambda_n(a)} o dL_a = Ad_{\Lambda_p(a)}\Theta_p \tag{2.16}$$

while we can combine (2.10) and (2.15) to obtain

$$Ad_{\Lambda_p(a)}J_p=J_p o Ad_a \quad \text{for} \quad a\varepsilon G_{\pi p}. \tag{2.17}$$

This last equation is due to Wang [49]. In the case that G acts transitively on X, giving an invariant J_p at one single p determines an invariant connection on P. Thus, the set of G invariant connections on P is parametized by the set of all J_p satisfying

$$J_{p|g_{\pi p}} = -\lambda_p \tag{2.11}$$

and

$$Ad_{\Lambda_p(a)}J_p=J_p Ad_a \qquad a\varepsilon G_{\pi p} \tag{2.17}$$

This is the theorem of Wang.

Let us now turn to the case where $X = M\times G/B$. We have constructed the bundle $Q\to M$ by identifying M with $M\times\{B\}$ and letting Q consist of all $p\in\pi^{-1}(M\times\{B\})$ with $\Lambda_p=\Lambda$, a fixed homorphism of $b\to K$. For any such p

$$L_a p = ap = p\Lambda(a) = R_{\Lambda(a)^{-1}} p$$

so

$$R_{\Lambda_p(a)} \circ L_a = \text{id} \quad \text{on } Q.$$

Thus (2.16) implies that

$\Theta = \Theta_{|Q}$ <u>takes values in the Lie algebra of</u> $\text{Cent}(\Lambda(B))$, <u>and hence</u> Θ <u>is a connection form on</u> $Q\to M$. <u>Similarly</u> $j=J_{|Q}$ <u>is a function from</u> $Q\to\text{Hom}(g,h)$ <u>satisfying</u>

$$J_{|b} = -\lambda \text{ and } \text{Ad}_{\Lambda(a)} J = j\text{Ad}_a \text{ for all } a\in B. \qquad (2.18)$$

Conversely any such pair Θ, j gives rise to a G invariant connection on P. Thus, the set of G-invariant connections, Θ, on P is parametized by the set of pairs (Θ,j), where Θ is a connection on Q (a Yang-Mills field) and j can be thought of as a Higgs field.

In case that G acts transitively on X, Wang shows that the curvature Θ at p is given by

$$F_p(\xi,\eta) = J_p[\xi,\eta] - [J_p(\xi), J_p(\eta)],$$

i.e. by the failure of J to be a Lie algebra homomorphism. Notice that this expression is quadratic in J. Thus, if (relative to an invariant bilinear form on k) we form $F\wedge *F$, this is of fourth order in

J. Under the assumption of G invariance, integration (relative to a G invariant measure on X) just gives some constant factor. Thus the pure Yang-Mills functional reduces to a quartic polynomial in the vector variable J. Similarly, as shown by Manton [24] and Harnad Shnider and Tafel [16] in the general case, the pure Yang-Mills functional on G invariant connections on P becomes a Yang-Mills-Higgs Lagrangian on Q, quartic in the Higgs field J. We shall give another, infinitesmal, derivation of the above construction, which has the advantage of being purely algebraic, and hence extending to superalgebras. The following is an extract from [40].

The idea is to express everything in terms of the right invariant vector fields on P. We can consider the right invariant vector fields on P as constituting sections of a vector bundle, $\underline{E} \to X$. The vertical right invariant vector fields are sections of a subbundle, $\underline{V}$ and so we have the exact sequence of vector bundles

$$0 \to \underline{V} \to \underline{E} \to TX \to 0.$$

A connection on E is then just a smooth splitting of this sequence. If G acts as automorphisms of P, then we get a map from its Lie algebra, g, into sections of E.

From this point of view, the starting point is an exact sequence of Lie algebra sheaves

$$V \to E \overset{\pi}{\to} M \to 0, \quad \text{with} \tag{2.18}$$

$$[v,e] \quad \in \quad V \quad \text{for } e \in E \text{ and } \quad v \in V \tag{2.19}$$

In the case that we have in mind, X will be a differentiable manifold, M a submanifold (a cross section for a group action) and M will be the sheaf of germs of sections of TX restricted to M. Similarly, $P \to X$ will be a principal bundle and E will be the sheaf of germs of sections of $\underline{E}$ restricted to M, where $\underline{E}$ is the bundle described above. Similarly, V will be the sheaf of germs of sections of V restricted to M. We will let g denote the locally constant sheaf associated with the Lie algebra g. We assume that the cross section M has been chosen so that the isotropy algebras at all points of M are the same Lie subalgebra, h, of g. The fact that we have G acting as automorphisms of P is then expressed by the condition that we have sheaf homomorphisms giving the diagram

$$\begin{array}{ccccccc} V & \to & E & \overset{\pi}{\to} & M & \to & 0 \\ & & \mu\uparrow & & \mathrm{id}\uparrow & & \\ & & g & \to & M & & \end{array} \qquad (2.20)$$

A connection is a splitting $\Theta: E \to V$ of (2.18). The connection Θ will be invariant if

$$\Theta[\xi,e] = [\xi,\Theta(e)] \quad \text{for } \xi \in \mu\,(g)\ , \text{ and } \ e \in E\ . \qquad (2.21)$$

We also want to consider the sheaf theoretical version of taking all germs of sections along M only. So we assume that we are given a sequence of sheaves

$$\nu_0 \to E_0 \overset{\pi^0}{\to} n_0 \to 0 \qquad (2.22)$$

where ν_0 is a sheaf of Lie algebra, E_0 is a sheaf of ν_0 modules with the action of elements of ν_0 on E_0 still denoted by [,] and with (2.19) satisfied with ν replaced by ν_0 and E replaced by E_0. We assume that there are "evaluation" sheaf maps, all denoted by ε making the diagram (2.23) commute, and

$$\begin{array}{ccccccc} \nu & \to & E & \to & M & \to & 0 \\ \varepsilon\downarrow & & \varepsilon\downarrow & & \varepsilon\downarrow & & \\ \nu_0 & \to & E_0 & \to & M_0 & \to & 0 \end{array} \tag{2.23}$$

which preserve the remaining algebraic structure, i.e. $\varepsilon : \nu \to \nu_0$ is a Lie algebra homomorphism and $\varepsilon: E \to E_0$ is consistent with the structures of E as a ν module and E_0 as ν_0 module. We let h denote the locally constant sheaf (along M) associated with the subalgebra h , and assume that we have the commutative sheaf diagram.

$$\begin{array}{ccccccc} \nu_0 & \to & E & \to & m & \to & 0 \\ \lambda\uparrow & & \mu_0\uparrow & & \uparrow & & \\ h & \to & g & \to & m_0 & & \end{array} \tag{2.24}$$

Let N_0 be a subsheaf of M with ι the inclusion map. Let us assume that there exists at least one connection $\Theta_0: E_0 \to \nu_0$ which satisfies

$$\Theta_0[\nu,e] = [\nu,\Theta_0(e)] \quad \text{for all } \nu \; \varepsilon \; \lambda(h) \tag{2.25}$$

(Notice that this would be a consequence of assuming the existence of a G invariant connection on the full bundle, i.e. of assuming (2.21)). Then we claim that the subsheaf, $\mathfrak{s}_0$, of ε_0 , consisting of all $\mathfrak{s}$ satisfying

$$\pi(\mathfrak{s}) \quad \varepsilon \quad \iota(N_0) \tag{2.26}$$

and

$$[\nu,\mathfrak{s}] = 0 \quad \text{for all} \quad \nu \; \varepsilon \; \lambda(h) \tag{2.27}$$

maps surjectively onto $\iota(N_0)$. Indeed, for any $e \in E$, let $f = e - \Theta(e)$ so $\pi(f) = \pi(e) - \pi\Theta(e) = \pi(e)$. Then

$$\begin{aligned}[\nu, f] &= [\nu, (e - \Theta(e))] \\ &= [\nu, e] - [\nu, \Theta(e)] \\ &= [\nu, e] - \Theta[\nu, e] \quad \text{by the invariance of } \Theta, \\ &= 0\end{aligned}$$

by (2.25) and the fact that Θ is the identity on V_0. Choosing e with $\pi(e) \in \iota(N_0)$ gives (2.26), and it is clear that we can span all of $\iota(N)$. We thus get the commutative diagram

$$\begin{array}{ccccccc} \mathrm{Cent}(\lambda(h)) & \to & F_0 & \to & N_0 & & \\ \downarrow & & \tau\downarrow & & \iota\downarrow & & \\ V_0 & \to & E_0 & \xrightarrow{\pi_0} & M_0 & \to & 0 \\ \lambda\uparrow & \Theta_0 & \mu_0\uparrow & & \uparrow & & \\ h & \to & g & \to & M_0 & & \end{array} \qquad (2.27)$$

The diagram (2.27) without the Θ_0 is given by the data (2.24) under the assumed hypotheses. The top row of (2.27) then constitutes the "reduced bundle". An invariant connection, Θ, then splits lower and upper corners of (2.27) and gives rise to a map

$$j = \Theta \bullet \mu$$

satisfying

$$j[\eta, \xi] = [\lambda(\eta), j(\xi)] \quad \text{for } \eta \text{ in } h \qquad (2.28)$$

and to a connection $\Theta: F_0 \to \mathrm{Cent}(\lambda(h))$. The curvature $K(\Theta)$ can be thought of as a map of $E \times E \to \nu$ given by

$$K(\Theta)(e_1,e_2) = \Theta([e_1-\Theta(e_1), e_2-\Theta(e_2)]). \qquad (2.29)$$

Let us write

$$e_i = \xi_i + f_i, \text{ where } \xi_i \in \mu(g) \text{ and } f_i \in \tau(F), \quad i = 1,2$$

Then

$$K(\Theta)(e_1,e_2) = \Theta([\xi_1-\Theta(\xi_1),\xi_2-\Theta(\xi_2)]) + \Theta([f_1-\Theta(f_1),f_2-\Theta(f_2)])$$

$$+\Theta([\xi_1-\Theta(\xi_1),f_2-\Theta(f_2)]) - \Theta([\xi_2-\Theta(\xi_2),f_1-\Theta(f_2)]). \qquad (2.30)$$

We can simplify the terms in this equation which involve ξ_i by using the invariance condition (2.21) and the fact that Θ is the identity on elements of ν. Thus, for example, in the first term we have the expression

$$-\Theta([\xi_1;\Theta(\xi_2)]) = -\Theta^2([\xi_1,\xi_2]) = \Theta([\xi_1,\xi_2])$$

and the entire first term simplifies to

$$[\Theta(\xi_1), \Theta(\xi_2)] - \Theta(\xi[_1,\xi_2]) .$$

If we define the map $Q(j) : \varepsilon \times \varepsilon \to$ by

$$Q(j)(\eta_1,\eta_2) = [j(\eta_1),j(\eta_2)] - j([\eta_1,\eta_2])$$

we see that the first term in (2.30) corresponds to $Q(j)$ (which is quadratic in j). The second term in (2.30) is clearly just $K(\nu)$ evaluated on f_1 and f_2. In the third term,

$\Theta([\xi_1,f_2]) - \Theta([\xi_1,\Theta(f_2)]) = 0$ and the remaining terms simplify to

$$[f_2-\Theta(f_2),\Theta(\xi_1)].$$

If $\xi_1 = \mu(\eta_1)$ this last expression is the covariant derivative of $j(\eta_1)$ in the direction $\pi'(f_2)$. The last term is the same with 1 and 2 interchanged.

Thus the curvature $K(\Theta)$ is expressed in terms of the data given by Θ and j (and is guadratic in j).

Absolutely all of the above sheaf theoretic constructions make sense with Lie algebras everywhere replaced by Lie superalgebras, giving a formulation of dimensional reduction for supersymmetry. It might thus be conjectured that our internal symmetry is derived via reduction from a larger orthosymplectic theory.

3. Internal Supersymmetry and Quantum Ghost Fields

In Particle Physics, if we require that the elements of the superalgebra arise from fields in ordinary space-time, the quantum statistics of End F (see section 1) become correlated to the Z(2) grading: End_0F carry Bose statistics (and by the spin-statistics theorem, integer spin J) and End_1F have Fermi statistics (and half-integer spin). This then also requires F_o and F_1 to have opposite statistics and univalence $\eta=(-1)^{2J}$. In the Bars-Iachello application to nuclei, End_1F indeed corresponds to the addition or removal of a proton, so that $\Delta J(End_1F)=\frac{1}{2}$, e.g. in going from ${}^{190}_{76}Os_{114}$ to ${}^{191}_{77}Ir_{114}$ etc... In both original applications of superalgebras to Particle Physics, Poincaré (or Conformal) Supersymmetry [Golfand-Likhtman 1971; Wess-Zumino 1974; Salam-Strathdee 1974] and Dual Model Supergauges [Aharonov-Casher-Susskind 1971; Ramond 1971; Neveu-Schwarz 1971], $\Delta J(End_1F)=\frac{1}{2}$ too.

In Internal supersymmetry, however, the superalgebra commutes with the Lorentz group, i.e. $\Delta J(End\ F)=o$ and the spin-statistics correlation is violated. One possible interpretation is that either F_o or F_1 then consist of unphysical "ghost" states, bosons with $\eta=-1$ or fermions with $\eta=1$, even though all other quantum numbers are still those of physical states. In the following, we discuss one such possible interpretation.

Ghosts were introduced in quantum field theory to restore unitarity [Feynman 1963; DeWitt 1965; Faddeev-Popov 1967], after gauge fixing. Gauge invariance (group G) requires the Yang-Mills potential A^i_μ (in the connection $\Theta|_\Sigma=A=A^i_\mu\lambda_i dx^\mu$, Σ a section, λ_i a basis of the Lie algebra of G) to be massless, and thus have only 2 physical components (the transverse ones). Loss of gauge invariance generates unphysical contributions by the longitudinal and time components of A^i_μ in closed loops. These are cancelled by the contributions of the J=o ghost X^i and antighost $\overline{X}_i$, with the Fermi statistics providing the necessary minus signs. The equations guaranteeing the cancellations and the resulting Unitarity of the quantized theory off mass-shell [Taylor 1971;

Slavnov 1972; Becchi et al 1976; Tyutin 1975, Curci-Ferrari 1975, 1976 a,b, 1978, Ojima 1980] were shown to coincide with the Maurer-Cartan structural equations of the Principal Bundle $P_2 \to M$, with structure group GXG, the fiber-product with itself of the gauge theory principal bundle $P \to M$ [Thierry-Mieg 1979; Ne'eman 1979a;Ne'eman-Thierry Mieg 1980a,b; Thierry-Mieg 1980a,b; Thierry-Mieg - Ne'eman 1982; Ne'eman 1982; Beaulieu- Thierry-Mieg 1982]

$$s A + DX = 0 \quad , \quad s X + \tfrac{1}{2}[X,X] = 0$$

$$\bar{s} A + D\bar{X} = 0 \quad , \quad \bar{s}\bar{X} + \tfrac{1}{2}[\bar{X},\bar{X}] = 0$$

where the exterior derivative $\tilde{d}$ and connection $\tilde{\Theta}$ on P_2 split, upon a choice of section Σ

$$\tilde{d} = d + s + \bar{s} \quad , \quad \tilde{\Theta} = A + X + \bar{X}$$

s, X in $G \varepsilon P$ and $\bar{s}$, $\bar{X}$ in $G=P_2/P$, X and $\bar{X}$ anticommutative (Fermionic) ghosts, and

$$\bar{s}\, X + s\, \bar{X} + [\bar{X},X] = 0$$

Introducing an auxiliary bosonic field b of dimension 2, we also have

$$\begin{cases} s\,\bar{X} =: -\tfrac{1}{2}[\bar{X},X] + b \\ \bar{s}\, X = -\tfrac{1}{2}[\bar{X},X] - b \\ s\, b = -\tfrac{1}{2}[X,b] - \frac{1}{8}[[X,X]\ ,\ \bar{X}] \\ \bar{s}\, b = -\tfrac{1}{2}[\bar{X},b] + \frac{1}{8}[[\bar{X},\bar{X}]\ ,\ X] \end{cases}$$

$$s^2 = \bar{s}^2 = s\bar{s} + \bar{s}s = 0$$

For matter fields ψ, we get a composite ghost state with identical quantum numbers:

$$s\, \psi = [X,\psi] := \psi_s$$

The ghost field X does not depend on the connection but does depend upon the choice of section. As it is given by $\Theta - \pi^*\Sigma^*\Theta$ it depends, at the point $\Sigma(x)$ only on the one jet of Σ at x. In fact, it really lives, as an invariant object, on $J_1(P)$ and is essentially the fundamental $V(P)$ valued one form defined on $J_1(P)$ where $V(P)$ denotes the vertical target bundle of P cf [14]. As such it is defined for any fibered manifold, not necessarily a principal bundle.

If G is a supergroup, with λ_o^i even, and λ_1^m odd parts of the superalgebra basis, the X_o^i are ghosts, but the X_1^m become bosonic scalar fields ϕ^m. It has been suggested that such "exorcized ghosts" ϕ^m be identified with Goldstone fields (and with Goldstone-Higgs fields if the odd part of the group be gauged) [Ne'eman 1979b; Ne'eman Thierry-Mieg 1980a,b]. In matter field representations of the supergroup, the ghost states will be interpreted as the ψ_s, with ψ corresponding to physical states in the statistics-conjugate representation. In this interpretation we thus double all matter representations by adjoining their ε^o-conjugate [Ne'eman 1979b; Balantekin-Bars 1981], where ε^o is an odd morphism inverting both statistics and chiralities in $SU(1/2)$ or $SU(1/7)$. This fits with equating the number of physical states to be identified phenomenologically with the dimensionality 2^{5+s} in $SU(1/5+s)$.

In the Weinberg-Salam Electroweak Theory, the gauge group $G{:}SU(2)_I \times U(1)_U$ is broken by an $I=\frac{1}{2}$ complex Goldstone-Higgs doublet ϕ with $U=\pm 1$,

$$\langle o \mid \phi_o \mid o \rangle \neq o$$

where ϕ_o is the $CP=1$ electrically neutral component. It was noticed [Ne'eman 1979b] that identification of G with the even subgroup of $SU(1/2)$ reproduces for one generation all (otherwise arbitrarily assigned) quantum numbers for the quarks [Ne'eman -Thierry Mieg 1980c] in the fundamental $\underset{\sim}{4}\ (=\sum_o^2 \Lambda_n)$; for integer electric charge Q, this decomposes into $\underset{\sim}{3}(=\sum_o^1\Lambda_n) + \underset{\sim}{1}(=\Lambda_2)$ precisely fitting the leptons. Using the ghost interpretation, we thus have e.g. for the leptons:

$\underset{\sim}{3}\ (\nu^{o}_{L}\ ,\ e^{-}_{L}\ /\ (e^{-}_{R})^{s}_{L}) \oplus \underset{\sim}{3}'\ ((\nu^{o}_{L})^{s}_{R}\ ,\ (e^{-}_{L})^{s}_{R}\ /\ e^{-}_{R})$ etc....

The 4 physical $W^{\pm}_{\mu}$, A_{μ} , Z^{o}_{μ} fields correspond to linear combinations of the gauge potentials V^{i}_{μ} of SU(2)xU(1). Their quantum ghosts X^{i} appear in the adjoint $\underset{\sim}{8}$ of SU(2/1) together with the Goldstone-Higgs ϕ^{m} , thus again constraining an otherwise arbitrary choice. Note that the statistics of the I=½ components of the octet are indeed required to be complimentary to those of the I=1,o X^{i}, which makes the ϕ become physical boson scalars.

All of the above ingredients thus correspond precisely to the Weinberg Salam theory, as treated quantum-mechanically by 't Hooft ['t Hooft 1971a,b]. There is, however, a new type of component, namely the ξ^{m}_{μ} isodoublet ghost (J=1 fermion) accompanying the W^{I}_{μ}, A_{μ}, Z^{o}_{μ} in $\underset{\sim}{8}$.

It is well-known that in a G=U(1) gauge theory, such as Quantum Electrodynamics, it is possible to renormalize the theory without introducing ghosts. Owing to the Abelian features of the gauge group, the unphysical components of V_{μ} (and the ghosts) simply decouple, so that Unitarity does not involve a cancellation by the contributions of X and $\overline{X}$. However, when the theory is coupled with Gravity, the ghosts become essential. The (non-Abelian) curvature of space-time interferes with the decoupling, and only cancellation by X, $\overline{X}$ can save Unitarity.

We conjecture that the J=1 , I=½ ghosts ξ^{m}_{μ}, $\overline{\xi}_{\mu m}$ were unnecessary in 't Hooft's flat space treatment, but become necessary when the theory is coupled to General Relativity. This would a priori not be happening in the "Covariant Quantization" treatment of Quantum Gravity ['t Hooft-Veltman 1974] in which the gravitational field is a perturbation of the Minkowski metric. In this treatment, the Hilbert space (and vacuum) correspond to a flat background. However, in a complete treatment of Quantum Gravity in curved space-time, the vacuum state itself is a functional of the gravitational field. For the vacuum to have an I=½ component, as required by $I \mid o > \neq o$ (i.e. spontaneous symmetry breakdown, leading to $< o \mid \phi_{o} \mid o > \neq o$) the graviational field has to generate a solution with I=½. Assuming that the contribution is due to a linear symmetry-breaking term in $g_{\mu\nu}(x)$, we write

$$g'_{\mu\nu}(x) = g_{\mu\nu}(x) + \delta g_{\mu\nu}(x)$$

where

$$[I, g_{\mu\nu}] = o \quad , \quad [I, \delta g_{\mu\nu}] \neq o \quad , \quad [I_3, \delta g_{\mu\nu}] = \tfrac{1}{2}\delta g_{\mu\nu}$$

The $\delta g_{\mu\nu}(x)$ field will have ξ^m_μ and $\bar{\xi}_{\mu m}$ as ghost and anti-ghost, in the same $\underset{\sim}{I}$,U direction as ϕ_o, just as $g_{\mu\nu}$ requires isoscalar ξ_μ and $\bar{\xi}_\mu$, in any Covariant Quantization of General Relativity with BRS equations [Delbourgo-Ramon Medrano 1976]. Since Covariant Quantization is at this stage the only relatively advanced technique for Quantum Gravity, we conjecture that a perturbed Covariant Quantization, with an $I=\tfrac{1}{2}$ increment $\delta g(x)$ added to $g_{\mu\nu}(x)$, is a good approximation of the complete theory of Quantum Gravity in the presence of matter undergoing a Spontaneously Broken Yang Mills interaction. We shall develop this model elsewhere.

Although the particular role of the ξ^m_μ, $\bar{\xi}_{\mu m}$ has thus not been completely elucidated, an Extended BRS algebra does exist for SU(2/1) and for any supergroup [Thierry-Mieg - Ne'eman 1982].

References

1. Y. Aharonov, A. Casher and L. Susskind, Phys. Letters B38, 512 (1971).

2. A. B. Balantekin and I. Bars, J. Math. Phys. 22, 1810 (1981).

3. I. Bars, in Group Theoretical Methods in Physics (Cocoyoc 1980 IX Intern. Conf). K. B. Wolf ed., Lecture Notes in Physics #135, Springer-Verlag, Berlin/Heidelberg/New York, pp. 319-332 (1980).
   F. Iachello, Phys. Rev. Letters 44, 772 (1980).
   A. B. Balantekin, I. Bars and F. Iachello, Yale report TYP 81.07.

4. L. Beaulieu and J. Thierry-Mieg, report CU-TP-196, Nucl. Phys. (1982).

5. C. Becchi, A. Rouet and R. Stora, Ann. Phys. (NY), 98, 287 (1976).

6. C. Bredon, Introduction to Compact Transformation Groups, New York Academic Press.

7. T. P. Cheng, E. Eichten and L. F. Li, Phys. Rev. D9, 2259 (1974) (Eight generations represent 16 flavors, and inserting the values of C_2 for $SU(3)_{color}$ fields, $4n_F < (11x6)$)

8. L. Corwin, Y. Ne'eman and S. Sternberg, Reviews of Modern Physics, 47, 573 (1975).

9. G. Curci and R. Ferrari, Nuovo Cim . 30A 155 (1975); ibid 32A, 151 (1976a); ibid 35A, 1, 273 (1976b); ibid 47A, 555 (1978).

10. R. Delbourgo and M. Ramon Medrano, Nucl. Phys. B110, 467 (1976).

11. B. S. DeWitt, Dynamical Theory of Groups and Fields, Gordon and Breach pub., New York/London/Paris (1965).

12. L. D. Faddeev and V. N. Popov, Phys. Lett. B25, 29 (1967).

13. R. P. Feynman, Acta Phys. Polon. 26, 697 (1963).

14. H. Goldschmidt, S. Sternberg, Ann. Inst. Four. 23, 203-267 (1973).

15. Yu. A. Golfand and E. P. Likhtman, JETP letters 13, 452 (1973).

16. J. Harnad, S. Shnider and J. Tafel, Letters on Mathematical Physics 4, 107-113 (1980).

17. G. 't Hooft, Nucl. Phys. B33, 173 (1971a).

18. G. 't Hooft, Nucl. Phys. B35, 167 (1971b).

19. G. 't Hooft and M. Veltman, Ann. Inst. Henri Poincare 20, 69 (1974).

20. R. Howe, Remarks on classical invariant theory. Preprint (1977).

21. V. Kac, Adv. Math. 26, 8 (1977).

22. M. Kashiwara and M. Vergne, Inventiones Math. 44, pp. 1-47 (1978).

23. D. Kazhdan, B. Kostant and S. Sternberg, Comm. Pure and App. Maths. 31, 481-507 (1978).

24. N. S. Manton, Nuclear Physics B158, 141-153.

25. G. Mostow, Ann. of Math. 65, 513-516 (1957).

26. A. Neveu and J. H. Schwarz, Nucl. Phys. B31, 86 (1971).

27. Y. Ne'eman, in Diff. Geom. Meth. in Math. Phys. (Proc. Aix-Salamanca (1979)), P. L. Garcia, A. Perez-Rendon, J. M. Souriau eds., Lecture Notes in Math. 836, Springer-Verlag, Berlin/Heidelberg/New York, pp. 318 ff. (1980c).

28. Y. Ne'eman, in High Energy Physics 1978 (XIX Intern. Conf., Tokyo), S. Homma, M. Kawaguchi and H. Miyazawa eds., Phys. Soc. Japan Pub., pp. 552-554 (1979a).

29. Y. Ne'eman, Phys. Letter 86B, 190 (1979b).

30. Y. Ne'eman, in High Energy Physics 1980 (XX Intern. Conf. Madison), L. Durand and L. G. Pondrom eds., A.I.P. Conf. Proc. 68, (P & F Subs. 22), New York, pp. 460-462 (1981).

31. Y. Ne'eman, in Differential Geometrical Methods in Math. Phys. (Proc. Clausthal 1980), M. Doebner ed., Lecture Notes in Mathematics 905, Springer-Verlag, Berlin/Heidelberg/New York (1982).

32. Y. Ne'eman, Proc. Nat. Acad. Sci. USA, 77, 720 (1980b).

33a. Y. Ne'eman and S. Sternberg, Proc. Nat. Acad. Sci., USA 77, 3217 (1980).

b. Y. Ne'eman and S. Sternberg, in High Energy Physics 1980 (XX Intern. Conf. Madison), L. Durand and G. Pomdrom eds., Ann. Inst. of Phys. Confer. Proceed. #68 (Particles and Fields subseries 22), New York, pp. 981-984 (1981).

c. Y. Ne'eman and S. Sternberg, in Group Theoretical Methods in Physics (Cocoyoc 1980 IX Intern. Conf.), K. B. Wolf ed., Lecture Notes in Physics 135, Springer-Verlag, Berlin/Heidelberg/New York, pp. 610-614 (1980).

34. Y. Ne'eman and J. Thierry-Mieg, in Group Theoretical Methods in Physics, (Kiryat Anavim 1979, VIII Conf.), Ann. Isr. Phys. Soc. 3, 100 (1980a).

35. Y. Ne'eman and J. Thierry-Mieg, Physics Letters B (to be pub.); available as report TAUP 143-81.

36. I. Ojima, Prog. Theoret. Phys. (Kyoto), 64, 625 (1980).

37. P. Ramond, Phys. Rev. D3, 2415 (1971).

38. A. Salam and J. Strathdee, Nucl. Phys. 76B, 477 (1974).

39. M. Scheunert, The Theory of Lie Superalgebras, Lecture Notes in Maths. 716, Springer-Verlag, Berlin/Heidelberg/New York.

40. S. Shnider and S. Sternberg, Dimensional Reduction from the Infinitesmal Viewpoint, to be pub. in Nuovo Cimento Letters.

41. A. A. Slavnov, Theo. Math. Phys. 10, 99 (1972).

42. J. C. Taylor, Nucl. Phys. B33, 436 (1971).

43. J. Thierry-Mieg, Thèse de Doctorat d'Etat (Paris-Sud) (1979).

44. J. Thierry-Mieg, J. Math. 21, 2834 (1980a).

45. J. Thierry-Mieg, Nuovo Cim. 56A, 396 (1980b).

46. J. Thierry-Mieg and Y. Ne'eman, Ann. of Phys. (NY), 123, pp. 247-273 (1979).

47. J. Thierry-Mieg and Y. Ne'eman, Phys. Rev. D, (1982).

48. I. V. Tyutin, report FIAN 39 (1975).

49. H. C. Wang, Nagoya Math. J. 13, 1-19 (1958).

50. J. Wess and B. Zumino, Nucl. Phys. 70B, 39 (1974).

51. A. R. White, CERN report Th. 3058, presented at the XVI Rencontre de Moriond.

II. Geometry of Classical Yang-Mills Fields

L e c t u r e s

on

Yang-Mills Fields and Vector Bundles

to be given at the

Poiana Brasov School on Gauge Theories 1981

Günther Trautmann

Most of the material presented in these lectures is contained in the work of Atiyah and co-authors on this subject. However I have tried to treat everything as elementary as possible and to follow a canonical way to derive the fields out of holomorphic bundles even so it might not be the most elegant. For example to follow the proof of the theorem giving the correspondence between bundles on P_3 and Yang-Mills fields, a lot of computations is necessary. These are more or less straightforward and omitted here. Instead I have shown in section 5 in the case of 1-instantons how the parameters of the moduli of the bundles convert into coefficients of the expressions for the fields. This does help to understand the theorem better then giving a complete proof of it. In section 6 some problems of the moduli of instantons are discussed.

1. Differentiable vectorbundles and connections.

1.1 In the sequal differentiable will always mean C^∞. Let us recall the definition of a differentiable (holomorphic) $\mathbb{C}$-vectorbundle of rank r on a differentiable (complex) manifold M of dimension n. By such a bundle we mean a differentiable (complex) manifold E together with a differentiable (holomorphic) map $E \underset{\pi}{\to} M$, such that

(i) for any $p \in M$ the fibre $E_p = \pi^{-1}(p)$ is a $\mathbb{C}$-vectorspace

(ii) the manifold is covered by open neighborhoods U such that for any such U there is a commutative diagram

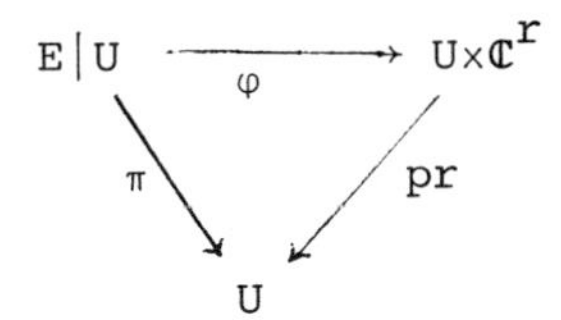

where $E|U$ is a notation for $\pi^{-1}(U)$, and for which

(a) φ is a diffeomorphism (biholomorphic map)

(b) for any $p \in U$ the induced map

$\varphi_p \colon E_p \to \{p\} \times \mathbb{C}^r \to \mathbb{C}^r$

is $\mathbb{C}$-linear.

A section of E over any open set $V \subset M$ is a differentiable (holomorphic) map $V \underset{s}{\to} E$ satisfying $\pi \circ s = \mathrm{id}$ or equivalently $s(p) \in E_p$ for any $p \in V$. The set $E(V)$ of all sections becomes a $\mathbb{C}$-vectorspace and a $C^\infty(V)$-module by defining addition and scalar multiplication pointwise and using (ii). The axiom (ii) implies that there is a covering of M by open sets U together with sections $e^1, \ldots, e^r \in E(U)$ such that for any $p \in U$ the vectors $e^1(p), \ldots, e^r(p) \in E_p$ form a basis of E_p. Such a system $(U, e^1, \ldots, e^r)$ is called a local frame of E.

1.2 Given any family $(E_p)_{p\in M}$ of $\mathbb{C}$-vectorspaces of dimension r and an open covering (U_i) of M together with functions $U_i \ni p \mapsto e_i^\rho(p) \in E_p$ for any i and $\rho = 1,\dots,r$, such that the vectors $e_i^1(p),\dots,e_i^r(p)$ span E_p for any $p\in U_i$ and such that on U_{ij} we have

$$e_i^\rho = \sum_\sigma g_{ij}^{\rho\sigma} e_j^\sigma$$

with differentiable (holomorphic) functions $g_{ij}^{\sigma\rho}$, then we obtain a differentiable (holomorphic) vectorbundle E having the E_p's as fibres and the systems $(U_i, e_i^1,\dots,e_i^r)$ as local frames. One only has to define E as the disjoint union of the E_p's, the projection $\pi: E \to M$ by sending all of E_p to the point p, and for any i a bijection

$\varphi_i: \pi^{-1}(U_i) \to U_i\times\mathbb{C}^r$, $\xi \to (p,c_1,\dots,c_r)$, when $\xi\in E_p$ and $\xi = \sum_\rho c_\rho e_i^\rho(p)$, and finally call $\Omega\subset E$ open, if for any i the set $\varphi_i(\Omega\cap\pi^{-1}(U_i))$ is open.

1.3 Examples are the tangent and cotangent bundles $T = TM$ and $T^* = T^*M$ respectively. Let T_pM resp. T_p^*M be the tangent resp. cotangent spaces of M at p. For any local coordinate system $(U,x_1,\dots,x_n)$ of M the derivations

$$\frac{\partial}{\partial x_1},\dots,\frac{\partial}{\partial x_n}$$

resp. the forms

$$dx_1,\dots,dx_n$$

provide local frames satisfying the conditions of 1.2, and therefore the bundles T and T^* are well defined.

1.4 Let E and F be differentiable (holomorphic) vectorbundles. Then the families (E_p^*), $(E_p\otimes F_p)$, $(\Lambda^k E_p)$ etc. are made differentiable (holomorphic) vectorbundles E^*, $E\otimes F$, $\Lambda^k E$ etc. by choosing the local frames as follows. If $(U,e^1,\dots,e^r)$ and

$(U,f^1,\dots,f^s)$ are local frames of E and F, let

$$e^{\rho *}(p)\in E_p^* \quad , \quad (e^\rho \otimes f^\sigma)(p)\in E_p\otimes F_p$$
$$(e^{i_1}\wedge\dots\wedge e^{i_k})(p) \quad {}^kE_p$$

be defined respectively as the dual basis, as $e^\rho(p)\otimes f^\sigma(p)$ or as $e^{i_1}(p)\wedge\dots\wedge e^{i_k}(p)$. Then the conditions of 1.2 are fulfilled and the bundles E^*, $E\otimes F$, $\wedge^k E$ etc. are defined having the above as local frames.

If for example $(U,x_1,\dots,x_n)$ is a coordinate neighborhood of M and if the bundle E has a local frame $e^1,\dots,e^r$ over U, then the sections

$$e^\sigma \otimes dx_{\nu_1}\wedge\dots\wedge dx_{\nu_k}$$

form a local frame of $E\otimes\wedge^k T$ over U, given by $p \to e^\sigma(p)\otimes dx_{\nu_1}(p)\wedge\dots\wedge dx_{\nu_k}(p)$.

1.5 Connections. Let E be a differentiable vectorbundle on the differentiable manifold M. A connection ∇ on E is an assignment of a section

$$\nabla(s)\in(T^*\otimes E)(U)$$

to any section $s\in E(U)$ for any open set $U\subset M$, satisfying : $\nabla(s|V) = \nabla(s)|V$ for $V\subset U$ and

$$\nabla(fs) = f\nabla(s) + df\otimes s,$$

whenever $f\in C^\infty(U)$, $s\in E(U)$.

If a connection ∇ on E is given then for any local frame $e^1,\dots,e^r$ of E over U we have

$$\nabla(e^\alpha) = \sum_\beta \omega^{\alpha\beta}\otimes e^\beta$$

with uniquely determined 1-forms $\omega^{\alpha\beta}\in T^*(U)$. The matrix $\omega = (\omega^{\alpha\beta})$ is called the connection matrix with respect to the frame $e^1,\dots,e^r$.

If over U two frames $e^1,\dots,e^r$ and $e'^1,\dots,e'^r$ of E are given, and if ω resp. ω' are the connection matrices defined by each frame and if the transition matrix $g = (g^{\alpha\beta})$ of differentiable functions is defined by

$$e'^{\alpha} = \sum_{\beta} g^{\alpha\beta} e^{\beta}$$

then we obtain the transformation law

$$\omega' = dg \circ g^{-1} + g \circ \omega \circ g^{-1}$$

in obvious matrix notation. Hence if M is covered by open sets U_i and if $E|U_i$ has local frames $e_i^1,\dots,e_i^r$ with transition matrices $g_{ij} = (g_{ij}^{\alpha\beta})$ defined over U_{ij} by

$$e_i^{\alpha} = \sum_{\beta} g_{ij}^{\alpha\beta} e_j^{\beta} ,$$

then the local connection matrices $\omega_i = (\omega_i^{\alpha\beta})$ satisfy on U_{ij} the equation

$$\omega_i = dg_{ij} \circ g_{ij}^{-1} + g_{ij} \circ \omega_j \circ g_{ij}^{-1} .$$

If conversely such a system ω_i of matrices of 1-forms is given, a unique connection ∇ on E is defined by

$$\nabla(e_i^{\alpha}) = \sum_{\beta} \omega_i^{\alpha\beta} \otimes e_i^{\beta} .$$

1.6 Curvature of a connection. Let ∇ be a connection on the differentiable vectorbundle E and let $e^1,\dots,e^r$ be a local frame. We assign to it the matrix

$$\Omega = (\Omega^{\alpha\beta})$$

of 2-forms given through the connection matrix $\omega = (\omega^{\alpha\beta})$ by

$$\Omega^{\alpha\beta} = d\omega^{\alpha\beta} - \sum_{\rho} \omega^{\alpha\rho} \wedge \omega^{\rho\beta}$$

or in obvious matrix notation by

$$\Omega = d\omega - \omega \wedge \omega .$$

This matrix is called the local curvature matrix of the connection ∇ with respect to the local frame. One verifies at once that

$$d\Omega = \omega\wedge\Omega - \Omega\wedge\omega \qquad \text{(Bianchi's identity)}.$$

The transformation law under transition from one frame to the other (with the same notation as in 1.5) becomes now

$$\Omega' = g\Omega g^{-1} ,$$

and if a covering U_i with local frames $e_i^1,\ldots,e_i^r$ and transition matrices g_{ij} is given, we have for the forms $\Omega_i = d\omega_i - \omega_i\wedge\omega_i$ over U_{ij}

$$\Omega_i = g_{ij}\Omega_j g_{ij} .$$

1.7 Connections compatible with a metric. Any differentiable $\mathbb{C}$-vectorbundle E on a differentiable manifold M admits a hermitian metric, i.e. for any $p\in M$ a hermitian metric $< , >_p$ on E_p such that for any two sections $s,t\in E(U)$ the function

$$<s,t>(p) = <s(p),t(p)>_p$$

is differentiable on U. The existence follows easily from the existence of a partian of unity, cf.f.e. [11]. Such a hermitian metric can be extended to the exterior algebra as follows. Given forms $\omega\in\Lambda^k T^*(U)$, $\omega'\in\Lambda^\ell T^*(U)$ and sections $s,s'\in E(U)$, the form

$$<\omega\otimes s,\ \omega'\otimes s'>: = <s,s'>\omega\wedge\overline{\omega'}$$

is well defined by bilinearity (and in the same way for any p). It is a section of $\Lambda^{k+\ell}T^*$ over U. Given now any two sections ξ resp. η over U of $\Lambda^k T^*\times E$ resp. $\Lambda^\ell T^*\times E$ we obtain a section

$$<\xi,\eta>$$

of $\Lambda^{k+\ell}T^*$ over U by linear extension. This bracket again behaves like a hermitian metric.

Also by using partion of unity, cf.[11], it can be shown that any differentiable vectorbundle E with a hermitian metric $< , >$ admits a connection ∇ satisfying the product rule

$$d<s,t> = <\nabla s,t> + <s,\nabla t>

for any two sections of E over an open subset $U \subset M$. Such a connection is called compatible with the metric or a metric connection.

Let now ∇ be a metric connection on E with hermitian metric $< , >$. If $e^1,\ldots,e^r$ is a local frame of $E|U$ we obtain the connection matrix

$$\omega = (\omega^{\alpha\beta}) \quad \text{by} \quad \nabla e^{\alpha} = \sum_{\beta} \omega^{\alpha\beta} \otimes e^{\beta}$$

and the hermitian matrix associated to the metric

$$h = (h^{\alpha\beta}) \quad \text{by} \quad h^{\alpha\beta} = \langle e^{\alpha}, e^{\beta} \rangle .$$

The compatibility of ∇ with $< , >$ is then expressed by

$$dh = \omega h + h\bar{\omega}^t$$

in obvious matrix notation. If especially $e^1,\ldots,e^r$ is an orthonormal frame (such frames always exist) then we obtain simply $\bar{\omega}^t = -\omega$.

1.8 G-bundles. Assume now that $G \subset GL(r,\mathbb{C})$ is a Lie subgroup and E has a G-structure. This means, there exists a covering (U_i) of M together with frames of $E|U_i$ such that their transition matrices have values in G. Then any local frame of $E|U$ for any open set $U \subset M$ is called a frame of the G-structure or G-frame, if its transition matrix with respect to any frame of the given ones has values in G. A connection ∇ on E is called a G-connection if for any G-frame its matrix ω has values in the Lie algebra $\mathfrak{g}$ of G. For example to any hermitian metric $< , >$ on E there corresponds an U(r)-structure on E given by the orthonormal frames and ∇ is a metric connection if and only if it is an U(r)-connection.

Again by using partition of unity one can easily prove that for any G-bundle there exists a G-connection on it.

1.8.1 Lemma. The differentiable vectorbundle E admits a SU(r)-structure if and only if $\Lambda^r E$ is trivial.

proof: If E has a SU(r)-structure one can choose the local frames such that their transition matrices g_{ij} have values in SU(r). Since $\wedge^r E$ has the transition functions $\det g_{ij} = 1$, it must be trivial. Let conversely $\wedge^r E$ be trivial. Since E admits a hermitian metric, it has a U(r)-structure given by the orthonormal frames. Let g_{ij} be the U(r)-transition matrices. Since $\wedge^r E$ is trivial, we can write $\det g_{ij} = a_i^{-1} a_j$ with nowhere vanishing differentiable functions a_i on U_i with $|a_i| = 1$. Let $b_i \in C^\infty(U_i)$ with $b_i^r = a_i$. Then $g'_{ij} = b_i g_{ij} b_j^{-1}$ are also transition functions for E with respect to the frames $b_i e_i^1, \ldots, b_i e_i^r$, which still are orthonormal. But now $\det g'_{ij} = 1$.

1.8.2 If ∇ is a SU(r)-connection of the SU(r)-bundle E and if $e^1, \ldots, e^r$ is an orthonormal frame (with respect to the metric corresponding to the underlying U(r)-structure) then the associated connection matrix ω satisfies

$$\omega + \bar{\omega}^t = 0$$

$$\mathrm{tr}(\omega) = 0.$$

Conversely, if this is fulfilled for any connection matrix with respect to an orthonormal frame, then ∇ is a SU(r)-connection.

2. Yang-Mills fields on S^4.

In the rest of these lectures we will only consider SU(2)-bundles on the manifold $M = S^4$. Note that any differentiable vectorbundle of rank 1 over S^4 is trivial, hence any differentiable vectorbundle over S^4 has a SU(r)-structure. We always have a metric on such a bundle corresponding to the SU(r)-structure.

2.1 If ∇ is a SU(2)-connection on the SU(2)-bundle E over S^4, and if ω is the connection matrix of ∇ with respect to an (orthonormal) SU(2)-frame over $U \cong \mathbb{R}^4$, then ω can be written a

$$\omega = A_1 dt_1 + A_2 dt_2 + A_3 dt_3 + A_4 dt_4$$

in the coordinates $t_1, \ldots, t_4$ of the chart U, such that the 2×2-matrices A_μ are differentiable mappings

$$A_\mu : \mathbb{R}^4 \to su(2)$$

i.e. satisfy

$$A_\mu + \bar{A}_\mu^t = o \quad \text{and} \quad \operatorname{tr} A_\mu = o.$$

For the curvature form $\Omega = d\omega - \omega\wedge\omega$ on U we can write in the same way

$$\Omega = F_{12} dt_1 \wedge dt_2 + \ldots + F_{34} dt_3 \wedge dt_4$$

with

$$F_{\mu\nu} = \frac{\partial A_\nu}{\partial t_\mu} - \frac{\partial A_\nu}{\partial t_\nu} - (A_\mu A_\nu - A_\nu A_\mu),$$

such that the $F_{\mu\nu}$ are also differentiable with values in su(2)

Now the second "Chern"-form $\sigma_2(\nabla)$ is defined by

$$\sigma_2(\nabla) = \frac{1}{(2\pi i)^2} \det(\Omega) = -\frac{1}{4\pi^2}(\Omega^{11}\wedge\Omega^{22} - \Omega^{21}\wedge\Omega^{12}) \ .$$

This 4-form is independent of the local curvature matrix Ω and is therefore defined globally on S^4. With respect to the above chart $U \cong \mathbb{R}^4$ one has

$$\sigma_2(\nabla) = \frac{1}{4\pi^2} \operatorname{tr}(F_{12}F_{34} - F_{13}F_{24} + F_{23}F_{14}) dt_1 \wedge \ldots \wedge dt_4 .$$

Now by the Chern-Weil-theory of characteristic classes, [6], it follows that the second Chern class $c_2 = c_2(E) \in \mathbb{Z}$ of E is given by

$$c_2 = \int_{S^4} \sigma_2(\nabla) = \frac{1}{4\pi^2} \int_{\mathbb{R}^4} \operatorname{tr}(F_{12}F_{34} - F_{13}F_{24} + F_{23}F_{14}) dt_1 \ldots dt_4$$

where we assume that $\mathbb{R}^4 = U = S^4 \setminus \{\text{point}\}$.

2.2 Duality. On the Riemannian manifold S^4 together with a fixed orientation the $*$-operator for differential forms is well defined. It is not difficult to check that for a chart U of S^4 obtained by stereographic projection with coordinates $t_1,\ldots,t_4$ we obtain

$$* \, dt_1 \wedge dt_2 = dt_3 \wedge dt_4$$
$$* \, dt_1 \wedge dt_3 = -dt_2 \wedge dt_4$$
$$* \, dt_2 \wedge dt_3 = dt_1 \wedge dt_4 .$$

Conversely using stereographic coordinates and the above formulas as local definitions of $*$, an operator $*$ for 2-forms on S^4 is invariantly defined. The SU(2)-connection ∇ of E is called <u>self-dual</u>, if $*\Omega = \Omega$ for any local curvature matrix Ω of ∇. In terms of stereographic voordinates this reads as

$$F_{12} = F_{34} , \; F_{13} = -F_{24} , \; F_{23} = F_{14} .$$

<u>2.3 Yang-Mills equations.</u>

Let E be a SU(2)-bundle on S^4 together with a SU(2)-connection ∇. ∇ is called a Yang-Mills potential, if for one (and then any) local curvature matrix Ω of ∇ the identity

$$d(*\Omega) = \omega \wedge (*\Omega) - (*\Omega) \wedge \omega$$

is satisfied. If ∇ is self-dual this equation is automatically satisfied because of Bianchi's identity. In terms of local stereographic coordinates the Yang-Mills equation is equivalent to the following system of nonlinear differential equations

$$\sum_{\lambda<\mu} \frac{\partial F_{\lambda\mu}}{\partial t_\lambda} - \sum_{\mu<\lambda} \frac{\partial F_{\mu\lambda}}{\partial t_\lambda} = \sum_{\lambda<\mu} [A_\lambda, F_{\lambda\mu}] - \sum_{\mu<\lambda} [A_\lambda, F_{\mu\lambda}]$$

for any fixed $1 \leq \mu \leq 4$.

<u>2.4 Remark:</u> The Yang-Mills equation for a potential ∇ as above can be derived by variation of the action integral (which is invariant)

$$|\Omega|^2 = -\int_{S^4} \mathrm{tr}\,\Omega \wedge *\Omega$$

$$\int_{\mathbb{R}^4} \Sigma\, |f_{\mu\nu\alpha\beta}|^2\, dt_1 \ldots dt_4$$

where $F_{\mu\nu} = (f_{\mu\nu\alpha\beta})$, and where the sum is over all indices with $\mu<\nu$.

2.5 The second Chern number $c_2 = c_2(E) = \int_{S^4} \sigma_2(\nabla)$

$$= \frac{1}{4\pi^2} \int_{\mathbb{R}^4} \mathrm{tr}(F_{12}F_{34} - F_{13}F_{24} + F_{23}F_{14})\, dt_1 \ldots dt_4$$

is called the instanton number of the "field" Ω. It is determined by the bundle $\underline{E}$. By decomposing $\Omega = \Omega^+ + \Omega^-$ with $\Omega^+ = \frac{1}{2}(\Omega + *\Omega)$ and $\Omega^- = \frac{1}{2}(\Omega - *\Omega)$ one easily obtains

$$|\Omega|^2 \geq 8\pi^2 |c_2| .$$

In this estimate equality holds iff $c_2 > o$ and $\Omega = *\Omega$ or $c_2 < o$ and $\Omega = -*\Omega$.

Therefore we consider the case $c_2 > o$ and $\Omega = *\Omega$ only, since the second case is obtained by changing the orientation of S^4.

3. Fibering $\mathbb{P}_3(\mathbb{C}) \to S^4$.

3.1 Let $x \neq y$ be points in $\mathbb{P}_3(\mathbb{C})$ with homogeneous coordinates $x_o,\dots,x_3$ resp. $y_o,\dots,y_3$. Then a point $p\in\mathbb{P}_5(\mathbb{C})$ with homogeneous coordinates

$$p_{ij} = x_i y_j - x_j y_i \quad \text{for } i<j$$

is uniquely determined by the line $\overline{x,y}$ through x and y and satisfies the equation

$$p_{o1}p_{23} - p_{o2}p_{13} + p_{o3}p_{12} = 0 .$$

By $\overline{x,y} \longmapsto p$ we thus obtain the Plücker-imbedding of the Graßmannian $Gr_1(\mathbb{P}_3)$ into $\mathbb{P}_5$ which identifies with the nonsingular quadric $Q\subset\mathbb{P}_5$ defined by the above equation. We write L_p for the line in $\mathbb{P}_3$ corresponding to $p\in Q$.

Now let us consider the homogeneous coordinates $x_o,\dots,x_5$ of $\mathbb{P}_5$ given by the transformation

$$x_o = p_{o1} + p_{23} \qquad x_2 = p_{13} + p_{o2} \qquad x_4 = p_{o3} - p_{12}$$

$$x_1 = p_{o1} - p_{23} \qquad x_3 = i(p_{13} - p_{o2}) \qquad x_5 = i(p_{o3} + p_{12}).$$

In these coordinates the equation for Q reads

$$x_o^2 = x_1^2 + \dots + x_5^2 .$$

By this we easily obtain the identification

$$S^4 = Q\cap\{[x]\in\mathbb{P}_5(\mathbb{C}),\ [x] = [\overline{x}]\}$$

and for $x\in S^4$ automatically $x_o \neq o$.

3.2 Let $\sigma\colon \mathbb{P}_3(\mathbb{C}) \to \mathbb{P}_3(\mathbb{C})$ be defined in homogeneous coordinates by $(z_o,\dots,z_3) \longmapsto (-\overline{z}_1,\overline{z}_o,\overline{z}_3,\overline{z}_2)$. Then σ is a diffeomorphism of $\mathbb{P}_3(\mathbb{C})$ without fix-points and $\sigma^2 = \text{id}$. Now one can check that for a point $p\in Q$ we have $p\in S^4$ if and only if L_p is a fix-line of σ, and that is the case if and only if for $z\in L_p$ also $\sigma z\in L_p$.

Thus a mapping

$$\mathbb{P}_3(\mathbb{C}) \xrightarrow[\pi]{} S^4$$

can be defined by assigning $z \in \mathbb{P}_3(\mathbb{C})$ the point p corresponding to the line through z and σz. This map turns out to be a differentiable fibre bundle with fibre $\mathbb{P}_1(\mathbb{C})$, and for $p \in S^4$ we have $L_p = \pi^{-1}(p)$. These lines will be called the real lines of $\mathbb{P}_3(\mathbb{C})$.

3.3 The open sets

$$V_1 = S^4 \cap \{p_{o1} \neq o\} \qquad \text{and} \qquad V_2 = S^4 \cap \{p_{23} \neq o\}$$

are coordinate neighborhoods of S^4 and cover S^4. If we choose the coordinate neighborhoods $U_i = \{z_i \neq o\}$ in $\mathbb{P}_3(\mathbb{C})$, then

$$\pi^{-1}V_1 = U_o \cup U_1 \qquad \text{and} \qquad \pi^{-1}V_2 = U_2 \cup U_3 \ .$$

Now on V_1 for example we obtain real coordinates by stereographic projection given by

$$t_\nu = \frac{x_{\nu+1}}{x_o + x_1} \qquad \text{for} \quad \nu = 1,\dots,4 \ .$$

Then for example $\pi|U_o$ is expressed in the coordinates t_μ and $z_\nu^* = \frac{z_\nu}{z_o}$ by

$$t_1 = \tfrac{1}{2}(p_{13}^* + p_{o2}^*) \qquad t_2 = \tfrac{i}{2}(p_{13}^* - p_{o2}^*)$$

$$t_3 = \tfrac{1}{2}(p_{o3}^* - p_{12}^*) \qquad t_4 = \tfrac{i}{2}(p_{o3}^* + p_{12}^*)$$

where $p_{ij}^* = p_{ij}(z)p_{o1}(z)^{-1}$ are functions in z_ν^*, $\bar{z}_\nu^*$ given by

$$\begin{aligned}
p_{o1}(z) &= |z_o|^2 + |z_1|^2 \\
p_{o2}(z) &= -z_o\bar{z}_3 + z_2\bar{z}_1 \\
p_{12}(z) &= -z_1\bar{z}_3 - z_2\bar{z}_o \\
p_{o3}(z) &= z_o\bar{z}_2 + z_3\bar{z}_1 \\
p_{13}(z) &= z_1\bar{z}_2 - z_3\bar{z}_o \\
p_{23}(z) &= |z_2|^2 + |z_3|^2 \ .
\end{aligned}$$

4. Instanton bundles on $\mathbb{P}_3(\mathbb{C})$.

It was the idea of Atiyah-Ward, [2], to use in addition to the holomorphic structure of vectorbundles on $\mathbb{P}_3(\mathbb{C})$ certain real and symplectic structures on such bundles in order to describe the property of the fields to be SU(2)-fields. Such a symplectic structure is by definition a holomorphic isomorphism $\tau: E \to \sigma^*\overline{E}$ with $\tau^2 = -1$, i.e. the bundle E is compatible with the symplectic involution σ induced from $\mathbb{C}^4$ on $\mathbb{P}_3(\mathbb{C})$. Let us describe such structures more closely.

4.1 Let $\sigma: \mathbb{P}_3 \to \mathbb{P}_3$ be the involution of section 3. If we put $\sigma(o)=1$, $\sigma(1)=o$, $\sigma(2)=3$, $\sigma(3)=2$ and $U_i = \{z_i \neq o\}$ we can write $\sigma^{-1}U_i = U_{\sigma(i)}$. If now (g_{ij}) is a holomorphic cocycle of the holomorphic bundle E on $\mathbb{P}_3$ with respect to the covering (U_i), then $\sigma^*\overline{E}$ has the cocycle (g^σ_{ij}) with

$$g^\sigma_{ij} = \overline{g}_{\sigma(i),\sigma(j)} \circ \sigma,$$

which is again holomorphic. Moreover there is a canonical antilinear isomorphism

$$\Gamma(\mathbb{P}_3, E) \longrightarrow \Gamma(\mathbb{P}_3, \sigma^*\overline{E})$$

described as follows. Any global section s of E can be given by a system (s_i) of holomorphic columns $s_i: U_i \to \mathbb{C}^r$ satisfying $s_i = g_{ij}s_j$ over U_{ij}. Now the above isomorphism is given by

$$(s_i) \longleftrightarrow (s^\sigma_i)$$

where $s^\sigma_i = \overline{s}_{\sigma(i)} \circ \sigma$. It is antilinear. A (holomorphic) <u>symplectic involution</u> on a holomorphic vectorbundle E on $\mathbb{P}_3(\mathbb{C})$ is now by definition a holomorphic isomorphism

$$E \xrightarrow[\tau]{} \sigma^*\overline{E}$$

satisfying $\tau^2 = -1$, i.e. the composition

$$E \xrightarrow[\tau]{} \sigma^*\overline{E} \xrightarrow[\sigma^*(\bar{\tau})]{} \sigma^* \overline{\sigma^*\overline{E}} = E$$

is -id. Any such involution induces now via

$$\Gamma(\mathbb{P}_3, E) \xrightarrow[\Gamma\tau]{} \Gamma(\mathbb{P}_3, \sigma^*\overline{E}) \xleftarrow[=]{} \Gamma(\mathbb{P}_3, E)$$

an antilinear involution (again denoted by τ)

$$\Gamma(\mathbb{P}_3, E) \xrightarrow[\tau]{} \Gamma(\mathbb{P}_3, E)$$

with $\tau^2 = -1$.

<u>4.2 Theorem</u> (Atiyah-Ward [2]) There is a 1-1 correspondence between

(I) the equivalence classes (under gauge transformation) of euclidean self-dual SU(2)-Yang Mills fields on S^4 and
(II) the isomorphism classes of holomorphic vectorbundles E of rank 2 on $\mathbb{P}_3(\mathbb{C})$ with the following properties:

(i) For any "real" line $L_p \subset \mathbb{P}_3(\mathbb{C})$ the bundle $E|L_p$ is holomorphically trivial

(ii) E has a (holomorphic) symplectic involution τ.

<u>Remark 1:</u> It is not difficult to show that bundles E of type (II) satisfy $c_1(E) = 0$ and $\Gamma(\mathbb{P}_3, E) = 0$

<u>Remark 2:</u> W. Barth has shown that bundles E of rank 2 with $c_1(E) = 0$ and $\Gamma(\mathbb{P}_3, E) = 0$ are "stable" and "simple", i.e. $\mathrm{Hom}(\mathbb{P}_3, E, E) = \mathbb{C}$.

<u>Remark 3:</u> The involution $\tau: E \to \sigma^*\overline{E}$ is unique up to a constant c with $|c| = 1$. For if τ_1 and τ_2 are two such involutions, $\tau_1^{-1}\tau_2$ has to be constant c as an endomorphism of E. By $\tau_i^2 = -1$ it follows from $\tau_2 = c\tau_1$ that $c\bar{c} = 1$.

<u>Remark 4:</u> (Vanishing theorem of Atiyah-Hitchin-Drinfeld-Manin) If E is a holomorphic vector bundle of type (II) then $H^1(\mathbb{P}_3, E(-2)) = 0$

and $H^1(\mathbb{P}_3, E(-2)\otimes E(-2)) = 0$. A proof has been explicated by Rawnsley [7] . Another proof is given by A. Douady in [5]. This theorem has as consequence that the space of coefficients of all possible selfdual SU(2)-Yang-Mills fields on S^4 with fixed instanton number c_2 is a real-analytic manifold of dimension $8c_2 - 3$.

4.3 The following terminology is now used: A holomorphic 2-bundle E of type (II) is called a real instanton bundle. A holomorphic 2-bundle E on $\mathbb{P}_3(\mathbb{C})$ with $c_1(E) = 0$, $c_2(E) > 0$, $\Gamma(\mathbb{P}_3, E) = 0$, $H^1(\mathbb{P}_3, E(-2)) = 0$ is called a mathematical instanton bundle.
In the description of mathematical instanton bundles certain matrices with complex entries are involved, cf. also the lectures of Barth. The condition (ii) of (II) exactly implies that these coefficients have to be real in a certain sense, cf. section 5 below.

Here I will briefly sketch the proof for the correspondence in the theorem, in the next section we shall see in the simplest case of instanton number $c_2 = 1$ how a Yang-Mills potential can be obtained explicitly from the data of the corresponding bundle on $\mathbb{P}_3$.

4.4 If the bundle F with ∇ is given, let $E = \pi^* F$. It turns out that the condition $*\Omega = \Omega$ for the curvature matrices Ω exactly implies that the differentiable bundle E has a holomorphic structure. Since F has a SU(2)-structure there exists an antilinear isomorphism $\tau : F \to \overline{F}$ such that $\tau^2 = -1$ and that it is compatible with the metric and the connection, i.e. satisfies $\langle \tau s, \tau t \rangle = \overline{\langle s,t \rangle}$ and $\nabla\tau = (\overline{\mathrm{id}}\otimes\tau)\nabla$.
By $E = \pi^* F \xrightarrow[\pi^*\tau]{} \pi^*\overline{F} = \sigma^*\pi^*\overline{F} = \sigma^*\overline{E}$ we obtain the corresponding isomorphism over $\mathbb{P}_3$. Again this turns out to be holomorphic. For details see [10].

4.5 If the bundle E on $\mathbb{P}_3(\mathbb{C})$ with $\tau\colon E \to \sigma^*\overline{E}$ is given, let for any $p \in S^4$ be

$$F_p = \Gamma(L_p, E|L_p).$$

Since $E|L_p$ is holomorphically trivial, $F_p \cong \mathbb{C}^2$ for any p. It turns out that the family of the F_p determines a real-analytic $\mathbb{C}$-vector bundle with $E = \pi^*F$. For this arguments in connection with Grauerts direct Image theorem are necessary. Now the involution τ induces by restriction to the lines L_p antilinear involutions

$$F_p \xrightarrow[\tau_p]{} F_p$$

with $\tau_p^2 = -1$, which fit together to an antilinear real-analytic bundle map

$$F \xrightarrow[\tau]{} F$$

with $\tau^2 = -1$. Furthermore, since $\wedge^2 F \cong C^\omega$, there is a nowhere degenerate symplectic form $(\ ,\ )$ on F. Since $\tau \wedge \tau$ induces an antilinear involution on C^ω and since any such must be multiplication with a constant c of modulus 1, we obtain with such a constant

$$(\tau s, \tau t) = c\overline{(s,t)}$$

for any two local sections of F. Since τ is antilinear, by

$$\langle s,t\rangle = \frac{1}{\sqrt{c}}(s,\tau t)$$

a hermitian metric is defined on the bundle F satisfying $\langle \tau s, \tau t\rangle = \overline{\langle s,t\rangle}$.

Since $\wedge^2 F$ is trivial, or by the existence of τ, we can assume that F has a $SU(2)$-structure. Now we can lift the metric $\langle\ ,\ \rangle$ to E by which we obtain a hermitian metric on E which is constant on the fibres. It is well known, cf. [11], that on E there exists a unique connection $\tilde{\nabla}$ which is compatible with the metric and such that any local connection matrix with respec to a holomorphic frame is of type (1,0). Now it can be verified, that $\tilde{\nabla} = \pi^*\nabla$ for a unique connection ∇ on F over S^4 and that ∇ is compatible with the metric and is a $SU(2)$-connection. For the last statement it is necessary to use the compatibility of τ with the metric and that τ is holomorphic. Moreover ∇ is self-dual which follows from the construction of $\tilde{\nabla}$, by which the local curvature matrices of $\tilde{\nabla}$ consist of forms of type (1,1). This is equivalent to the self-duality on S^4 by the special type of the fibering $\mathbb{P}_3(\mathbb{C}) \to S^4$.

5. How to derive explicit expressions for the potentials.

Some of the facts concerning holomorphic vectorbundles on $\mathbb{P}_3(\mathbb{C})$ which we will use now, can also be found in the lecture notes of W. Barth.

5.1 Let us first consider the bundle Ω^1 of holomorphic 1-forms over $\mathbb{P}_3(\mathbb{C})$. There is a natural sequence

$$0 \longrightarrow \Omega^1 \longrightarrow 4\mathcal{O}(-1) \xrightarrow[Z]{} \mathcal{O} \longrightarrow 0$$

where the homomorphism Z can be described by a column of linear forms

$$Z = \begin{pmatrix} z_0 \\ z_1 \\ z_2 \\ z_3 \end{pmatrix}$$

sending locally a row f to $f \circ \frac{1}{z_i} Z$. The linear forms z_i can be thought of as homogeneous coordinates of $\mathbb{P}_3(\mathbb{C})$. The sheaf Ω^1 does not have sections but dim $\Gamma(\mathbb{P}_3, \Omega^1(2)) = 6$ and this space is spanned by the vectors

$$\omega_{01} = (-z_1, z_0, 0, 0)$$
$$\vdots$$
$$\omega_{23} = (0, 0, -z_3, z_2)$$

of $4\,\Gamma\mathcal{O}(1)$.

5.2 Let now E be a mathematical instanton bundle on $\mathbb{P}_3(\mathbb{C})$. By tensoring the sequence 5.1 with E(-1) and taking cohomology we obtain the exact sequence

$$\begin{array}{ccccccc} 4H^1E(-2) & \longrightarrow & H^1E(-1) & \xrightarrow{\delta} & H^2\Omega^1 \otimes E(-1) & \longrightarrow & 4H^2E(-2) \\ \| & & & & & & \| \\ 0 & & & & & & 0 \end{array}$$

Since by self-duality also $H^2E(-2) = 0$, it follows that δ is an isomorphism. From this fact we can define a bilinear operation $\omega\otimes\xi \overset{\phi}{\longmapsto} \omega(\xi)$

$$\begin{array}{ccc} \Gamma\Omega^1(2)\otimes H^2E(-3) & \xrightarrow{\ \phi\ } & H^1E(-1) \\ & \text{cup} \searrow \quad \swarrow \delta & \\ & H^2\Omega^1\otimes E(-1) & \end{array}$$

by $\phi = \delta^{-1}\circ \text{cup}$, where cup is as usual. Especially for any $\omega_{ij}\in\Gamma\Omega^1(2)$ we obtain a $\mathbb{C}$-linear map

$$\omega_{ij}\colon\ H^2E(-3) \longrightarrow H^1E(-1)$$

defined by $\omega_{ij}(\xi) = \phi(\omega_{ij}\otimes\xi)$.

5.3 Let now E be a mathematical instanton bundle on $\mathbb{P}_3(\mathbb{C})$ with $c_2 = 1$ (a so-called null-correlation bundle). Then by the Riemann-Roch-formula and Serre-duality we obtain $\dim H^2E(-3)=1$, $\dim H^1E(-1)=1$. If $\xi\in H^1E(-1)$ and $\eta\in H^2E(-3)$ are basis vectores, then the operations ω_{ij} determine coefficients a_{ij} by

$$\omega_{ij}(\eta) = a_{ij}\xi.$$

By looking at some more canonical operations like the above, one finds the condition

$$a_{o1}a_{23} - a_{o2}a_{13} + a_{o3}a_{12} \neq 0 .$$

If $A = (a_{ij})$ is the column with these coefficients we can define the sheaf E(A) as the kernel of the homomorphism

$$0 \longrightarrow E(A) \longrightarrow 6\mathcal{O}(1) \xrightarrow{\begin{pmatrix} a_{o1} & -z_1 & z_o & 0 & 0 \\ \vdots & & \vdots & & \vdots \\ a_{23} & 0 & 0 & -z_3 & z_2 \end{pmatrix}} \mathcal{O}(1)\oplus 4\mathcal{O}(2) .$$

It can be checked that E(A) is locally-free and that $E \cong E(A)$, cf.[10]. Moreover if A' is another column as the above, then $E(A) \cong E(A')$ if and only if $A = \lambda A'$ for some $\lambda \neq 0$. Therefore the set of isomorphism classes of instanton bundles with $c_2=1$ is identified with $\mathbb{P}_5(\mathbb{C}) \setminus Q$, where $Q \subset \mathbb{P}_5(\mathbb{C})$ is the quadric of section 3. The columns $A = (a_{ij})$ are now parameters for those instanton bundles.

Remark: The last facts can be generalized to instanton bundles of arbitrary $c_2>0$, but the matrices having E as a kernel are more complicated. There is a connection between these matrices and the monads in Barth's lecture, which I will mention in the last section.

5.4 Let $E = E(A)$ be as in 5.3 and let $p \in Q$ some point with Plücker coordinates $p_{01},\ldots,p_{23}$. Using the representation of E(A) one can derive that

$$E|L_p \cong 2\mathcal{O}_{L_p}$$

if and only if the expression

$$\delta_A(p) = a_{01}p_{23}-a_{02}p_{13}+a_{03}p_{12}+a_{12}p_{03}-a_{13}p_{02}+a_{23}p_{01} \neq 0 ,$$

and that in the case $\delta_A(p) = 0$ we have

$$E|L_p \cong \mathcal{O}_{L_p}(-1)\oplus\mathcal{O}_{L_p}(1)$$

Remark: $\delta_A(p) = 0$ is the equation of the polar hyperplane of A with respect to Q. Statements like this can also be made for the case of higher c_2, see last section.

5.5 Let E now be a real instanton bundle on $\mathbb{P}_3(\mathbb{C})$ with $c_2=1$, and let as before $E = E(A)$. We want to derive conditions for the parameters a_{ij} which correspond to (II), (i), (ii). These are

5.5.1 $a_{o1}a_{23}-a_{o2}a_{13}+a_{o3}a_{12} \neq 0$

5.5.2 $\delta_A(p) \neq 0$ for $p \in S^4 \subset Q$

5.5.3 $a_{o1} = \bar{a}_{o1},\ a_{o2} = \bar{a}_{13},\ a_{12} = -\bar{a}_{o3},\ a_{23} = \bar{a}_{23}$.

The first condition was already derived. The second follows from (II), (i) by 5.4. The third condition will now be a consequence of the symplectic involution $\tau\colon E \to \sigma^*\bar{E}$. First note that τ induces in a natural way antilinear involutions

$$\tau^1\colon H^1E(-1) \to H^1E(-1)$$
$$\tau^2\colon H^2E(-3) \to H^2E(-3)$$

as was shown for ΓE. Here however $(\tau^1)^2 = \mathrm{id}$ and $(\tau^2)^2 = \mathrm{id}$. (The sign of $(\tau^\nu)^2$ depends on the twisting degree d of E(d)). The vectorspaces being 1-dimensional, we can choose the bases η resp. ξ to be real, i.e. invariant under τ^2 resp. τ^1. Furthermore on $\Gamma\Omega^1(2)$ there is also a natural antilinear involution Σ which can be induced by σ via the sequence in 5.1. Explicitly

$$\Gamma\Omega^1(2) \xrightarrow{\Sigma} \Gamma\Omega^1(2)$$

can be calculated to have the form (for details see [10])

$$\begin{aligned}
\omega_{o1} &\longrightarrow \omega_{o1}\\
\omega_{o2} &\longrightarrow \omega_{13}\\
\omega_{12} &\longrightarrow -\omega_{o3}\\
\omega_{o3} &\longrightarrow -\omega_{12}\\
\omega_{13} &\longrightarrow \omega_{o2}\\
\omega_{23} &\longrightarrow \omega_{23}\ .
\end{aligned}$$

By canonical arguments the following diagram is commutative

$$\begin{array}{ccc}
\Gamma\Omega^1(2)\otimes H^2E(-3) & \xrightarrow{\phi} & H^1E(-1)\\
\big\downarrow \Sigma\otimes\tau^2 & & \big\downarrow \tau^1\\
\Gamma\Omega^1(2)\otimes H^2E(-3) & \xrightarrow[\phi]{} & H^1E(-1)
\end{array}\ .$$

This implies $\pm a_{k\ell}\xi = \bar{a}_{ij}\xi$ if $\pm\omega_{k\ell} = \Sigma\omega_{ij}$, since ξ and η are invariant. This proves 5.5.3.

5.6 Remark: Starting with 5.5.1, 5.5.2, 5.5.3 the bundle E(A) is a real instanton bundle, because 5.5.3 allows us to construct $\tau: E \to \sigma^*\bar{E}$ by using Σ.

5.7 Using 5.5.3 one immediately verifies that $d := a_{01}a_{23}-a_{02}a_{13}+a_{03}a_{12}>0$ and that also $\delta_A(p)>0$ for any $p\in S^4$. Now we calculate the potential corresponding to the bundle $E = E(A)$ by using the advice of 4.5.

It is not hard to find a holomorphic trivialization

$$U_o\times\mathbb{C}^2 \xleftarrow{\varphi_o} E|U_o$$

and a differentiable trivialisation

$$E|U_o\cup U_1 \xrightarrow{\psi} (U_o\cup U_1)\times\mathbb{C}^2$$

such that $h = \varphi_o\,\psi^{-1}$ can be explicitly given as

$$h = \sqrt{\delta^*}\begin{pmatrix} h_{11} & h_{12} \\ h_{21} & h_{22} \end{pmatrix}$$

where δ^*, $h_{11},\dots,h_{22}$ are the following functions:

$$\delta^* = \delta^*(p^*) = \frac{\delta_A(p)}{p_{01}} = a_{01}p^*_{23}+\dots+a_{23}> 0$$

$$h_{11}= \frac{1}{\delta^*}(a_{12}-a_{02}z^*_1+a_{01}z^*_2)(p^*_{03}a_{01}-a_{03}) - a_{01}$$

$$h_{21}= \frac{1}{\delta^*}(a_{12}-a_{02}z^*_1+a_{01}z^*_2)(a_{02}-p^*_{02}a_{01})$$

$$h_{12}= \frac{1}{\delta^*}(-a_{13}+a_{03}z^*_i-a_{01}z^*_3)(p^*_{03}a_{01}-a_{03})$$

$$h_{22}= \frac{1}{\delta^*}(-a_{13}+a_{03}z^*_1-a_{01}z^*_3)(a_{02}-p^*_{02}a_{01})+a_{01} \ .$$

Here we use the local coordinates $z_i^* = \frac{z_i}{z_o}$ of $U_o \subset \mathbb{P}_3$ and $p_{ij}^* = \frac{p_{ij}}{p_{o1}}$ of $V_1 = S^4 \cap \{p_{o1} \neq 0\}$, where $\pi^{-1} V_1 = U_o \cup U_1$. Note that p_{ij}^* are functions in z_i^* describing $\pi : \mathbb{P}_3(\mathbb{C}) \to S^4$. (To get φ_o, we simply use the representation of 5.3. To get ψ, we construct a section $s_p \in \Gamma(L_p, E|L_p) = F_p$ also using 5.3, such that $p \xrightarrow[s]{} s_p$ becomes a section of F. Then $s_1 = s$ and $s_2 = \tau s$ are orthogonal and yield a trivialization of $F|V_1$ or after lifting a trivialization of $E|U_o \cup U_1$) .

5.8 Now the connection matrix $\tilde{\omega}$ of $\tilde{\nabla}$ of E with respect to ψ is given by

$\tilde{\omega} = \tilde{\omega}' + \tilde{\omega}''$, $\tilde{\omega}' = -\overline{\tilde{\omega}''^t}$, $\tilde{\omega}''^t = h^{-1}\bar{\partial} h$,

cf. [11]. After calculating we obtain

$$\tilde{\omega} = \begin{pmatrix} \Theta_{11} & \Theta_{12} \\ \Theta_{21} & \Theta_{22} \end{pmatrix}$$

with

$$\Theta_{11} = \frac{1}{2\delta^*}[(p_{13}^* a_{o1} - a_{13}) dp_{o2}^* - (p_{o3}^* a_{o1} - a_{o3}) dp_{12}^* + (p_{12}^* a_{o1} - a_{12}) dp_{o3}^* - (p_{o2}^* a_{o1} - a_{o2}) dp_{13}^*]$$

$$\Theta_{12} = \frac{1}{\delta^*}[(p_{12}^* a_{o1} - a_{12}) dp_{o2}^* - (p_{o2}^* a_{o1} - a_{o2}) dp_{12}^*]$$

$$\Theta_{22} = -\Theta_{11} \quad , \quad \Theta_{21} = -\overline{\Theta}_{12} \quad .$$

For more details see [10].

By this one immediately has the result that $\tilde{\omega} = \pi^* \omega$ for some matrix of forms on $V_1 \subset S^4$: One has only to forget that p_{ij}^* are functions on $U_o \cup U_1$ and consider them as coordinate functions on V_1, cf. 3. Thus we have already obtained an explicit expression of the connection matrix ω on V_1 of ∇ on F over S^4 with $\tilde{\omega} = \pi^*\omega$. We can write

$$\omega = \begin{pmatrix} \theta_{11} & \theta_{12} \\ \theta_{21} & \theta_{22} \end{pmatrix}.$$

Note that ω has already values in su(2) and that its curvatur is self-dual.

5.9 It is convenient to use other coordinates $t_1,\dots,t_4$ on V_1 instead of the p^*_{ij}, which are given by

$$t_1 - it_2 = p^*_{13} \qquad t_3 - it_4 = p^*_{o3}$$

$$t_1 + it_2 = p^*_{o2} \qquad t_3 + it_4 = -p^*_{12}\text{ , cf. section 3.}$$

If we use new parameters $a_1,\dots,a_4$, d (where we assume $a_{o1}=1$; note that $a_{o1}a_{23} > 0$) by

$$a_1 - ia_2 = a_{13} \qquad a_3 - ia_4 = a_{o3}$$

$$a_1 + ia_2 = a_{o2} \qquad a_3 + ia_4 = -a_{12}$$

$$d = a_{23} - a_{o2}a_{13} + a_{o3}a_{12}$$

we obtain the following expressions for ω:

$$\omega = A_1(t)dt_1 + \dots + A_4(t)dt_4$$

with

$$A_1(t) = \frac{1}{d+\|t-a\|^2}\begin{pmatrix} -i(t_2-a_2) & -(t_3-a_3)-i(t_4-a_4) \\ (t_3-a_3)-i(t_4-a_4) & i(t_2-a_2) \end{pmatrix},$$

$$A_2(t) = \frac{1}{d+\|t-a\|^2}\begin{pmatrix} i(t_1-a_1) & (t_4-a_4)-i(t_3-a_3) \\ -(t_4-a_4)-i(t_3-a_3) & -i(t_1-a_1) \end{pmatrix}$$

$$A_3(t) = \frac{1}{d+\|t-a\|^2}\begin{pmatrix} -i(t_4-a_4) & (t_1-a_1)+i(t_2-a_2) \\ -(t_1-a_1)+i(t_2-a_2) & i(t_4-a_4) \end{pmatrix}$$

$$A_4(t) = \frac{1}{d+\|t-a\|^2}\begin{pmatrix} i(t_3-a_3) & -(t_2-a_2)+i(t_1-a_1) \\ (t_2-a_2)+i(t_1-a_1) & -i(t_3-a_3) \end{pmatrix}$$

Since we started with a general real instanton bundle with $c_2=1$, this is exactly the general form of a Yang-Mills potential with instanton number 1 in the self-dual euclidean case. Note that $d+\|t-a\|^2$ is nothing else then the function δ^*, which by the above description has a geometrical meaning. Furthermore by the above transformation it follows that $a_1,\dots,a_4$ are real parameters. The parameter space of the potentials above is thus the open upper halfspace $\mathbb{R}^5_+ = \{(d,a_1,\dots,a_4)\in\mathbb{R}^5 \mid d>0\}$.

6. Some remarks on moduli of instanton bundles.

6.1 Let us consider a $k\times k$-matrix $A=(a^{\mu\nu})$ of 2-forms $a^{\mu\nu}\in\Lambda^2\mathbb{C}^4$ and assume that it is symmetric. It defines an operator A from $k\mathbb{C}^4$ to $k\Lambda^3\mathbb{C}^4$ by

$$\begin{pmatrix} a^{11} & \cdots & a^{1k} \\ \vdots & & \vdots \\ a^{k1} & \cdots & a^{kk} \end{pmatrix} \wedge \begin{pmatrix} x^1 \\ \vdots \\ x^k \end{pmatrix}$$

where each x^ν denotes a vector of $\mathbb{C}^4$. We can choose a matrix

$$B = \begin{pmatrix} b^{11} & \cdots & b^{1\ell} \\ \vdots & & \vdots \\ b^{k1} & \cdots & b^{k\ell} \end{pmatrix}$$

of vectors $b^{\kappa\lambda}$ of $\mathbb{C}^4$ such that the ℓ columns of B are a basis of the kernel of A . Let us write KA = B. Choosing in the same way a basis of the kernel of $(KA)\wedge$, we obtain a matrix $K^2A=K(KA)$, and continuing in this way, a sequence of matrices

$$A,\ KA,\ K^2A,\ \ldots$$

is constructed. The entries of each $K^\mu A$ are vectors of $\mathbb{C}^4$ and obviously the rank of $K^\mu A$ (as operator $K^\mu A\wedge$) is independent of the choices made. Now let $M(k)$ be the set of such matrices satisfying the two conditions

(i) rank $A = 2k + 2$
(ii) $K^m A = 0$ for some natural number m.

Two matrices $A, B\in M(k)$ are said to be equivalent if there is a matrix $S\in GL(k,\mathbb{C})$ with

$$b^{\mu\nu} = \sum_{\kappa,\lambda} s^{\mu\kappa} b^{\kappa\lambda} s^{\nu\lambda}$$

or $B = SAS^t$ for short.

6.2 Given a matrix $A \in M(k)$ one can construct a mathematical instanton bundle $E = E(A)$ with $c_2 E = k$, cf. [10], and such that

$$h^1 E(d) = \text{dim kernel of } K^d A \wedge$$

for $d \geq 0$. It follows that for $d \geq 0$

$$h^1 E(d) = 4h^1 E(d-1) - \text{rank}(K^d A) .$$

Moreover one can find a basis $\xi_1, \ldots, \xi_k \in H^1 E(-1)$ and the Serre-dual basis $\eta_1, \ldots, \eta_k \in H^2 E(-3)$ such that the coefficients $a_{ij}^{\mu\nu}$ of the form

$$a^{\mu\nu} = \Sigma a_{ij}^{\mu\nu} e_i \wedge e_j$$

occur as the coefficients of the operators ω_{ij}, cf. 5.2, via

$$\omega_{ij} \eta_\mu = \sum_\nu a_{ij}^{\mu\nu} \xi_\nu .$$

Given any mathematical instanton bundle E we obtain such a matrix A representing the operators ω_{ij} such that $E \cong E(A)$. Moreover $E(A) \cong E(B)$ if and only if A and B are equivalent, i.e. $B = SAS^t$. Equivalence just means transformation of the basis of $H^1 E(-1)$. By these results $M(k)/GL(k)$ is the set of isomorphism classes of instanton bundles.

6.3 One can determine the jumping lines of $E = E(A)$ by the matrix A; cf. Barth's script. Let as in 5.4

$$\delta^{\mu\nu}(p) = a_{o1}^{\mu\nu} p_{23} - a_{o2}^{\mu\nu} p_{13} + \ldots$$

be the equation of the polar hypersurface of $a^{\mu\nu}$ as a point of $\mathbb{P}_5$ with respect to the quadric Q. It turns out that

$$E|L_p = \mathcal{O}_{L_p}(-d) \oplus \mathcal{O}_{L_p}(d)$$

if and only if

$$\text{rank}(\delta^{\mu\nu}(p)) = k-d.$$

6.4 Given a 2-form $a = \Sigma a_{ij} e_i \wedge e_j$ in $\Lambda^2 \mathbb{C}^4$ resp. a point $z \in \mathbb{P}_3$ with homogeneous coordinates $z_o, \dots, z_3$ let us denote by $A_{1,1}$ resp. Z_o resp. Z_3 the matrices

$$A_{1,1} = \begin{pmatrix} -a_{12} & a_{o2} & -a_{o1} & 0 \\ -a_{13} & a_{o3} & 0 & -a_{o1} \\ -a_{23} & 0 & a_{o3} & -a_{o2} \\ 0 & -a_{23} & a_{13} & -a_{12} \end{pmatrix} \qquad Z_o = \begin{pmatrix} z_o \\ z_1 \\ z_2 \\ z_3 \end{pmatrix} \qquad Z_3 = (-z_3 z_2 - z_1 z_o).$$

One can prove that a symmetric matrix $A = (a^{\mu\nu})$ of 2-forms with condition (i) satisfies condition (ii) if and only if for any $z \in \mathbb{P}_3$

$$\text{rank} \begin{pmatrix} A^{11}_{1,1} Z_o & \cdots & A^{1k}_{1,1} Z_o \\ \vdots & & \vdots \\ A^{k1}_{1,1} Z_o & \cdots & A^{kk}_{1,1} Z_o \end{pmatrix} = k.$$

6.5 Relations between monads and matrices. Let us for somplicity forget the symmetry condition for the matrices of $M(k)$. Then (ii) has to be fulfilled by A and A^t simultaneously (A^t here is the transposed with respect to the upper indices of the 2-forms). This is equivalent to rank $(Z_3 A^{\mu\nu}_{1,1}) = k$ and rank $(A^{\mu\nu}_{1,1} Z_o) = k$ for any z. It is shown by Barth in his lectures that any mathematical instanton bundle is the cohomology of a monad

$$k\mathcal{O}(-1) \xrightarrow[a]{} (2k+2)\mathcal{O} \xrightarrow[b]{} k\mathcal{O}(1),$$

where a is a monomorphism of bundles, b is an epimorphism and $a \circ b = 0$ (as matrices). Such a monad is not necessarily selfdual. We can define now the constant matrices A and B by

$$a = \begin{bmatrix} Z_3 & & \\ & \ddots & \\ & & Z_3 \end{bmatrix} \circ A \quad \text{and} \quad B \circ \begin{bmatrix} Z_o & & \\ & \ddots & \\ & & Z_o \end{bmatrix} = b$$

such that A is of type $4k\times(2k+2)$ and B of type $(2k+2)\times 4k$. If we subdivide the product $A\circ B$ into 4×4 blocks $A^{\mu\nu}$, i.e. $A\circ B = (A^{\mu\nu})$, $1\leq\mu,\nu\leq k$, then by the relation $a\circ b = o$ it follows that

$$Z_3 A^{\mu\nu} Z_o = 0 .$$

This condition implies that each $A^{\mu\nu}$ is of the form $A^{\mu\nu}=A^{\mu\nu}_{1,1}$ coming from a 2-form $a^{\mu\nu}$. The matrix $A\circ B$ satisfies (i) and (ii) and thus we have obtained a morhism from the space Mon(k) of not necessarily selfdual monads to $M(k)$. It was shown by W. Böhmer that this morphism is a fibre bundle with fibre $GL(2k+2,\mathbb{C})$. The same is true if we restrict ourselves to selfdual monads resp. symmetric matrices. In this case the fibre becomes $Sp(k+1)$.

References

[1] Atiyah, Geometry of Yang-Mills Fields, Lezioni Fermiane, Pisa 1979

[2] Atiyah-Ward, Instantons and Algebraic Geometry, Commun. Math. Phys. 55, 117-124 (1977)

[3] Atiyah-Hitchin-Drinfeld-Manin, Construction of Instantons, Phys. Letters 65 A, 185-187 (1978)

[4] Barth-Hulek, Monads and moduli of vector bundles, manuscr. math. 25, 323-347 (1978)

[5] Douady - Verdier, Les équations de Yang-Mills, Seminaire E.N.S. Paris 1977-78, astérisque 71-72 (1980)

[6] Milnor-Stasheff, Characteristic classes, Princeton University Press 1974

[7] Rawnsley, On the Atiyah-Hitchin-Drinfeld-Manin vanishing theorem for cohomology groups of instanton bundles, Math. Ann. 241, 43-56 (1979)

[8] Rawnsley, Self-dual Yang-Mills fields, manuscript.

[9] Okonek-Schneider-Spindler, Vectorbundles on complex projective space, Progr. in Mathem. 3, Birkhäuser Boston 1980

[10] Trautmann, Zur Berechnung von Yang-Mills-Potentialen durch holomorphe Vektorbündel, Proc. Nice conference 1979, Progr. in Mathem. 7, Birkhäuser Boston 1980

[10'] Trautmann, Moduli for vectorbundles on $\mathbb{P}_n(\mathbb{C})$, Math. Ann. 23 167-186 (1978)

[11] Wells, Differential analysis on complex manifolds, Prentice Hall 1973

LECTURES ON MATHEMATICAL INSTANTON BUNDLES

to be given at the

POIANA BRASOV SCHOOL ON GAUGE THEORIES

Wolf Barth

1. Vector bundles on $\mathbb{P}_n$.

The complex-projective n-space $\mathbb{P}_n = \mathbb{P}_n(\mathbb{C})$ is the basic space of global algebraic geometry. Although algebraic geometry is the study of sub-varieties of $\mathbb{P}_n$, an important tool is the notion of vector bundle, which is an object lying over $\mathbb{P}_n$.

Points in $\mathbb{P}_n$ are given by homogeneous coordinates $(z_1:\ldots:z_{n+1})$, where $z_i \in \mathbb{C}$. It is understood that for $o \neq \lambda \in \mathbb{C}$ we have $(\lambda z_1:\ldots:\lambda z_{n+1}) = (z_1:\ldots:z_{n+1})$, and the z_i cannot vanish simultaneously. The sets $U_i = \{z_i \neq o\}$ are open and dense in $\mathbb{P}_n$. The quotients $\frac{z_j}{z_i}$, $j \neq i$, are complex-valued functions on U_i. Using them as coordinates one may identify U_i with $\mathbb{C}^n$.

A vector bundle of rank r over $\mathbb{P}_n$ is given by matrix-valued functions

$$c_{ij}: U_i \cap U_j \to GL(r,\mathbb{C}), \quad i,j = 1,\ldots,n+1$$

satisfying the cocycle conditions

$$\begin{aligned} c_{ii} &= 1_r && \text{(unit matrix)}, \\ c_{ij}\cdot c_{ji} &= 1_r && \text{on } U_i \cap U_j, \\ c_{ij}\cdot c_{jk}\cdot c_{ki} &= 1_r && \text{on } U_i \cap U_j \cap U_k. \end{aligned}$$

The total space V of the bundle is obtained by glueing the disjoint union $\bigcup_{i=1}^{n+1} U_i \times \mathbb{C}^r$ identifying $u \times v \in U_i \times \mathbb{C}^r$ and $u' \times v' \in U_j \times \mathbb{C}^r$, if $u=u' \in \mathbb{P}_n$ and $v = c_{ij}(u)v'$. The following examples are the most basic ones:

The line bundles $\mathcal{O}_{\mathbb{P}_n}(k)$, $k \in \mathbb{Z}$: Here $r = 1$ and

$$c_{ij} = \left(\frac{z_j}{z_i}\right)^k .$$

The tangent bundle $T_{\mathbb{P}}$: Here $r = n$ and c_{ij} is the functional matrix obtained by differentiating the coordinates $\frac{z_k}{z_i}$ on U_i with respect to the coordinates $\frac{z_\ell}{z_j}$ on U_j. If $i < j$, then

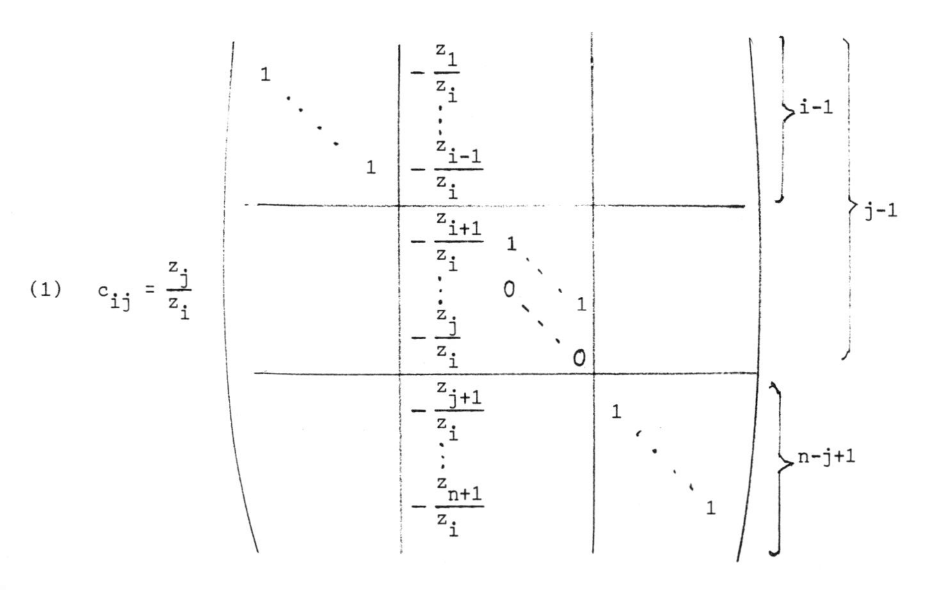

The cotangent bundle $\Omega_{\mathbb{P}}$: Again $r = n$, but c_{ij} is the transposed inverse of the matrix in the preceding example.

The bundle V is called algebraic, if the r^2 entries in the matrix c_{ij} are polynomials in the coordinates $\frac{z_k}{z_i}$ and $\frac{z_k}{z_j}$, as in the examples above. There is a natural projection $\pi: V \to \mathbb{P}_n$, which on $U_i \times \mathbb{C}^r$ is given by $(u,v) \to u$. The fibres $V(x) = \pi^{-1}x$, $x \in \mathbb{P}_n$, are r-dimensional vector spaces. A section in the algebraic bundle V is a map $\sigma: \mathbb{P}_n \to V$ with $\pi \circ \sigma = id_{\mathbb{P}}$, such that $\sigma|U_i$ is defined by a map $U_i \to \mathbb{C}^r$ consisting of r polynomials in the coordinates $\frac{z_k}{z_i}$, $k \neq i$. Every bundle admits the zero-section, on U_i given by $u \to u \times 0$, but in general there need not be other sections. All sections form a vector space, denoted by $\Gamma(V)$, which for nontrivial reasons is finite dimensional. Consider the example:

Sections in $\Gamma\mathcal{O}_{\mathbb{P}}(k)$: Every $\sigma \in \Gamma\mathcal{O}_{\mathbb{P}}(k)$ is defined by $\mathbb{C}$-valued functions σ_i, $i=1,\dots,n+1$, which are polynomials in $\frac{z_k}{z_i}$, $k \neq i$, and satisfy

$$\sigma_i = \left(\frac{z_j}{z_i}\right)^k \sigma_j \qquad \text{on } U_i \cap U_j.$$

Since σ_i stays bounded when $z_j \to o$ on U_i, we see that σ_j must be a polynomial $p(\frac{z_1}{z_j},\ldots,\frac{z_{n+1}}{z_j})$ of degree $\leq k$. If we write it homogeneously $P(z_1,\ldots,z_{n+1}) = z_j^k\, p$, then for all i we have $\sigma_i = \frac{1}{z_i^k} P$. In this way $\Gamma\mathcal{O}_{\mathbb{P}_n}(k)$ is identified with the $\binom{n+k}{k}$-dimensional vector space of homogeneous polynomials in $z_1,\ldots,z_{n+1}$ of degree k.

By performing on the cocycle matrices all the usual tensor operations, one defines the tensor bundles associated with a given bundle V, resp. with two given bundles V and W: The direct sum $V \oplus W$, tensor product $V \otimes W$, the tensor powers $\otimes^p V$, exterior and symmetric powers $\Lambda^p V$ and $S^p V$, and the dual bundle V^*. For $V \otimes \mathcal{O}_{\mathbb{P}}(k)$ one uses the abbreviation $V(k)$.

In particular, the fibres of $\mathrm{Hom}(V,W) = V^* \otimes W$ are the vector spaces $V(x)^* \otimes W(x)$ of linear maps from $V(x)$ to $W(x)$. So sections in $V^* \otimes W$ define <u>bundle morphisms</u> $\gamma: V \to W$, i.e., algebraic maps which preserve the fibres and act linearly on them. Such a <u>bundle morphism</u> γ is called injective (surjective) if $\gamma(x) : V(x) \to W(x)$ is so for all $x \in \mathbb{P}_n$. Concretely, if V is defined by a cocycle c_{ij}, and W by b_{ij}, then γ is given by polynomial maps γ_i on U_i with $b_{ij}\gamma_j = \gamma_i c_{ij}$.

Consider e.g. the $n \times (n+1)$ - matrices

$$\gamma_j = \begin{pmatrix} 1 & & & -\frac{z_1}{z_j} & & & \\ & \ddots & & \vdots & & & \\ & & 1 & -\frac{z_{j-1}}{z_j} & & & \\ & & & -\frac{z_{j+1}}{z_j} & 1 & & \\ & & & \vdots & & \ddots & \\ & & & -\frac{z_{n+1}}{z_j} & & & 1 \end{pmatrix}$$

With c_{ij} the cocycle given above for $T_{\mathbb{P}}$, one easily checks that $c_{ij}\gamma_j = \gamma_i \frac{z_j}{z_i}$. So the γ_j define a morphism

$$\gamma: (n+1)\mathcal{O}_{\mathbb{P}}(1) \rightarrow T_{\mathbb{P}}$$

(here and in the sequel we abbreviate the m-fold direct sum $\mathcal{O}_{\mathbb{P}}(k)\oplus\ldots\oplus\mathcal{O}_{\mathbb{P}}(k)$ by $m\cdot\mathcal{O}_{\mathbb{P}}(k)$). Since rank $\gamma_j = n$ in all points, the morphism γ is surjective. The vector

$$\beta_j : \left(\frac{z_1}{z_j},\ldots,\frac{z_{j-1}}{z_j},1,\frac{z_{j+1}}{z_j},\ldots,\frac{z_{n+1}}{z_j}\right)^t$$

spans the kernel of γ_j in all points, hence defines a morphism

$$\beta: \mathcal{O}_{\mathbb{P}} \rightarrow (n+1)\,\mathcal{O}_{\mathbb{P}}(1)$$

of the trivial bundle onto ker γ. One expresses this fact by saying the sequence

$$(2) \qquad 0 \longrightarrow \mathcal{O}_{\mathbb{P}} \xrightarrow{\beta^t} (n+1)\,\mathcal{O}_{\mathbb{P}}(1) \xrightarrow{\gamma} T_{\mathbb{P}} \longrightarrow 0$$

is <u>exact</u>. This sequence, usually called the Euler sequence and derived in a coordinate free way, gives a description of the tangent bundle which is used much more frequently than the definition of $T_{\mathbb{P}}$ by a cocycle. Dualizing it, one obtains the description

$$(3) \qquad 0 \longrightarrow \Omega_{\mathbb{P}} \xrightarrow{\gamma^t} (n+1)\,\mathcal{O}_{\mathbb{P}}(-1) \xrightarrow{\beta} \mathcal{O}_{\mathbb{P}} \longrightarrow 0$$

of the cotangent bundle by an exact sequence.

Finally, one must mention that the description of bundles by cocycles is very far from being unique! Two cocycles c_{ij} and b_{ij} define the same bundle (more correctly: two isomorhic bundles) if there is a bijective bundle map between their bundles. That means, there is a relation $c_{ij}\gamma_j = \gamma_i b_{ij}$. Two cocycles satisfying such a relation (hence defining the same bundle) are called cohomologous.

Consider e.g. $\Lambda^n T_{\mathbb{P}}$. Forming the determinants of the matrices (1), we obtain for this rank-1 bundle the cocycle $c_{ij} = (-1)^{j-i-1}\left(\frac{z_j}{z_i}\right)^{n+1}$, $i < j$. Putting $\gamma_j = \sqrt{-1}\cdot(-1)^j$, we find it is cohomologous with $\left(\frac{z_j}{z_i}\right)^{n+1}$, i.e., we have

$\Lambda^n T_{\mathbb{P}_n} = \mathcal{O}_{\mathbb{P}_n}(n+1)$. The dual bundle

$$\Omega^n_{\mathbb{P}_n} = \Lambda^n \Omega_{\mathbb{P}_n} = \mathcal{O}_{\mathbb{P}_n}(-n-1)$$

is called the <u>canonical bundle</u> on $\mathbb{P}_n$.

Since the bundles $\mathcal{O}_{\mathbb{P}_n}(k)$ for distinct k are distinguished by the dimensions of their (resp. their duals') spaces of sections, they are not isomorphic. It is a non-trivial fact however (essentially equivalent with the vanishing of $H^1(\mathcal{O}_{\mathbb{P}_n})$, see 2.4 below) that every rank-1 bundle (= <u>line bundle</u>) on $\mathbb{P}_n$ is isomorphic with one of the $\mathcal{O}_{\mathbb{P}_n}(k)$. So for every rank-r bundle V there is some $k \in \mathbb{Z}$ with $\Lambda^r V = \mathcal{O}_{\mathbb{P}_n}(k)$. One calls this k the <u>first Chern class</u> $c_1(V)$ of V.

2. Cohomology.

Fix some bundle V on $\mathbb{P}_n$ with cocycle c_{ij}. We define (o,q)-forms ω with values in V. Such a form is given on U_i by an expression

$$\omega_i = \sum_{1\le k_1<\ldots<k_q\le n} f^{(i)}_{k_1\ldots k_q}\, d\bar{u}_{k_1}\wedge\ldots\wedge d\bar{u}_{k_q},$$

where for simplicity we write $u_1,\ldots,u_n$ instead of $\frac{z_k}{z_i}$, $k\neq i$, and where the coefficients f are $\mathcal{C}^\infty$-maps $U_i\to\mathbb{C}^r$. To define a global form, these ω_i must satisfy the patching conditions $\omega_i = c_{ij}\omega_j$ with multiplication by c_{ij} exerted on the coefficients f. Since different sets of coordinates are used on U_i and U_j, the explicit formula is complicated.

Anyway, the c_{ij} are polynomial functions, hence holomorphic on $U_i\cap U_j$ and $\bar{\partial}\, c_{ij} = o$ therefore. This implies that $\bar{\partial}\omega_i$ patches with $\bar{\partial}\omega_j$, and the form $\bar{\partial}\omega$ of type $(o,q+1)$ with values in V is well-defined. As usual one has $\overline{\partial\partial}\omega = o$ and defines the (Dolbeault-) cohomology groups

$$H^q(V) = \frac{\bar{\partial}\text{-closed } \omega \text{ of type } (o,q)}{\bar{\partial}\text{ (all } (o,q-1)\text{-forms)}}.$$

We shall need the following basic facts:

2.1 Finiteness theorem (Cartan-Serre, [6,p. 202]: *All cohomology groups $H^q(V)$ are $\mathbb{C}$-vector spaces of finite dimension.*

2.2 *There is a natural isomorphism* $\Gamma(V)\to H^o(V)$. This map is described easily: Locally, i.e., on each U_i a section $\sigma\in\Gamma(V)$ is given by a polynomial map σ_i, which is holomorphic, hence a $\bar{\partial}$-closed 0-form. The converse, a weak form of Serre's Gaga theorem [11] is less trivial.

2.3 Serre duality [10]: *For every V and q there is a pairing*

$$H^q(V)\otimes_{\mathbb{C}} H^{n-q}(V^*\otimes\Omega^n_{\mathbb{P}_n})\to\mathbb{C}$$

making one of these vector spaces the dual of the other.

2.4 <u>For the line bundles</u> $\mathcal{O}_{\mathbb{P}_n}(k)$ <u>one has</u>

$$\dim_{\mathbb{C}} H^q(\mathcal{O}_{\mathbb{P}_n}(k)) = \begin{cases} \binom{k+n}{n} & q = 0, \quad k \geq 0 \\ 0 & q = 0, \quad k \leq -1 \\ 0 & q = 1,\dots,n-1, \quad \text{all } k \\ 0 & q = n, \quad k \geq -n \\ \binom{-k-1}{n} & q = n, \quad k \leq -n-1 \end{cases} .$$

The computation of the dimensions for the H^0 is nothing but counting homogeneous polynomials. The computation of the other dimensions is still elementary. In algebraic context this is done e.g. in [9, section 65].

The machinery of cohomology theory, although its ingredients are fairly abstract, is basically simple: Each morphism $\gamma: V \to W$ between vector bundles induces (by letting it act on differential forms) a morphism $H^q(\gamma) : H^q(V) \to H^q(W)$ for all q. Everything is governed by the <u>long exact cohomology sequence</u>:

2.5. <u>Given a short exact sequence</u>

$$0 \to U \to V \to W \to 0$$

<u>of vector bundles, there are morphisms</u>

$$\delta^q : H^q(W) \to H^{q+1}(U) , \quad q = 0,1,\dots$$

<u>such that the sequence</u>

$$0 \to H^0(U) \to H^0(V) \to H^0(W) \xrightarrow{\delta^0}$$
$$\xrightarrow{\delta^0} H^1(U) \to H^1(V) \to H^1(W) \xrightarrow{\delta^1} \dots$$

<u>of vector spaces is exact</u> (which means the kernel of one arrow is the image of the preceding one). Although the proof of all the exactness assertions is cumbersome, it is not hard. The definition of $\delta^q: H^q(W) \to H^{q+1}(U)$ is as follows: Represent the class $[\omega] \in H^q(W)$ by the $\bar{\partial}$-closed $(0,q)$-form ω. Then one easily finds $(0,q)$-forms φ with values in V which are mapped onto ω. These φ however need not be $\bar{\partial}$-closed! Then $\bar{\partial}\varphi$ is a $\bar{\partial}$-closed $(0,q+1)$-form with

values in U representing $\delta^q[\omega]$

Consider e.g. the description of Ω by (3). It leads to a long exact sequence

$$o \to H^o(\Omega_{\mathbb{P}}) \to (n+1)\, H^o(\mathcal{O}_{\mathbb{P}}(-1)) \to H^o(\mathcal{O}_{\mathbb{P}}) \xrightarrow{\delta^o}$$

$$\to H^1(\Omega_{\mathbb{P}}) \to (n+1)\, H^1(\mathcal{O}_{\mathbb{P}}(-1)) \to H^1(\mathcal{O}_{\mathbb{P}}) \to$$

$$\to H^2(\Omega_{\mathbb{P}}) \to \dots$$

Using 2.4 one finds $H^q(\Omega) = o$ for all $q \neq 1$, but $H^1(\Omega_{\mathbb{P}}) = \delta^o\, H^o(\mathcal{O}_{\mathbb{P}}) = \mathbb{C}$.

We shall need a converse of 2.4:

2.6 Theorem [8]: Let V on $\mathbb{P}_n$ be a vector bundle with

$$H^q(V(k)) = o \quad \text{for } 1 \le q \le n-1, \quad \text{all } k,$$

then V is a direct sum of line bundles $\mathcal{O}_{\mathbb{P}_n}(\ell)$.

We shall prove this by induction on n. The hardest part (to be left as an exercise) is the case $n = 1$. Explicitly the assertion is this: Consider some $r \times r$ matrix $c(z)$ the entries of which are polynomials in z and $\frac{1}{z}$ (here $z = \frac{z_2}{z_1}$ is the inhomogeneous coordinate).
If rank $c(z) = r$ for all z, then there are matrices γ_1, γ_2 with entries polynomials in z, resp. $\frac{1}{z}$ such that

$$\gamma_1(z)\, c(z)\, \gamma_2(z) = \begin{pmatrix} z^{k_1} & & o \\ & \ddots & \\ o & & z^{k_r} \end{pmatrix}, \quad k_1, \dots, k_r \in \mathbb{Z}.$$

Number theorists claim to have known this already a hundred years ago. At the beginning of the century Birkhoff [4] proved it. In 1956 it appeared in algebraic geometry [7] somewhat generalized, and is called Grothendieck's theorem since then.

To perform the induction step, assume $n \ge 2$ and consider $\mathbb{P}_{n-1} \subset \mathbb{P}_n$ embedded as the subset $z_{n+1} = o$. This polynomial $z_{n+1} \in \Gamma\mathcal{O}_{\mathbb{P}_n}(1)$ defines a morphism $z_{n+1} : \mathcal{O}_{\mathbb{P}_n} \to \mathcal{O}_{\mathbb{P}_n}(1)$, which is bijective outside of $\mathbb{P}_{n-1}$, but vanishes on $\mathbb{P}_{n-1}$. The quotient $\mathcal{O}_{\mathbb{P}_n}(1)/z_{n+1}\mathcal{O}_{\mathbb{P}_n}$ cannot be a vector bundle, because it vanishes outside of $\mathbb{P}_{n-1}$, whereas on this subspace it looks like $\mathcal{O}_{\mathbb{P}_{n-1}}(1)$. In fact, it is a <u>sheaf</u> on $\mathbb{P}_n$, which may be identified with $\mathcal{O}_{\mathbb{P}_{n-1}}(1)$. Then we have the <u>exact sequence of sheaves</u>

$$o \to \mathcal{O}_{\mathbb{P}_n} \xrightarrow{z_{n+1}} \mathcal{O}_{\mathbb{P}_n}(1) \longrightarrow \mathcal{O}_{\mathbb{P}_{n-1}}(1) \to o \quad .$$

Given the bundle V on $\mathbb{P}_n$, we denote by $V|\mathbb{P}_{n-1}$ its restriction, which is a rank-r bundle on $\mathbb{P}_{n-1}$. Then the sequence above "tensored with $V(k)$" gives an exact sequence

$$o \to V(k) \xrightarrow{z_{n+1}} V(k+1) \longrightarrow V(k+1) \mid \mathbb{P}_{n-1} \to o \ .$$

Here the long exact cohomology sequence comes in (it also holds for sequences of sheaves) and tells us that $V|\mathbb{P}_{n-1}$ satisfies on $\mathbb{P}_{n-1}$ the assumptions in theorem 2.6. So by induction there are $k_1,\dots,k_r \in \mathbb{Z}$ with

$$V|\mathbb{P}_{n-1} = \mathcal{O}_{\mathbb{P}_{n-1}}(k_1)\oplus\dots\oplus\mathcal{O}_{\mathbb{P}_{n-1}}(k_r) \quad .$$

Denote by S the direct sum $\mathcal{O}_{\mathbb{P}_n}(k_1)\oplus\dots\oplus\mathcal{O}_{\mathbb{P}_n}(k_r)$ on $\mathbb{P}_n$. Then we have so far an isomorphism

$$h: S \mid \mathbb{P}_{n-1} \to V \mid \mathbb{P}_{n-1}$$

of restricted bundles, and we want to extend this $h \in \Gamma(S^* \otimes V|\mathbb{P}_{n-1})$ to $\mathbb{P}_n$. The exact cohomology sequence for

$$o \to S^* \otimes V(-1) \to S^* \otimes V \to S^* \otimes V|\mathbb{P}_{n-1} \to o$$

tells us that this is possible, if $H^1(S^* \otimes V(-1)) = o$. But $S^* \otimes V(-1)$ is the direct sum of bundles $V(-k_i-1)$, for which we assumed the vanishing of their H^1. So we have a morphism $S \to V$, which we denote by h again, restricting on $\mathbb{P}_{n-1}$ to an isomorphism. Of course h is bijective, where $\det h: \Lambda^r S \to \Lambda^r V$ does not vanish. Since $\Lambda^r S = \Lambda^r V = \mathcal{O}_{\mathbb{P}_n}(k_1+\dots+k_r)$, this $\det h$ is a section in $(\Lambda^r S)^* \otimes \Lambda^r V = \mathcal{O}_{\mathbb{P}}$, which does not vanish on $\mathbb{P}_{n-1}$, i.e. $\det h$ is a non zero constant.

3. <u>Mathematical instanton bundles on</u> $\mathbb{P}_3$.

In these lectures a <u>mathematical instanton bundle</u> will mean: a vector bundle V on $\mathbb{P}_3$ with

$$\boxed{\begin{array}{l} \text{rank}\,(V) = 2 \\ c_1(V) = o \\ H^o(V) = o \\ H^1(V(-2)) = o \end{array}}$$

having additionally the property that

$$V|L = \mathcal{O}_L \oplus \mathcal{O}_L$$

for some line $L \subset \mathbb{P}_3$ (notice that a line is a copy of $\mathbb{P}_1$). This last property is not independent, it can be deduced from the first three ones, but this does not matter at the moment (cf. section 4).

Vector bundles are shy creatures, hard to approach. The safest thing to do is to compute as much of their cohomology as possible. So we collect cohomology information on mathematical instanton bundles V:

i) $H^o(V(k)) = o$ <u>for all</u> $k \leq o$. If $k = o$, this is one of our assumptions. If $k < o$, take an arbitrary homogeneous polynomial $f(z_1,\ldots,z_4) \not\equiv o$ of degree $-k$. It can be viewed as a nontrivial map $\mathcal{O}_{\mathbb{P}}(k) \to \mathcal{O}$ inducing a map $V(k) \to V$ which is an isomorphism on the dense set where $f \not\equiv o$. This makes $H^o(V(k))$ a subspace of $H^o(V) = o$.

ii) $H^1(V(k)) = o$ <u>for all</u> $k \leq -2$. For $k = -2$ this is assumed. For $k < -2$ the reasoning is similar to the preceding one, but more subtle: Fix a linear polynomial $f(z_1,\ldots,z_4)$ and denote by $\mathbb{P}_2$ the plane where it vanishes. Then we have the exact sheaf sequence (used in the proof of 2.6 already)

$$o \to V(k) \xrightarrow{f} V(k+1) \to V(k+1) \mid \mathbb{P}_2 \to o$$

for all k, and the exact piece

$$H^o(V(k+1) \mid \mathbb{P}_2) \to H^1(V(k)) \to H^1(V(k+1))$$

of its cohomology sequence. If we know $H^o(V(k+1)\,|\,\mathbb{P}_2) = o$ for all $k \leq -3$, then the map $H^o(V(k)) \to H^o(V(k+1))$ would be injective, and the assertion would follow by decreasing induction on k. But we have the freedom to choose f such that the plane $\mathbb{P}_2$ contains a line L on which $V|L = \mathcal{O}_L \oplus \mathcal{O}_L$. So for $k \leq -1$, any section $\sigma \in \Gamma(V(k)|\mathbb{P}_2)$ would vanish along L. If $g(z_1,z_2,z_3) = o$ is a linear equation for L, this implies that $\frac{1}{g}\sigma$ is a section in $V(k-1)|\mathbb{P}_2$, which again vanishes along L. We may repeat this reasoning arbitrarily often and find, that the Taylor expansion of σ in all points $x \in L$ vanishes, i.e. $\sigma = o$.

iii) V <u>is selfdual</u>, i.e., $V = V^*$. Indeed, by assumption $\Lambda^2 V = \mathcal{O}_{\mathbb{P}_3}$. Take some section $\varepsilon \neq o$ in $\Lambda^2 V$. In each point $x \in \mathbb{P}_3$ this is an alternating bilinear form on $V(x)$, which may be viewed as an anti-selfadjoint map $V(x) \to V^*(x)$. So ε defines an anti-selfadjoint morphism $V \to V^*$ which in all x is bijective.

iv) $H^2(V(k)) = o$ <u>for all</u> $k \geq -2$,

$H^3(V(k)) = o$ <u>for all</u> $k \geq -4$.

Since $\Omega^3_{\mathbb{P}_3} = \mathcal{O}_{\mathbb{P}_3}(-4)$ by section 1, Serre-duality 2.3, combined with iii) shows $H^q(V(k)) = H^{3-q}(V^*(-4-k))$. The assertions therefore are dual to i) and ii).

v) <u>For all</u> $k \geq o$ <u>the multiplication map</u>

$$H^1(V(k-1)) \otimes_{\mathbb{C}} H^o(\mathcal{O}_{\mathbb{P}_3}(1)) \to H^1(V(k))$$

<u>is surjective</u>. In plain language this means the following: Every linear polynomial f defines a map $V(k-1) \to V(k)$, hence a map $H^1(V(k-1)) \to H^1(V(k))$. If we denote by $f\cdot\varphi$ the image of $\varphi \in H^1(V(k-1))$, then the assertion is that $H^1(V(k))$ is spanned by these products $f\cdot\varphi$, $f \in H^o(\mathcal{O}_{\mathbb{P}_3}(1))$, $\varphi \in H^1(V(k-1))$.

To prove v) we fix a line $L \subset \mathbb{P}_3$ on which $V|L = \mathcal{O}_L \oplus \mathcal{O}_L$. W.l.o.g. we assume L has equations $z_1 = z_2 = o$. This implies that the map

$$\mathcal{O}_{\mathbb{P}_3}(-1) \oplus \mathcal{O}_{\mathbb{P}_3}(-1) \xrightarrow{(z_1,z_2)} \mathcal{O}_{\mathbb{P}_3}$$

is surjective away from L, but vanishes on L. Just as in the proof of 2.6 one may identify the quotient of $\mathcal{O}_{\mathbb{P}_3}$ by the image of this map with $\mathcal{O}_L$. One obtains an exact sequence

$$o \to \mathcal{O}_{\mathbb{P}_3}(-2) \xrightarrow{\begin{pmatrix}-z_2\\ z_1\end{pmatrix}} \mathcal{O}_{\mathbb{P}_3}(-1) \oplus \mathcal{O}_{\mathbb{P}_3}(-1) \xrightarrow{(z_1,z_2)} \mathcal{O}_{\mathbb{P}_3} \to \mathcal{O}_L \to o$$

of sheaves. Tensored with $V(k)$ it becomes

$$o \to V(k-2) \to V(k-1) \oplus V(k-1) \xrightarrow{(z_1,z_2)} V(k) \to V(k)|L \to o$$
$$V(k)|L = \mathcal{O}_L(k)\oplus\mathcal{O}_L(k)$$

Such a sequence may be split in two:

$$o \to V(k-2) \to V(k-1) \oplus V(k-1) \xrightarrow{r} J \to o$$
$$o \to J \xrightarrow{s} V(k) \to \mathcal{O}_L(k) \oplus \mathcal{O}_L(k) \to o.$$

Here J is some unknown sheaf and $s\circ r = (z_1,z_2)$. The cohomology sequence of the first row and iv) imply for $k \geq o$:

$$H^1(r) : H^1(V(k-1)) \oplus H^1(V(k-1)) \to H^1(J)$$

is surjective. Formula 2.4 on $L = \mathbb{P}_1$ shows for these k:

$$H^1(s) : H^1(J) \to H^1(V(k))$$

is surjective too. Composing both maps we see

$$(z_1,z_2) : H^1(V(k-1)) \oplus H^1(V(k-1)) \to H^1(V(k))$$

is surjective, i.e., $H^1(V(k))$ is spanned by products $z_1\cdot\varphi$ and $z_2\cdot\varphi$ already. To apply our knowledge of H^1, we must use a more concrete interpretation of these groups:

An <u>extension</u> of V is an exact sequence

$$o \to V \to W \to \mathcal{O}_{\mathbb{P}_3} \to o$$

with some unknown bundle W. The associated cohomology sequence induces a map

$$\delta^o : H^o(\mathcal{O}_{\mathbb{P}_3}) \to H^1(V) ,$$

and the image under this map of the section $1 \in H^o(\mathcal{O}_{\mathbb{P}_3})$ is a distinguished element $\varphi \in H^1(V)$. It is a theorem from homological algebra that this correspondence between extensions and elements $\varphi \in H^1(V)$ is one to one. So every class in $H^1(V)$ comes from an extension as above, and every k-tuple $\varphi_1,\ldots,\varphi_k \in H^1(V)$ from a larger extension

(4) $$o \to V \to W \to k\cdot\mathcal{O}_{\mathbb{P}_3} \to o.$$

We apply this to $V(-1)$ instead of V. Put

$$\boxed{k = \dim H^1(V(-1))}$$

and choose a basis $\varphi_1,\ldots,\varphi_k \in H^1(V(-1))$. With these elements consider the extension (4). The induced map δ^o maps the k unit sections in the direct sum $k\,\mathcal{O}_{\mathbb{P}_3}$ onto the k generators of $H^1(V(-1))$. By exactness, the map $H^1(V(-1)) \to$ $\to H^1(W)$ must vanish. Since $H^1(\mathcal{O}_{\mathbb{P}_3}) = o$ too by 2.4, the exact cohomology sequence shows $H^1(W) = o$. For $k < o$, by ii) above, $H^1(V(k))$ vanishes and $H^1(W(k)) = o$ again. For $k > o$ and $f \in \Gamma\mathcal{O}_{\mathbb{P}}(1)$ there is a commutative diagram

$$\begin{array}{ccc} H^1(V(k-1)) & \to & H^1(W(k-1)) \\ f\downarrow & & \downarrow f \\ H^1(V(k)) & \to & H^1(W(k)) \end{array}$$

and v) above shows that $H^1(V(k)) \to H^1(W(k))$ is the zero-morphism, hence $H^1(W(k)) = o$ in this range too. We conclude:

$$H^1(W(k)) = o \qquad \text{for all } k \in \mathbb{Z}.$$

This is half of the assumption in theorem 2.6!

To get our hand on the other half, we dualize (4) and tensor by $\mathcal{O}_{\mathbb{P}_3}(-1)$:

$$o \to k\mathcal{O}_{\mathbb{P}}(-1) \to W^*(-1) \to V^* \to o.$$

This sequence now induces an isomorphism $H^1(W^*(-2)) \to H^1(V^*(-1))$, and together with the isomorphism $V = V^*$ this suffices to lift the extension (4) to $W^*(-1)$. We obtain a big diagram

(5) 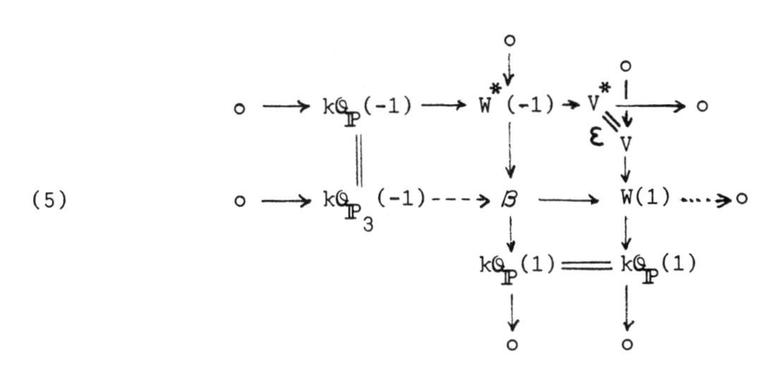

where the columns and first row are exact by construction and the second row (dotted arrows) is exact by inspection. The essential point is, that the vanishing of all $H^1(W(k))$, and dually of all $H^2(W^*(k))$, implies for all k:

$$H^1(\mathcal{B}(k)) = H^2(\mathcal{B}(k)) = o.$$

So by theorem 2.6, this $\mathcal{B}$ decomposes in a direct sum of line bundles. Since rank $\mathcal{B} = 2k + 2$, there must be $2k + 2$ such summands. What are they ?

Notice first, that diagram (5) is self-dual. If we dualize everything and simultaneously switch the upper left with the lower right corner, we get the same diagram again - except for the isomorphism $\varepsilon: V^* \to V$, which changed its sign. This shows $\mathcal{B}^* = \mathcal{B}$, and upon futher inspection this isomorphism turns out anti-self adjoint (for this and further information see e.g. [3]).

Now let $\mathcal{O}_{\mathbb{P}_3}(\ell) \subset \mathcal{B}$ be one of its summands. If $\ell \geq 2$, then the image of this bundle in $k\mathcal{O}_{\mathbb{P}}(1)$ would be trivial, hence $\mathcal{O}_{\mathbb{P}_3}(\ell)$ would come from $W^*(-1)$, showing $H^o(W^*(-1)) \neq o$. But this would contradict $H^o(V^*) = H^o(V) = o$. If $\ell = 1$, similar reasoning shows that its image in $k\mathcal{O}_{\mathbb{P}}(1)$ cannot be zero. It then must be a direct summand here. But this implies that in (4) one of the direct summands in $k\mathcal{O}_{\mathbb{P}_3}$ comes from W. Then $\dim \delta^o H^o(k\mathcal{O}_{\mathbb{P}_3}) < k$, a contradiction again. The only possibility is $k \leq o$, but because of $\mathcal{B} = \mathcal{B}^*$, $k \geq o$ at the same time. Hence $\mathcal{B} = (2k + 2)\mathcal{O}_{\mathbb{P}_3}$, and we arrive at

3.1 Monad theorem: <u>For every mathematical instanton bundle there is a map</u>

$$\alpha: k\mathcal{O}_{\mathbb{P}_3}(-1) \to (2k+2)\mathcal{O}_{\mathbb{P}_3}$$

and a nondegenerate symplectic form on $(2k+2)\mathcal{O}_{\mathbb{P}_3}$, *such that for* α^t, *the adjoint of* α *with respect to this form, we have* $\alpha^t\alpha = 0$ *and*

$$V = \ker \alpha^t \ / \ \mathrm{im}\ \alpha .$$

In shorter terms: V is the cohomology of the selfdual *monad*

$$k\mathcal{O}_{\mathbb{P}_3}(-1) \xrightarrow{\alpha} (2k+2)\mathcal{O}_{\mathbb{P}_3} \xrightarrow{\alpha^t} k\mathcal{O}_{\mathbb{P}_3}(1) .$$

The technique of describing bundles by monads is due to Horrocks, who never published it, because he thought it was useless.

4. Properties of monads.

It is time now to emphasize that a monad, although it looks impressingly abstract, consists of three very concrete pieces of information:

- the integer $k = \dim H^1(V(-1))$, which (if one applies some rules of Chern class algebra) turns out to be the second Chern class $c_2(V)$.
- the symplectic form on $(2k+2)\mathcal{O}_{\mathbb{P}}$. Being an isomorphism of a trivial bundle with itself it is given by a constant $(2k+2) \times (2k+2)$ matrix. For our purposes, we usually can assume this matrix to be

$$(6) \qquad \begin{pmatrix} 0 & -1_{k+1} \\ 1_{k+1} & 0 \end{pmatrix}$$

- the morphism $\alpha: k\mathcal{O}_{\mathbb{P}_3}(-1) \to (2k+2)\mathcal{O}_{\mathbb{P}}$. It can be considered as some $(2k+2)\times k$ - matrix $\alpha_{\mu\nu}$ of morphisms $\alpha_{\mu\nu}: \mathcal{O}_{\mathbb{P}_3}(-1) \to \mathcal{O}_{\mathbb{P}}$. Such a morphism $\alpha_{\mu\nu}$ however is nothing but a section in $\mathcal{O}(1)$, i.e., a linear polynomial (2.4). If we write $\alpha_{\mu\nu} = \sum_1^4 (\alpha_{\mu\nu})_i z_i$, then

$$\alpha = \alpha(z) = \sum_1^4 \alpha_i \, z_i$$

with four constant $(2k+2) \times k$ matrices $\alpha_i = (\alpha_{\mu\nu})_i$.

These four matrices are the essential ingredients in a monad. Each 4-tuple of matrices α_i, constitutes a morphism $\alpha = \sum \alpha_i z_i$, but to define a monad, this α must satisfy

(7) $\alpha^t(z)\alpha(z) = 0$ identically in z, where $\alpha^t = \sum \alpha_i^t \, z_i$, and the α_i^t are the transposed matrices with respect to (6).

(8) $\alpha(z) : \mathbb{C}^k \to \mathbb{C}^{2k+2}$ is injective for all z.

Starting with α one may define a new $k\times k$-matrix

$$Q(z,z') = \alpha^t(z')\alpha(z) = \sum_{i,j} \alpha_i^t \, \alpha_j \, z_i' z_j$$

depending linearly on two sets of coordinates z, z'. Since by (7)

$$Q(z,z) = Q(z',z') = Q(z+z',\ z+z') = o,$$

we find

$$Q(z,z') = -\ Q(z',z)\ .$$

And since $\alpha^{tt} = -\ \alpha$,

$$Q^t(z,z') = \alpha^t(z)\ \alpha^{tt}(z') = -\ Q(z',z) \ = \ Q(z,z'),$$

i.e., the matrix Q is symmetric.

The examples discussed in section 5 will be presented in monad form. So we first should answer the question: When is the bundle V defined by a monad, a mathematical instanton bundle?

4.1 Proposition: Let $\alpha\colon k\mathcal{O}_{\mathbb{P}_3}(-1) \to (2k+2)\mathcal{O}_{\mathbb{P}_3}$ be a morphism satisfying (7) and (8). Then $V = \ker \alpha^t\ /\ \mathrm{im}\ \alpha$ is a mathematical instanton bundle if and only if

(9) the induced map $H^o(\alpha^t) : H^o((2k+2)\mathcal{O}_{\mathbb{P}}) \to H^o(k\mathcal{O}_{\mathbb{P}}(1))$ is injective.

Proof: Given α there is a diagram

$$(10)\qquad \begin{array}{ccccccccc}
 & & & & o & & o & & \\
 & & & & \downarrow & & \downarrow & & \\
o & \to & k\mathcal{O}_{\mathbb{P}}(-1) & \longrightarrow & \mathcal{C}^* & \longrightarrow & V & \longrightarrow & o \\
 & & \| & & \downarrow & & \downarrow & & \\
o & \to & k\mathcal{O}_{\mathbb{P}}(-1) & \xrightarrow{\alpha} & (2k+2)\mathcal{O}_{\mathbb{P}} & \longrightarrow & \mathcal{C} & \longrightarrow & o \\
 & & & & \downarrow{\scriptstyle \alpha^t} & & \downarrow & & \\
 & & & & k\mathcal{O}_{\mathbb{P}}(1) & = & k\mathcal{O}_{\mathbb{P}}(1) & & \\
 & & & & \downarrow & & \downarrow & & \\
 & & & & o & & o & &
\end{array}$$

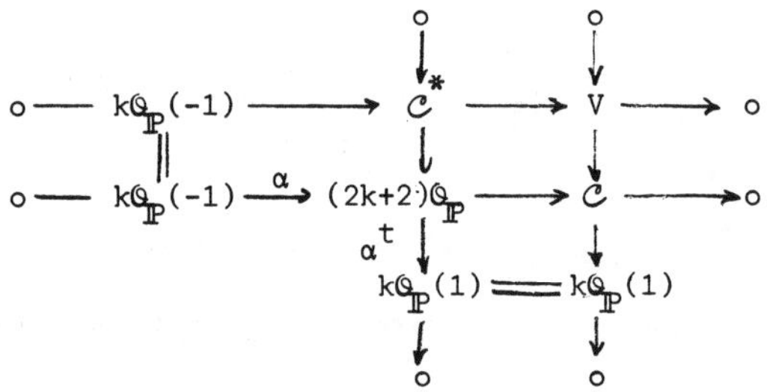

Similar to (5). Its middle column and first row show

$$\ker H^o(\alpha^t) = H^o(\mathcal{C}^*) = H^o(V)\ ,$$

so $H^o(V) = o$ is equivalent with (9). This proves the only if part. To prove the converse, assume (9). We verify the properties listed at the beginning of section 3. We know $\mathrm{rank}\ V = 2$ and $H^o(V) = o$ already. Also $c_1(V) = o$ is obvious, because $V = V^*$ implies $\Lambda^2 V = \mathcal{O}_{\mathbb{P}}$. Furthermore, using 2.4 and the diagram above we see

$$H^1(V(-2)) = H^1(\mathcal{C}^*(-2)) = o.$$

It remains to find a line $L \subset \mathbb{P}_3$ with $V|L = 2\mathcal{O}_L$.

Unfortunately, I only have a nontrivial proof for the existence of such a line [1, theorem of Grauert-Mülich]. This proof works in a far more general situation. For mathematical instanton bundles, however, there should be an easier one. Although I cannot give such a proof, I want to discuss here the behaviour of V on lines $L \subset \mathbb{P}_3$, because this will be of use later:

By Grothendieck's theorem (see 2.6) for every line $L \subset \mathbb{P}_3$ we have

$$V|L = \mathcal{O}_L(d_1) \oplus \mathcal{O}_L(d_2), \qquad d_1, d_2 \in \mathbb{Z} \quad .$$

Since

$$\mathcal{O}_L(d_1+d_2) = \Lambda^2(V|L) = (\Lambda^2 V)|L = \mathcal{O}_L ,$$

necessarily $d_1 + d_2 = o$.

4.2 Proposition: Let $V = \ker \alpha^t / \operatorname{im} \alpha$ be the cohomology bundle of a monad as above. Let the line $L \subset \mathbb{P}_3$ be spanned by the points $x \neq y \in \mathbb{P}_3$. Then $V|L = \mathcal{O}_L(d) \oplus \mathcal{O}_L(-d)$ with d the dimension of the null-space of the $k \times k$-matrix $Q(x,y)$.

Proof: The space of sections in $V|L$ vanishing at one point of L, say at x, has dimension d, Diagram (10) shows that this dimension coincides with the dimension of the space of sections $\sigma \in H^o((2k+2)\mathcal{O}_L)$ satisfying

i) $\sigma(x) \in \operatorname{im} \alpha(x)$,

ii) $\alpha^t(p)\sigma(p) = o$ for all $p \in L$.

But a section $\sigma \in H^o((2k+2)\mathcal{O}_L)$ is constant, determined by a vector $v \in \mathbb{C}^{2k+2}$. Condition i) means $v \in \operatorname{im} \alpha(x)$. Since $\alpha^t(p)$ depends linearly on the homogeneous coordinates of $p \in L$, property ii) holds if $\alpha(x)^t v = o$ (true for v, because $\alpha(x)^t \alpha(x) = o$) and $\alpha(y)^t v = o$. So our space of vectors v is just $\{\operatorname{im} \alpha(x)\} \cap \{\ker \alpha^t(y)\}$ and its dimension is the nullity of $Q(x,y)$, because $\alpha(x)$ is injective.

Assuming that there is some line L spanned by x,y with rank $Q(x,y) = k$, we see that the lines for which this is not the case form a closed thin subset. To be explicit, write for $x = (x_1:\ldots:x_4)$, $y = (y_1:\ldots:y_4)$

$$Q(x,y) = \sum_{i,j} Q_{ij}\, x_i y_j = \sum_{i<j} Q_{ij}(x_i y_j - x_j y_i) \quad .$$

Here the six combinations

$$x_i y_j - x_j y_i = z_i \wedge z_j(x,y) = z_{ij}(L)$$

are the Plücker coordinates of the line L. So we may write $Q = \sum_{i<j} Q_{ij}\, z_{ij}$. The set $\{\mathrm{rank}\; Q < k\}$ can be described by a $k\times k$ determinant, hence is a hyper-surface of degree k in Plücker's $\mathbb{P}_5$ with homogeneous coordinates z_{ij}. Exactly for the lines L parametrized by points on this hypersurface we have $V|L \nsubseteq 2\mathcal{O}_L$. These lines are called the <u>jumping lines</u> of V.

5. t'Hooft's examples.

The problem to find rank-2 bundles on $\mathbb{P}_3$, which are not the direct sums of two line bundles, is not totally trivial. The oldest example known is the null-correlation bundle [8, section 10]. It is described as follows: Choose some nondegenerate skew-symmetric 4x4-natrix $M = (m_{\mu\nu})_{\mu,\nu=1,\dots,4}$. It defines a null-correlation (i.e., a map $\mathbb{P}_3 \to \mathbb{P}_3^*$ with all points x contained in their image plane E_x) by associating with $x = (x_1:\dots:x_4)$ the plane E_x of equation $\sum m_{\mu\nu} x_\mu z_\nu = o$. The null-correlation bundle for M is the rank-2 subbundle $N \subset T_{\mathbb{P}_3}(-1)$ of tangent vectors which in x are tangent to E_x. To describe $T_{\mathbb{P}}(-1)$ we tensor (2) with $\mathcal{O}_{\mathbb{P}}(-1)$, where we put $\alpha = \beta^t$:

$$o \to \mathcal{O}_{\mathbb{P}_3}(-1) \xrightarrow{\alpha} 4\mathcal{O}_{\mathbb{P}_3} \to T_{\mathbb{P}_3}(-1) \to o \quad .$$

Then we consider M as symplectic form on $\mathbb{C}^4$, and on the bundle $4\mathcal{O}_{\mathbb{P}_3}$. Denoting by α^t the transpose with respect to this form we have $\alpha^t\alpha = 0$ and a diagram

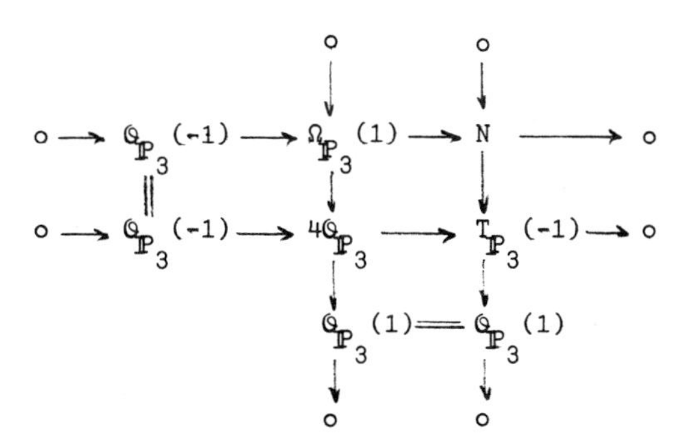

defining N. Thus N even comes as cohomology of a monad

$$\mathcal{O}_{\mathbb{P}_3}(-1) \xrightarrow{\alpha} 4\mathcal{O}_{\mathbb{P}_3} \xrightarrow{\alpha^t} \mathcal{O}_{\mathbb{P}_3}(1)$$

with $k = 1$, as it should be. Since $\alpha(z) = z$ the four image points $\alpha(1:o:o:o),\dots,\alpha(o:o:o:1)$ span $\mathbb{C}^4$. So condition (9) is verified. Also jumping lines are determined easily. Since $Q(x,y) = \sum m_{\mu\nu}x_\mu y_\nu$, by 4.2 the jumping lines through x are precisely the lines in E_x. In particular there are lines L with $N|L = 2\mathcal{O}_L$, and N is a mathematical instanton bundle.

A whole series of examples is due to t'Hooft: Write as above $\alpha(z) = \sum_1^4 \alpha_i z_i$ and fix

$$(11) \qquad \alpha_3 = \begin{pmatrix} 1_k \\ o \\ \hline 0 \end{pmatrix} \begin{matrix} k+1 \\ \\ k+1 \end{matrix} \qquad \alpha_4 = \begin{pmatrix} 0 \\ \hline 1_k \\ o \end{pmatrix} \begin{matrix} k+1 \\ \\ k+1 \end{matrix}$$

as well as the symplectic form (6) an $\mathbb{C}^{2k+2}$. Then

$$(12) \qquad \alpha_1 = \begin{pmatrix} A_1 \\ a_1 \\ \\ B_1 \\ b_1 \end{pmatrix} \begin{matrix} k \\ 1 \\ \\ k \\ 1 \end{matrix} \qquad \alpha_2 = \begin{pmatrix} A_2 \\ a_2 \\ \\ B_2 \\ b_2 \end{pmatrix} \begin{matrix} k \\ 1 \\ \\ k \\ 1 \end{matrix}$$

with $k \times k$-blocks A_i, B_i and row vectors a_i, b_i. Condition (7) for such an α reads:

$$(13) \qquad \alpha_j^t \, \alpha_i + \alpha_i^t \, \alpha_j = o \qquad \text{for all } i,j.$$

Putting $j = 4$, resp. 3 here, the equations for $i=1,2$ are equivalent with the symmetry of the matrices A_i, B_i. For <u>symmetric</u> A_i, B_i the remaining equations $(i,j = 1,2)$ are:

$i=j=1,2$: $\qquad - B_i A_i - b_i^t a_i + A_i B_i + a_i^t b_i = o$, i.e.

$$(14) \qquad [A_i, B_i] + a_i \wedge b_i = o$$

$i=1,\ j=2$: $\qquad - B_2 A_1 - b_2^t a_1 + A_2 B_1 + a_2^t b_1 - B_1 A_2 - b_1^t a_2 + A_1 B_2 + a_1^t b_2 = o$, i.e.

$$[A_1, B_2] + [A_2, B_1] + a_1 \wedge b_2 + a_2 \wedge b_1 = o \quad .$$

In this generality the equations will be needed in section 6. For t'Hooft's examples one takes the A_i and B_i diagonal and the vector products zero:

$A_1 = \text{diag}\,(a_1,\dots,a_k)$

$B_1 = \text{diag}\,(b_1,\dots,b_k)$

$A_2 = \text{diag}\,(c_1,\dots,c_k)$

$B_2 = \text{diag}\,(d_1,\dots,d_k)$

$a_1 = b_2 = (\lambda_1,\dots,\lambda_k)$

$a_2 = b_1 = 0$

(Please don't get confused: the vectors a_i and b_i have nothing to do with the entries a_i, b_i in the matrices.)

Next we have to discuss condition (8). To do this define $k+1$ lines $L_1,\dots,L_{k+1} \subset \mathbb{P}_3$ by

$$L_\mu = \{a_\mu z_1 + c_\mu z_2 + z_3 = b_\mu z_1 + d_\mu z_2 + z_4 = o\}, \quad \mu = 1,\dots,k$$

$$L_{k+1} = \{\dot{z}_1 = z_2 = o\} \quad .$$

5.1 Proposition: <u>The map</u> α <u>defines a monad (i.e., condition</u> (8) <u>is met) if and only if</u>

i) $\lambda_\mu \neq o$ <u>for all</u> $\mu = 1,\dots,k$

ii) <u>the lines</u> $L_1,\dots,L_{k+1}$ <u>are disjoint.</u>

Proof: Assume first that (8) holds, i.e., for all $o \neq v \in \mathbb{C}^k$ and $(z_1:\dots:z_4)$ we have $\sum_1^4 \alpha_i z_i v \neq o$. If then $\lambda_\mu = o$ for some μ, put $v = (o,\dots,o,1,o,\dots,o) = v_\mu$, the μ-th coordinate vector. Then $\sum \alpha_i z_i v$ has the two entries $a_\mu z_1 + b_\mu z_2 + z_3$ and $b_\mu z_1 + d_\mu z_2 + z_4$, hence vanishes on the line L_μ. If two lines meet, say L_μ and L_ν, then obviously $\mu \neq k+1 \neq \nu$. Put $v = -\lambda_\nu v_\mu + \lambda_\mu v_\nu$, then one easily checks $\sum \alpha_i z_i v = o$ in the intersection $L_\mu \cap L_\nu$.

Then assume that i) and ii) hold. If there is some $o \neq v = \sum t_\mu v_\mu$ with $\sum \alpha_i z_i v = o$, because of i) at least two entries in v will be nonzero, say t_μ and t_ν. Then $\sum \alpha_i z_i v$ contains the four entries

$$a_\mu z_1 t_\mu + c_\mu z_2 t_\mu + z_3 t_\mu$$
$$b_\mu z_1 t_\mu + d_\mu z_2 t_\mu + z_4 t_\mu$$
$$a_\nu z_1 t_\nu + c_\nu z_2 t_\nu + z_3 t_\nu$$
$$b_\nu z_1 t_\nu + d_\nu z_2 t_\nu + z_4 t_\nu$$

This is a system of linear equations in $z_1,\ldots,z_4$. Up to the factors t_μ, $t_\nu \neq o$, the rows of the coefficient matrix are the equations defining L_μ and L_ν. So because of $L_\mu \cap L_\nu \neq o$, there is no nontrivial solution $z_1,\ldots,z_4$.

If the conditions i) and ii) in proposition 5.1 are met, then α defines a monad. By 4.1 the corresponding bundle V is a mathematical instanton bundle if

$$\text{rank } (\alpha_1,\alpha_2,\alpha_3,\alpha_4) = 2k + 2 \quad .$$

This is obviously the case if i) holds.

For t'Hooft's example the variety of jumping lines can be easily written down. (Doing this we can avoid to apply the part of the proof of 5.1 which was not given there.) In fact we have

$$Q_{12} = -B_2A_1 + A_2B_1 - b_2^t a_1 = \text{diag}(b_\mu c_\mu - a_\mu d_\mu) - (\lambda_\mu \lambda_\nu) \ ,$$
$$Q_{13} = B_1, \ Q_{14} = -A_1, \ Q_{23} = B_2, \ Q_{24} = -A_2, \ Q_{34} = -1,$$
$$\sum_{i<j} Q_{ij}z_{ij} = \text{diag } \{-(a_\mu d_\mu - b_\mu c_\mu)z_{12} + b_\mu z_{13} - a_\mu z_{14} + d_\mu z_{23} - c_\mu z_{24} - z_{34}\} - (\lambda_\mu \lambda_\nu) z_{12},$$

$$(-1)^k \det(\sum_{i<j} Q_{ij}z_{ij}) = \prod_1^k \ell_\mu(z_{ij}) + \sum_1^k \lambda_\mu^2 z_{12} \prod_{\nu \neq \mu} \ell_\mu(z_{ij}) \ ,$$

with $\ell_\mu(z_{ij}) = (a_\mu d_\mu - b_\mu c_\mu)z_{12} + a_\mu z_{14} - b_\mu z_{13} + c_\mu z_{24} - d_\mu z_{23} + z_{34}$.

The hyperplane $\ell_\mu(z_{ij}) = o$ parametrizes in Plücker's $\mathbb{P}_5$ the lines intersecting L_μ. If we include $\ell_{k+1}(z_{ij}) = z_{12}$ and $\lambda_{k+1}^2 = 1$, then the equation for jumping lines has the fairly simple form

$$\left(\sum_1^{k+1} \frac{\lambda_\mu^2}{\ell_\mu} \right) \ell_1 \cdot \ldots \cdot \ell_{k+1} = o.$$

In particular, none of the lines L_μ is a jumping line, but all the lines meeting two of the L_μ are.

6. Commutator equations.

The final section gives a short introduction into classification questions for mathematical instanton bundles. Only few answers are known. Classification of bundles is essentially classification of monads (essentially means: up to the natural $GL_k \times Sp_{2k+2}$ action). So let us restrict our attention to monads, even to monads coming from morphisms $\sum \alpha_i z_i$ with α_3 and α_4 fixed as in the preceding section, see (11). This last condition is not really very restrictive, it just means the line $z_1 = z_2 = o$ is not a jumping line. Then we are dealing only with the two matrices α_1, α_2 as in (12), containing four symmetric blocks A_1, B_1, A_2, B_2 and four row vectors a_1, b_1, a_2, b_2. The essential requirement is (7), which in section 5 was explicitely written down (14), (15) in terms of A_i, B_i, a_i, b_i. These somewhat confusing commutator equations can be cast in a more compact form, if we introduce new homogeneous coordinates $(u_1 : u_2)$. If you want, they can be interpreted as the coordinates of planes through the line $z_1 = z_2 = o$, so they may be thought of as other names for z_1, z_2. Then (14) and (15) are equivalent with

$$(16) \qquad [u_1A_1 + u_2A_2, u_1B_1 + u_2B_2] + (u_1a_1 + u_2a_2) \wedge (u_1b_1 + u_2b_2) = o$$

holding identically in u_1, u_2 (check it !). Writing down all solutions $A_1, \ldots, b_2$ of this equation is the basic problem.

Formally, (16) is a one-variable analog of the equation

$$(17) \qquad [A,B] + a \wedge b = o,$$

which plays the central role in classifying "stable" rank-2 bundles on the projective plane $\mathbb{P}_2$. Compared with (16), equation (17) is easy to solve: Project the space of variables A,B,a,b onto the smaller space of triples A,a,b, and ask for solutions B of (17) when A,a,b are given. Denote by S_k the space of complex symmetric $k \times k$-matrices A. It contains an open dense subset of matrices A having a minimal polynomial of degree k. (The typical case is a diagonal matrix, where all entries are pairwise distinct.) Le me call such a

atrix regular. It is easy to see that A is regular if and only if its cen-
ralizer

$$Z_A = \{B \in S_k : [A,B] = o\}$$

as dimension k, or what is the same, the map

$$S_k \to \Lambda_k, \quad B \to [A,B]$$

s surjective, where Λ_k is the space of alternating $k \times k$-matrices. So for
egular A and arbitrary a,b there is a k-dimensional affine space of so-
utions B for (17). Thus we found a set of solutions of dimension

$$\begin{array}{ll} \frac{1}{2}k(k+1) \quad = \dim S_k & \text{(for the choice of } A\text{)} \\ + \; 2k & (\text{" " " " } a,b) \\ + \; k & (\text{" " " " } B) \\ \hline = \frac{1}{2}k^2 + \frac{7}{2}k & \end{array}$$

Subtracting $\dim O_k + \dim SL_2$, one obtains $4k-3$, the right dimension of stab-
e bundles on $\mathbb{P}_2$ with $c_1 = o$, $c_2 = k$.) But there might be other solutions
f course, coming from A which are not regular. It is not hard to see that
uch solutions form a variety of dimension smaller than the dimension just
omputed, so they do not play an important role.

ow let us return to (16). We shall abbreviate

$$u_1A_1 + u_2A_2 = \mathcal{A}(u) \qquad u_1B_1 + u_2B_2 = \mathcal{B}(u)$$

$$u_1a_1 + u_2a_2 = a(u) \qquad u_1b_1 + u_2b_2 = b(u) \; ,$$

nd call the objects $\mathcal{A}, \mathcal{B}$ pencils (of quadrics). In fact A_i determines
quadric $xA_ix^t = o$ in $\mathbb{C}^k$ (or $\mathbb{P}_{k-1}$), and a one-dimensional linear system
f quadrics is called a pencil. Copying the procedure described above, we con-
ider the map

$$\mathcal{B} \to [\mathcal{A},\mathcal{B}] \; .$$

t can be considered as morphism of vector bundles: Let $S_k(1) = \frac{1}{2}k\,(k+1)\mathcal{O}_{\mathbb{P}_1}(1)$
e the bundle $S_k \otimes \mathcal{O}_{\mathbb{P}}(1)$ of symmetric $k \times k$-matrices depending linearly on

u_1, u_2. Then $\mathcal{A}, \mathcal{B} \in H^o(S_k(1))$. Similarly let $\Lambda_k(2) = \frac{1}{2}k(k-1)\mathcal{O}_{\mathbb{P}_1}(2)$ be the bundle of alternating $k \times k$-matrices depending quadratically on u_1, u_2. Then

$$[\mathcal{A},-] : B \to [\mathcal{A}(u), B], \quad S_n(1) \to \Lambda_n(2)$$

is indeed a vector bundle morphism. Since all matrices $A \in S_k$ are regular, except for a (closed algebraic) subvariety of codimension 2, the general pencil $\mathcal{A}$ will contain only regular matrices $\mathcal{A}(u)$. We know that $[\mathcal{A}(u), -]$ is surjective then with its kernel $Z_{\mathcal{A}(u)}$ spanned by $1, \mathcal{A}(u), \mathcal{A}^2(u), \ldots \mathcal{A}^{k-1}(u)$. In other words, we have an exact sequence

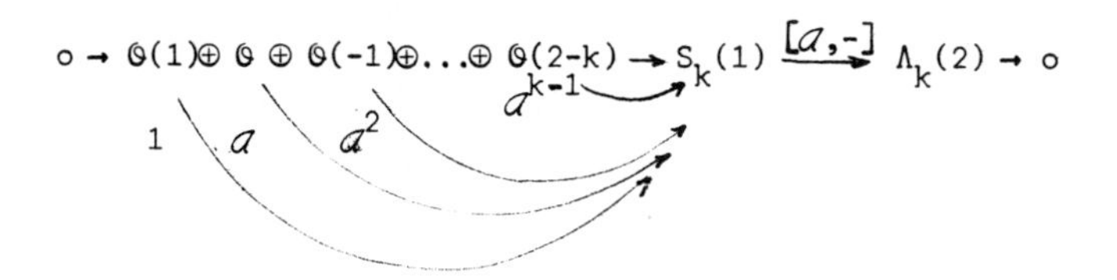

and an exact cohomology sequence

$$H^o(S_k(1)) \to H^o(\Lambda_k(2)) \to H^1(\mathcal{O}_{\mathbb{P}_1}(1) \oplus \mathcal{O}_{\mathbb{P}_1} \oplus \ldots \oplus \mathcal{O}_{\mathbb{P}_1}(2-k)) \quad .$$

By 2.4 the H^1-space has dimension $1 + \ldots + k-3 = \frac{1}{2}(k-3)(k-2)$. For $k \geq 4$ this does not vanish. Unfortunately, because classes in this space are obstructions against solving the equation $[\mathcal{A}, \mathcal{B}] = a \wedge b$ when $\mathcal{A}, a, b$ are given. Only for $k = 2,3$ the obstruction vanishes, and (16) can be solved easily in the same way as (17). The procedure even works (with much more complications) for $k = 4$, [2], because the obstruction there is small (1-dimensional). But with k increasing, the obstruction grows quadratically, whereas $4k$ the number of parameters in a and b grows only linearly. For $k \geq 13$ it seems that this method will definitely fail, because from $k = 13$ on

$$\frac{1}{2}(k-3)(k-2) > 4k.$$

So for $k = 1,2,3$ one knows that the space of mathematical instanton bundles is smooth, connected and rational. (This means essentially its elements can be parametrized by polynomial functions with parameters forming a vector space.)

In fact indirect methods have led to these results earlier, cf. [5]. For $k = 4$ one knows its irreducibility. And for $k \geq 5$ one only knows it is not empty of dimension $\geq 8k - 3$ and containing at least one component of dimension equal to $8k - 3$.

Recently I was informed that J. Le Potier has shown the smoothness for $k = 4$.

References

1. Barth,W.: Some properties of stable rank-2 vector bundles on $\mathbb{P}_n$, Math.Ann. 226, 125-150 (1977).

2. Barth,W.: Mathematical instanton bundles of rank 2 with $c_2 = 4$. To appear.

3. Barth,W. and K.Hulek: Monads and moduli of vector bundles. manuscr.math. 25, 323-347 (1978).

4. Birkhoff,G.D.: A theorem on matrices of analytic functions. Math.Ann. 74, 122-133 (1913).

5. Ellingsrud,G. and S.A.Strømme: Stable rank-2 vector bundles on $\mathbb{P}^3$ with $c_1 = o$ and $c_2 = 3$, Math.Ann. 255, 123-135 (1981).

6. Grauert,H. and R.Remmert: Theory of Stein Spaces. Grundlehren Band 236, Springer.

7. Grothendieck,A.: Sur la classification des fibrés holomorphes sur la sphère de Riemann, Amer.J.Math. 79, 121-138 (1956).

8. Horrocks,G.: Vector bundles on the punctured spectrum of a local ring. Proc. London Math.Soc. (3), 14, 689-713 (1964).

9. Serre,J.P.: Faisceaux algébriques coherents, Ann.Math. 61, 197-278 (19

10. Serre,J.P.: Un théorème de dualité, Comm. Math.Helv. 29, 9-26 (1955).

11. Serre,J.P.: Géomètrie algebrique et géométrie analytique, Ann.Inst. Fourier 6, 1-42 (1956).

Wolf Barth
Mathematisches Institut
Universität Erlangen-Nürnberg
Bismarckstr. 1 1/2
D-8520 ERLANGEN
West-Germany

Solutions of Classical Equations

M.F. Atiyah

§1. Generalities

The purpose of these lectures is to explain some of the new methods that have successfully been applied to obtain explicit solutions of some of the non-linear partial differential equations arising in gauge theories.

To set these methods in a proper context let me begin by reviewing briefly the various standard methods that are used to solve differential equations. For constant coefficient linear equations there is of course the powerful technique of the Fourier Transform, which reduces the problem to one of algebra. In a different direction, equations which possess symmetries lend themselves to group-theoretical methods. For example spherically symmetric solutions of appropriate equations reduce to ordinary differential equations involving only functions of the radius. Within the past decade or so there has been very extensive work on a large class of non-linear equations in two dimensions (one space and one time) involving inverse scattering theory. This can be viewed as some kind of non-linear Fourier transform and it is extremely effective for the class of problems of so-called "Lax type".

For fully four-dimensional problems, which are the ones of main physical interest, the "Penrose Transform" is a new technique which has been remarkably effective for certain problems. These problems are of very restricted type mathematically but they are also the ones of main interest in physics. The Penrose transform is much less universal in its applicability than the other transforms we have mentioned but it is able to deal with genuinely four-dimensional situations of real interest.

The Penrose transform converts certain types of differential equations into problems of complex analysis or algebraic geometry and one must then apply the appropriate techniques in these fields to exploit the method.

Since complex analytic geometry differs in some important

respects from differential geometry, which is more familiar to physicists working on gauge theories, I shall next make some general remarks comparing the two types of geometry

In differential geometry the basic ideas of connection and curvature (or potential and field in physical terminology) are local. This does not mean that global questions in differential geometry do not exist or are not interesting: on the contrary they can be very interesting, particularly in relation to topology. But the basic starting point lies in the local information provided by the curvature. This is true both for Riemannian manifolds (for Gravitation) and for fibre bundles (for Gauge-theories).

In complex analytic geometry on the other hand there is no local information. Any complex manifold looks locally like C^n, with no special features, and any holomorphic fibre bundle is locally an analytic product. All the information now lies in the global aspect. This is familiar in elementary complex analysis where the singularities of analytic functions play such a dominant role, especially in connection with the theory of residues. In higher dimensions the systematic generalization of residue theory requires the machinery of sheaf cohomology groups. Although these are formally modelled on the familiar homology and cohomology groups of ordinary topology, they are much more refined because they encode the global complex analytic information, not just topological information.

The Penrose transform converts problems from differential geometry into problems of complex analytic geometry. Like the Fourier transform it is non-local, so that local curvature information is coded into global holomorphic information. The fundamental reason for this is that to each point in Minkowski space we have the 2-sphere of light-cone directions which is a global complex-analytic object. Perhaps I could add one further comment respecting the non-locality of the Penrose transform. In four-dimensions there are two types of spinors, left or right-handed. The Penrose theory treats these asymmetrically, dealing locally with one and non-locally with the other (of course one can switch between left and right to derive the "opposite" Penrose transform).

The simplest example of the way the Penrose transform works is provided by the Laplace equation $\Delta f = 0$ in R^4. Solutions of this equation, defined in some open set U, correspond by the Penrose transform to elements of a sheaf-cohomology group $H^1(V, \mathcal{O}(-2))$. Here

V is an open set determined by U in complex projective 3-space, and $\mathcal{O}(-2)$ is the sheaf of local holomorphic functions on C^4 which are homogeneous of degree -2. If U is a small neighbourhood of a point p in R^4, then V is a small neighbourhood of a line L_p in $P_3(C)$. Restricting to L_p an element of $H^1(L_p, \mathcal{O}(-2))$ can be represented by a holomorphic function ϕ (or differential) in an annulus on the complex line L_p and the value $f(p)$ is then given by a contour integral of ϕ around the unit circle.

A more sophisticated application of the Penrose transform is to the problem of instantons. These are self-dual solutions of the Yang-Mills equations on R^4 with finite action. They correspond to holomorphic bundles on the whole of the complex projective 3-space $P_3(C)$ and so are algebraic. Moreover they can all be constructed by fairly simple algebraic methods. For further details I refer to [1] and also to the lectures of Barth in this summer school.

The self-duality equations also turn up in a different physical situation, in connection with static magnetic monopoles. These then correspond to holomorphic bundles on a 2-dimensional complex manifold. This case is in some ways more elementary than the instanton problem because the time independence reduces the number of independent variables. On the other hand there are other features which make it more difficult and progress has been made only very recently. For these two reasons I will denote the rest of my lectures to the detailed investigation of magnetic monopoles. On the one hand the Penrose transform in this case is much easier to explain, while on the other hand I can report on new results. In particular I shall describe some very recent work of N. Hitchin [3].

§2. The Bogomolny Equations

Consider on R^3 an SU(2)-potential A, together with a Higgs field Φ in the adjoint representation. As usual we introduce the covariant derivative

$$\nabla_\mu = \partial_\mu + A_\mu$$

and the curvature

$$F_{\mu\nu} = [\nabla_\mu, \nabla_\nu] = [A_\mu, A_\nu] + \partial_\mu A_\nu - \partial_\nu A_\mu .$$

The Bogomolny equations are then:

$$\nabla\Phi = *F$$

where $*F$ is the dual of F, so that both sides of this equation are 1-forms on R^3 with values in the adjoint representation of SU(2).

We shall be interested in solutions of this equation which satisfy the condition:

$$|\Phi| \to 1 \quad \text{at} \quad \infty$$

More precisely we shall assume some uniform estimate of the form

$$(1.1) \qquad |\Phi| \sim 1 - \frac{m}{r} + O(r^{-2}) \quad \text{as} \quad r \to \infty$$

where m is a constant. Up to a universal constant m is also the magnetic or topological charge k defined as the degree of the map $S^2 \to S^2$ given by Φ on a large sphere.

The Bogomolny equations (which are first-order) imply the second-order Yang-Mills-Higgs equations, and their solutions are interpreted as non-abelian magnetic monopoles. For comments on their physical significance I refer to the lectures of D.Olive.

The mathematical problem here is how to find all solutions of the Bogomolny equations for all k. If $k = 0$ the only solution is the trivial solution $F = 0$, Φ constant. For $k = 1$, there is spherically symmetric solution, usually referred to as the Bogomolny-Prasad-Sommerfield monopole. For a long time the existence of solutions for $k > 1$ was an open problem. Recently however there has been rapid progress. An explicit solution for $k = 2$ was found by R. Ward [4] using the Penrose transform, and this has been followed up by others. On the other hand C. Taubes [6] has proved the existence of solutions for all k by methods of functional analysis. Moreover E. Weinberg [5] has computed the number of parameters on which the solution depends and found the number $4k - 1$. Thus solutions exist, there are many of them and it would be desirable to obtain more explicit information about them, comparable perhaps to what has been done for instantons.

§3. The Penrose-Weierstrass transform

For this 3-dimensional problem the Penrose transform actually goes back to 1866 when Weierstrass used it to study the minimal surface equation in R^3. The basic geometry is very simply described.

We consider an oriented line L in R^3. Such a line can be parametrized by a pair of vectors (u,v) in R^3, where u is the unit vector in the direction ℓ and v is the perpendicular distance of ℓ from the origin.

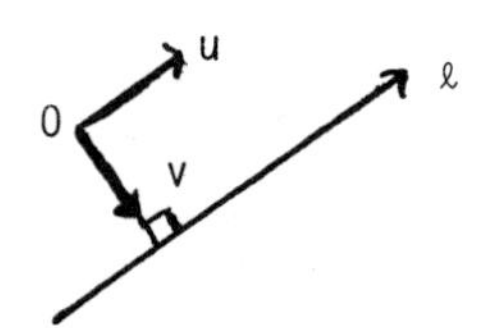

Thus $|u| = 1$ and $u \cdot v = 0$. Such a pair describes a point of the tangent bundle T to the 2-sphere. Identify the 2-sphere with the complex sphere (or projective line): we see that T becomes naturally a 2-dimensional complex manifold. We can cover T explicitly by two complex coordinate systems (ξ,η) and (ξ',η') where in the overlap $\xi' = \frac{1}{\xi}$ is the coordinate on the 2-sphere and $\eta' = -\xi^{-2}\eta$ is the fibre coordinate. This formula arises by considering the tangent vector field

$$\eta \frac{\partial}{\partial \xi} = \eta' \frac{\partial}{\partial \xi'}$$

Thus T parametrizes the oriented lines of R^3. Conversely a point p of R^3 gives rise to a holomorphic section of T (i.e. a tangent vector field on the 2-sphere) corresponding to all lines through p. There are in fact a 3-complex parameter family of such sections. The real ones must satisfy the further condition of being invariant under the anti-holomorphic involution $\sigma : T \to T$ induced by the anti-podal map of the 2-sphere ($\sigma(\ell)$ is the same line with the opposite orientation). This exemplifies the fact that we must study T not only holomorphically but also with respect to the real structure given by σ.

The idea now is to use the correspondence between R^3 and T to convert problems from R^3 to T. Naturally the hope is that certain problems then become simpler.

Suppose then that (A,Φ) are a solution of the Bogomolny equations. We shall associate to this some geometric data on T as follows. For each oriented line L in R^3 consider the ordinary differential equation:

$$(3.1) \quad (\nabla_\ell - i\Phi)s = 0$$

where ∇_ℓ denotes the covariant derivative (given by A) along ℓ and s is a function $\ell \to C^2$. This differential equation has a 2-dimensional space of solutions which we denote by E_ℓ. By varying ℓ we then get a 2-dimensional complex vector bundle E over T. Using A we can endow E with a natural operator $d''E$ of type (0,1) satisfying the derivation rule

$$(3.2) \quad d''_E(fs) = fd''_E s + sd''f$$

where s is a section of E, f is a function on T and $d''f$ is the $(0,1)$ component of df (so that $d''f = 0$ gives the Cauchy-Riemann equations for a holomorphic function). An elementary computation then shows that the Bogomolny equations are essentially equivalent to the commutation formula

$$[d''_E, d''_E] = 0.$$

To see that this is plausible note that if (x,y,z) are R^3-coordinates the Bogomolny equations imply the commutation formula

$$[\nabla_t - i\Phi,\ \nabla_x + i\nabla_y] = 0,$$

and are equivalent to this if we permute the coordinates. Thus E acquires a <u>holomorphic structure</u>, i.e. a local section s of E over T is called holomorphic if it satisfies the equation

$$d''_E s = 0.$$

The commutation relation above is just the integrability condition which ensures that this equation has plenty of local solutions. Note that (3.2) implies that fs is again holomorphic if f and s are holomorphic. Thus we can always choose a local holomorphic gauge or basis for E, and any two differ by a holomorphic (non-singular) matrix-valued function.

The holomorphic bundle has some further properties. These are as follows:

(i) symplectic structure (i.e. its group is $SL(2,C)$)

(ii) real structure (i.e. σ on T lifts to an anti-holomorphic involution on E).

(iii) E is trivial when restricted to any real holomorphic section of T.

Conversely Hitchin proves that a holomorphic bundle E with these properties uniquely determines a solution (A,Φ) of the Bogomolny equations (up to equivalence). This correspondence does not so far involve the asymptotic conditions on Φ which we must next proceed to study.

§4. Asymptotic Conditions

Before proceeding further we should comment that Hitchin's re-interpretation of the Bogomolny equations applies with only minor changes to any Lie group. In particular consider the abelian case of $U(1)$. There is the trivial solution $F = 0, \Phi = i$ of the Bogomolny equations, and this therefore corresponds to a holomorphic line-bundle L on T. This bundle can be given, relative to our coordinate covering of T, by the single transition function $\exp(\eta/\xi)$. This "exponential" line-bundle will play an important role for the $SU(2)$-solutions. In a sense it will represent the only non-algebraic part of the solution. Modulo this exponential the entire theory will be described in terms of the classical algebraic geometry of curves.

Returning now to the $SU(2)$-case let us consider the solutions of (3.1) which tend to zero at $+\infty$ on the line ℓ. If t is the parameter on ℓ the asymptotic condition (1.1) on Φ implies that these solutions decay like $t^k \exp(-t)$ where k is the magnetic charge. These decaying solutions give a one-dimensional subspace of E and one can show that these form a <u>holomorphic</u> sub-bundle of E over T. Moreover this bundle is isomorphic to $L^*(-k)$, where L is the "exponential" line-bundle introduced above, L^* its dual and

$$L^*(-k) = L^* \otimes H^{-k}$$

where H is the standard Hopf line-bundle on P_1 pulled up to T. Thus $L^*(-k)$ can be given by the transition function

$$\xi^{-k}\exp(-\eta/\xi).$$

Since E is an $SL(2,C)$-bundle it follows that it is described by an exact sequence of the form:

$$(4.1) \quad 0 \to L^*(-k) \to E \to L(k) \to 0.$$

Such extensions are classified by elements β of the sheaf cohomology group

$$H^1(T, L^{-2}(-2k))$$

In more concrete terms this means that the transition matrix of E can be put into the triangular form:

$$\begin{pmatrix} a & b \\ 0 & a^{-1} \end{pmatrix}$$

where $a = \xi^{-k}\exp(^{-\eta}/_{\xi})$ and b is the cocycle representing the element β of $H^1(T,L^{-2}(-2k))$.

We note that a triangular matrix cannot also be unitary (unless $b = 0$) so that this description of E does not so far incorporate the reality (or unitary) constraint. To get this we note that we could equally well have considered the solutions of (3.1) which tended to zero at $-\infty$ on ℓ. Since reversing orientation of ℓ gives the real structure one can check that we now get another description of E in terms of an exact sequence:

$$(4.2) \quad 0 \to L(-k) \to E \to L^*(k)$$

which differs from (4.1) in that L is replaced by L^*. The reality constraint is essentially expressed by the fact that E has simultaneously the forms (4.1) and (4.2): they are complex-conjugates of one another.

Combining (4.1) and (4.2) we can consider the composite homomorphism

$$\begin{array}{lll} (4.1) & L(-k) \to & E \\ & & \| \\ (4.2) & & E \to L(k) \end{array}$$

which gives a section of H^{2k} over T, i.e. a polynomial of the form

$$(4.3) \quad s = \eta^k + a_1(\xi)\eta^{k-1} + \ldots + a_k(\xi)$$

where $a_i(\xi)$ is a polynomial of deg $2i$ in ξ. The equation $s = o$ defines an algebraic curve S on T which Hitchin calls the <u>spectral curve</u>. Points of S correspond to points ℓ of T at which the homomorphism $L(-k) \to L(k)$ is zero or equivalently where the fibres of the sub-bundles of (4.1) and (4.2) coincide.

In other words the equation (3.1) on the line ℓ has a solution which tends to zero at $\pm\infty$: since the decay is exponential the solution is then square-integrable. This explains the terminology of spectral curve: regarding (3.1) as a differential equation depending non-linearly on the parameter ℓ, the spectrum corresponds to values of the parameter for which L^2-solutions exist.

The reality conditions imply that the polynomial (4.3) has to be invariant under

$$\sigma : (\xi, \eta) \to (-\bar{\xi}^{-1}, \bar{\eta}/\bar{\xi}^{2}).$$

Moreover on S we must have $L \cong L^*$, by comparing (4.1) and (4.2). Hence on S the line-bundle L satisfies $L^2 = 1$. This imposes further conditions on the curve S. The number of conditions is given by the genus, which can be computed to be $(k-1)^2$. Hence the number of free parameters left to describe S is

$$(k+1)^2 - 1 - (k-1)^2 = 4k - 1$$

as predicted in [5].

Finally Hitchin shows that the solution of the Bogomolny equations is (up to equivalence) uniquely determined by the curve S. This is proved as follows. The equation for S leads to an exact sequence

$$0 \to \mathcal{O}_T(-2k) \to \mathcal{O}_T \to \mathcal{O}_S \to 0.$$

Tensoring by L^{-2} we get

$$0 \to L^{-2}(-2k) \to L^{-2} \to L_S^{-2} \to 0.$$

The cohomology sequence of this gives

$$\to H^0(S, L_S^{-2}) \to H^1(T, L^{-2}(-2k)) \to$$

Now on S we know that L^{-2} is trivial so that $H^0(S, L_S^{-2}) \cong C$ has a basic element 1. Taking $\delta(1)$ we get an element

$$\beta \in H^1(T, L^{-2}(-2k))$$

which determines the extension (4.1) and so the bundle E.

For example when $k = 1$, the spectral curve S is just the section corresponding to a point in R^3. This point is the "source" of the monopole.

Let me sum up the situation, and explain how Hitchin's work sheds light on the prior work of Ward [4] and Corrigan-Goddard [2]. In these papers an Ansatz is used which is essentially equivalent to assuming the exact sequences (4.1) and (4.2). Moreover the polynomial (4.3) turns up in the calculations. Hitchin's approach proves that this Ansatz gives all solutions and explains the geometric significance of (4.3). There are still problems involved which have not been completely resolved, notably concerning the condition that L^2 is trivial on S. There are countably many solutions, involving discrete parameters, and it is not clear which of these choices lead to non-singular solutions of the Bogomolny equations.

To derive really explicit solutions one has in general to deal with the algebraic function theory of the curve S. For special cases, when S decomposes into linear factors, this is elementary. For general irreducible S of high genus the theory is likely to be quite intricate.

There is a remarkable analogy between the role of this spectral curve for the Bogomolny equations and the role of algebraic curves in the theory of finite-zone potentials for KdV-type equations. This analogy may repay further investigation.

Finally let me recall that Weierstrass used the transform between R^3 and T in the study of minimal surfaces. In fact the curve S defines a minimal surface in R^3, so that any magnetic monopole defines such a minimal surface which characterizes it. It is not clear whether this has any real physical significance but it is intriguing.

References

1. M.F. Atiyah, Geometry of Yang-Mills Fields, Fermi Lectures, Scuola Normale Pisa 1979.
2. E. Corrigan and P. Goddard, Commun. Math. Phys. 86 (1981), 575-587.
3. N.J. Hitchin, Monopoles and Geodesics, Commun. Math. Phys. (to appear).
4. R.S. Ward, Commun. Math. Phys. 79 (1981), 317-325.
5. E. Weinberg, Phys. Rev. D20(1979), 936.
6. C. Taubes, The Existence of Multimonopole Solutions to the Non-Abelian Yang-Mills Equations for Arbitrary Simple Groups, Commun. Math. Phys. 80, 343-376 (1981).

THE SELF DUALITY PROBLEM FOR GAUGE THEORIES

Arthur Jaffe

1. PURE YANG-MILLS THEORY

The pure Yang-Mills action functional on $\mathbb{R}^d$ is

$$\mathcal{A}_{YM}(A) = \|F_A\|^2 = \operatorname{tr}\int_{\mathbb{R}^d} F_A \wedge *F_A \quad . \tag{1}$$

Here A is a Lie algebra valued connection, on a principal G-bundle over $\mathbb{R}^d$. The gauge field $F_A = dA + A\wedge A$ is the curvature two form, and $*$ denotes the Hodge dual. Every such field satisfies the Bianchi identity

$$d_A F_A = 0 \ , \tag{2}$$

where d_A denotes covariant differentiation. The Yang-Mills equations

$$d_A * F_A = 0 \tag{3}$$

are the variational equations for the functional (1). Thus critical points of $\mathcal{A}_{YM}$ are harmonic in the sense that $\Delta_A F_A = 0$, where $\Delta_A = d_A d_A^* + d_A^* d_A$ is the covariant Laplacian. The system of equations (3) for A is second order.

On a compact 4-manifold M, the topological invariant

$$\operatorname{tr}\int_M F_A \wedge F_A = 8\pi^2 k \ , \quad k \ \text{integer}, \tag{4}$$

divides gauge fields into classes. Critical points of (1) on $\mathbb{R}^4$ also have integer k, as was proved by Uhlenbeck [22]. It is conjectured that k is integer whenever $\mathcal{A}_{YM} < \infty$. In any case, for $d = 4$,

$$\mathcal{A}_{YM}(A) = \frac{1}{2}\|F_A \mp *F_A\|^2 \mp 8\pi^2 k \quad , \tag{5}$$

and minima of $\mathcal{A}_{YM}(A)$ satisfy the first order system of equations

$$F_A = *F_A, \quad \text{if} \quad k>0; \qquad \text{or} \qquad F_A = -*F_A, \quad \text{if} \quad k<0 \ . \tag{6}$$

Solutions to (6) are called *self dual* or *instantons*. Clearly every minimum of $\mathcal{A}$ is a critical point, and the Yang-Mills equations (3) hold as a consequence of the Bianchi identities (2). All minima on $\mathbb{R}^4$ have been characterized by Atiyah, Hitchin, Drinfeld and Manin [2].

2. MONOPOLE EQUATIONS

The $SU(2)$ monopole equations arise from coupling a pure $SU(2)$ Yang-Mills theory on $\mathbb{R}^3$ to a scalar (Higgs) field Φ. The Yang-Mills-Higgs functional is

$$\mathcal{A}_{YMH}(A,\Phi) = \mathcal{A}_{YM}(A) + \|d_A\Phi\|^2 + V(\Phi) \quad . \tag{7}$$

For $d=3$, the Yang-Mills functional reduces to the square of the magnetic field, $\mathcal{A}_{YM}(A) = \|F\|^2$, where $F = dA + A \wedge A$. It is also standard to take the Higgs field Φ to be Lie algebra valued, with G acting by the adjoint action on Φ. The covariant derivative d_A is defined on the associated vector vector bundle by $d_A\Phi = d\Phi + [A,\Phi]$. The potential function $V(\Phi)$ is a function of the gauge invariant $|\Phi| = (\mathrm{tr}\Phi^*\Phi)^{1/2}$. The simplest case is $V=0$, which is sometimes called the Prasad-Sommerfield limit. This case is also equivalent to studying time independent solutions to (3). Furthermore in this case every critical point (A,Φ) of (7) has the property

$$\lim_{|x|\to\infty} |\Phi(x)| = \text{constant} \ , \tag{8}$$

with the constant zero only for pure gauges [11]. Thus it is convenient to normalize (A,Φ) so that $\text{constant} = 1$.

We consider $V=0$. Let $B = *F$. Then

$$\mathcal{A}_{YMH}(A,\Phi) = \|B\|^2 + \|d_A\Phi\|^2 \ , \qquad |\Phi| \to 1 \quad . \tag{9}$$

The critical points of (9) satisfy the second order system of differential equations:

$$d_A B = *[d_A\Phi,\Phi] \ , \qquad \Delta_A\Phi = 0 \quad . \tag{10}$$

There is also a first order system for the minima of $\mathcal{A}_{YMH}$. In fact

$$\mathcal{A}_{YMH}(A,\Phi) = \|B \mp d_A\Phi\|^2 \mp 8\pi k \quad , \tag{11}$$

where

$$k = \frac{1}{4\pi}\int_{S^2_\infty} \mathrm{tr}(\Phi * B) = \frac{1}{4\pi}\int_{\mathbb{R}^3} \mathrm{tr}(d_A\Phi \wedge *B) \tag{12}$$

is the monopole number. For any critical point (A,Φ) of (9), one can show that k is integer [29,28,11,19]; again k is conjectured to be integer whenever $\mathcal{A}_{YMH} < \infty$. In any case the minima of (9) are critical points which satisfy the Bogomol'nyi equations:

$$B = d_A\Phi, \quad \text{if } k > 0; \quad \text{or} \quad B = -d_A\Phi, \quad \text{if } k < 0. \tag{13}$$

Solutions to (10) or (13) can be interpreted as having magnetic charge k. In fact $|\Phi| \to 1$, so (12) is the gauge invariant contribution of the magnetic flux, calculated on the two-sphere at infinity. For this reason, solutions to the equations are thought of as being composed of magnetic monopoles. For other gauge groups G, several charges characterize each solution, see [11,19]. They are theoretically nicer than Dirac monopoles, since they are *finite energy*, time independent solutions to the Yang-Mills-Higgs equations. Experimental physicists are now searching for such particles in nature!

Explicit solutions in closed form to (13) were found by Prasad and Sommerfield [15] and Ward [24,25]. A $4|k| - 1$ parameter family of proposed solutions, has been given by Corrigan and Goddard [6], whose work is based on the Ward correspondence. Important earlier work includes Atiyah and Ward [3], Ward [26], Prasad [13], and Prasad and Rossi [14]. What remains to be proved is the regularity of the configurations (A,Φ) constructed by Corrigan and Goddard. In fact it is known that every critical point of $\mathcal{A}_{YMH}$ must be real analytic [11]. Once regularity is established for the Corrigan-Goddard proposed fields, the results of Hitchin [9,1] show that all solutions to (13) would be obtained by this construction.

Explicit formulas aside, an existence theorem has been established by Taubes [11,18] for another class of solutions: those with arbitrary k, but which can be thought of as composed of $|k|$ individual constituent monópoles located sufficiently far from each other.

I should mention that whether one can characterize a general monopole solution in terms of a local monopole particle structure is an open problem. One conjectures that for monopole number k, $\Phi(x)$ has $|k|$ zeros which indicate individual monopole centers. Furthermore, one expects that the winding number of Φ on a sphere enclosing n of these zeros is n. The remaining $|k|-1$ parameters (of the $4|k|-1$ total) should be associated with a $U(1)$ group for each monopole center--modulo a global $U(1)$ gauge transformation. For groups G other than $SU(2)$, such a classification presumably involves both G and the asymptotics of $\Phi(x)$ as $|x| \to \infty$.

3. VORTICES

On $\mathbb{R}^2$, there is a third related gauge theory--another Yang-Mills-Higgs theory. In this case we study the $U(1)$ gauge group and choose Φ to be complex-valued. The group $U(1)$ is associated to act on Φ by left multiplication.

This model is also called the Landau-Ginzburg theory. It is used as a semiclassical description of superconductors (with z-independent solutions). The relevant functional is

$$\mathcal{A}_{GL}(A,\Phi) = \mathcal{A}_{YMH}(A,\Phi) = \|B\|^2 + \|d_A\Phi\|^2 + \frac{\lambda}{4}\left\| |\Phi|^2 - 1 \right\|^2 . \tag{14}$$

Here $d_A\Phi = d\Phi - iA\Phi$ gives the "minimal" coupling. The model with $\lambda \geqslant 1$ describes type II superconductors. For $\lambda \leqslant 1$ the model applies to type I superconductors (which exhibit the Meissner effect). The intermediate case, $\lambda = 1$, has properties similar to the two examples above, and we now specialize to that model. The variational equations of (14) for a critical point are the second order system

$$*d_A B = \operatorname{Im} \Phi \overline{d_A\Phi} , \qquad \nabla_A^2 \Phi = \frac{1}{2}(1-|\Phi|^2)\Phi . \tag{15}$$

Again, for this system there is a topological integer; in this case the magnetic flux. (In two dimensions the magnetic field has only one component, which "points" in the z direction of homogeneity.) The field is $B = *F$, and the flux

$$k = \int \frac{1}{2\pi} F \tag{16}$$

is the vortex number. Any critical point of (15) has integer k (due to the decay of F at infinity), see [11]. Thus $\mathfrak{A}_{GL}$ can be written

$$\mathfrak{A}_{GL}(A,\Phi) = \left\| B \mp \frac{1}{2}(1-|\Phi|^2)\right\|^2 + \frac{1}{2}\|d_A\Phi \mp i*d_A\Phi\|^2 \pm 2\pi k \quad . \tag{17}$$

Accordingly the minima (A,Φ) of $\mathfrak{A}_{GL}$ satisfy the first order system of equations

$$B = \frac{1}{2}(1-|\Phi|^2), \qquad d_A\Phi = i*d_A\Phi, \qquad \text{if} \quad k>0 \; ; \tag{18a}$$

$$B = -\frac{1}{2}(1-|\Phi|^2), \qquad d_A\Phi = -i*d_A\Phi, \qquad \text{if} \quad k<0 \; . \tag{18b}$$

Solutions to (18) are not known in closed form. They have been completely characterized by the zeros of Φ, see [11]. In fact given m points $\{x_1,\ldots,x_m\} \in (\mathbb{R}^2)^m$, there exists a solution to (18a) and to (18b) with these points as the zero set of $\Phi(x)$ (counting multiplicities). For these solutions $|k| = m$. Furthermore this solution is unique modulo gauge transformations. The vortex nature of the solution is the property that in the neighborhood of a zero $a = (a^1,a^2) \in \mathbb{R}^2$ of multiplicity n,

$$\Phi(x) \sim (x^1-a^1 \pm i(x^2-a^2))^n \; h(x) \; , \tag{19}$$

where $h(x) \in C^\infty$ and $h(a) \neq 0$. Here the choice of $\pm$ sign coincides with the choice of (18)a or b.

4. PARTICLE INTERPRETATION

The first order equations (13) and (18) for monopoles and vortices are called *self dual*, in analogy with the equations (6) for Yang-Mills fields. In each case we have seen that minima of $\mathfrak{A}$ have the property

$$\mathfrak{A} = \text{constant} \times \text{integer} = \text{constant}|k| \tag{20}$$

where k is a topological invariant. For this reason the solutions can be interpreted as composed of $|k|$ noninteracting particles. The total value of the energy functional $\mathfrak{A}$ depends only on the number of particles (and is linear in this number), but it does not depend on the relative positions, orientations, etc. of the particles. Hence the interaction energy is zero.

Furthermore two basic forms of each particle exist (particle/antiparticle) corresponding to k being positive or negative. Note further that we can express $k = \int k(x)d(\text{volume})$, as the integral of a charge density defined by (4), (12) or (16). In each case solutions of (6), (13) or (18), when inserted into the definitions of $k(x)$ above have the general property that

$$\begin{aligned} & k(x) > 0, \quad \text{if} \quad k > 0 , \\ \text{or} \quad & \\ & k(x) < 0, \quad \text{if} \quad k < 0 . \end{aligned} \tag{21}$$

(This is immediate for instantons and monopoles. For vortices, $B = \pm \frac{1}{2}(1-|\Phi|^2)$ has a constant sign as a consequence of $|\Phi| < 1$ established in (23) below.) Thus solutions to the self dual (minimum) equations are made up purely of particles, if $k > 0$, or of antiparticles, if $k < 0$.

Given his picture of the local charge density $k(x)$, one might speculate that other, nonminimal solutions exist which describe mixtures of particles and antiparticles, or which may even have some other physical interpretation. For such solutions, $k(x)$ would be both positive and negative. Before discussing that question further, we remark that only for special values of the field interaction potentials V can minima of $\mathbb{A}$ be interpreted in terms of free particles. In general, one might expect that the constituent particles in a solution either attract or repel, and thus are not in general stable. (Furthermore, particles which repel each other to infinity might be expected to yield trivial solutions to the field equations, unless, for example, the equations are conformally invariant.) The choices of V which we have made here ensure that the particle-particle forces vanish. However the particle-antiparticle forces do not vanish, and leads one to suspect that particle-antiparticle configurations are unstable.

5. THE SELF DUALITY PROBLEM

For the reasons discussed above, it has been conjectured that every critical point of $\mathbb{A}$ is a minimum (self dual). In other words, for $\mathbb{A} < \infty$, one conjectured that

(i) every solution to (15) satisfies (18),

(ii) every solution to (10) satisfies (13),

(iii) every solution to (3) satisfies (6).

Theorem [17,11]. *The conjecture holds in case* (i).

We outline the five steps in the proof. Assume that (A,Φ) is a critical point of $\mathfrak{A}_{GL}$ defined in (17). Let $w = \frac{1}{2}(1-|\Phi|^2)$.

Step 1.
$$\|B\| = \|w\| \quad . \tag{22}$$

For $0<\alpha$, define a one parameter family $(A_\alpha(x),\Phi(x)) = (\alpha^{-1}A(x/\alpha), \Phi(x/\alpha))$. Then $\mathfrak{A}_{GL}(A_\alpha,\Phi_\alpha) = \alpha^{-2}\|B\|^2 + \|d_A\Phi\|^2 + \alpha^2\|w\|^2$. Since $\alpha = 1$ is a critical point, the vanishing first variation ensures (22).

Step 2.
$$\text{Either} \quad |\Phi(x)| < 1 \quad \text{or} \quad |\Phi(x)| \equiv 1 \ . \tag{23}$$

This follows from the maximum principle. In fact, differentiating w and using (15) shows that
$$\Delta w = -|d_A\Phi|^2 + w|\Phi|^2 \ . \tag{24}$$

Thus $\Delta w \leqslant |\Phi|^2 w$. From $\mathfrak{A} < \infty$, we infer $w \in L_2(\mathbb{R}^2)$. Thus the maximum principle applies and either $0 < w(x)$ or $w(x) \equiv 0$, proving (23).

Step 3.
$$\text{Either} \quad |B(x)| < w(x) \quad \text{or} \quad \pm B(x) = w(x) \ . \tag{25}$$

Again we use the maximum principle. Differentiating (15) yields
$$\Delta B = -(d_A\Phi, i{*}d_A\Phi) + B|\Phi|^2 \ . \tag{26}$$

Adding (24), (26) we find
$$\Delta(w \pm B) = (w \pm B)|\Phi|^2 - |d_A\Phi|^2 \mp (d_A\Phi, i{*}d_A\Phi) \ . \tag{27}$$

By the Schwarz inequality, $\Delta(w \pm B) \leqslant |\Phi|^2(w \pm B)$. From $\mathfrak{A} < \infty$, we infer that $w, B \in L_2(\mathbb{R}^2)$. Thus the maximum principle ensures (25).

Remark. The inequality (25) let us estimate the magnitude of the magnetic field in terms of the magnitude of Φ. In other words Φ is an *order parameter*. Since $|\Phi| \leqslant 1$,
$$|B(x)| \leqslant 1 - |\Phi| \ , \tag{28}$$

and B vanishes as $\Phi \to 1$.

Step 4.
$$w(x) = \pm B(x) \ . \tag{29}$$

By (22), $0 = \int(w^2 - B^2) = \int(w+B)(w-B)$. But by (25), the integrand is either strictly positive or identically zero. Hence it must vanish.

Continuity ensures (29) which is one of the equations (18).

Step 5. $$d_A\Phi = \pm i{*}d_A\Phi \quad . \tag{30}$$

Inserting (29) in (27) yields $|d_A\Phi|^2 = (d_A\Phi, \pm i{*}d_A\Phi)$. Since $\pm i{*}$ is a unitary operator, we have (30), to complete the proof.

The self duality conjecture was generally believed to be true for cases (ii) and (iii) as well. However, Taubes has recently proved the existence of a non-self-dual solution in case (ii), the monopole equations [21]. (This is unrelated to the question of gauge groups of higher rank [7].) His proof is closely related to Morse theory. In fact the truth of the conjecture would indicate that the functional $\mathfrak{A}$ is an unsuitable Morse function in the study of the space of gauge fields. The existence of nonminimal solutions to the Yang-Mills equations is an encouraging step toward further developing the Morse theory.

The recent work of Taubes also throws doubt on the self duality conjecture for pure Yang-Mills fields (case iii). Technically the conformally invariant and $d = 4$ cases are quite different, and many interesting open problems remain.

REFERENCES

1. M.F. Atiyah, this volume.
2. M.F. Atiyah, N. Hitchin, V. Drinfeld and Yu. Manin, Construction of instantons, *Phys. Lett.* 65A, 185 (1978).
3. M.F. Atiyah and R. Ward, Instantons and algebraic geometry, *Commun. Math. Phys.* 55, 117 (1977).
4. E.B. Bogomol'nyi, The stability of classical solutions, *Sovt. J. Nucl. Phys.* 24, 449 (1976).
5. J.P. Bourguinon and B. Lawson, Stability and isolation phenomena for Yang-Mills fields, *Commun. Math. Phys.* 79, 189 (1980).
6. J.P. Bourguinon, B. Lawson and J. Simons, Stability and gap phenomena for Yang-Mills fields, Proc. Nat. Acad. Sci. U.S.A. 76, 1550 (1979).
7. J. Burzlaff, A finite energy SU(3) solution which does not satisfy the Bogomol'nyi equations, to appear.
8. E. Corrigan and P. Goddard, An n-monopole solution with 4n-1 degrees of freedom, *Commun. Math. Phys.* 80, 575 (1981).

9. N. Hitchen, Monopoles and geodesics, *Commun. Math. Phys.* 83, 579 (1982).

10. A. Jaffe, New results for classical gauge theories: Qualitative and quantitative, Berlin Conference 1981. To be published in Springer Lecture Notes in Physics.

11. A. Jaffe and C. Taubes, *Vortices and Monopoles*, Birkhäuser, Boston, 1980.

12. T. Parker, Gauge theories on 4-dimensional, Riemannian manifolds, *Commun. Math. Phys.*, to appear.

13. M. Prasad, Yang-Mills-Higgs monopole solutions of arbitrary topological charge, *Commun. Math. Phys.* 80, 137 (1981).

14. M. Prasad and P. Rossi, Construction of exact multimonopole solutions, *Phys. Rev.* D24, 2182 (1981).

15. M. Prasad and C. Sommerfield, Exact classical solution for the 't Hooft monopole and the Julia-Zee dyon, *Phys. Rev. Lett.* 35, 760 (1975).

16. S. Sedlacek, A direct method for minimizing the Yang-Mills functional over 4-manifolds, *Commun. Math. Phys.*, to appear.

17. C. Taubes, On the equivalence of first and second order equations for gauge theories, *Commun. Math. Phys.* 75, 207 (1980).

18. C. Taubes, The existence of multi-monopole solutions to the non-abelian, Yang-Mills-Higgs equations for arbitrary simple gauge groups, *Commun. Math. Phys.* 80, 343 (1981).

19. C. Taubes, Surface integrals and monopole charges in nonabelian gauge theories, *Commun. Math. Phys.* 81, 299 (1981).

20. C. Taubes, Self-dual Yang-Mills fields on non-self-dual 4-manifolds, *J. Diff. Geom.*, to appear.

21. C. Taubes, The existence of a non-minimal solution to the SU(2) Yang-Mills-Higgs equations on $\mathbb{R}^3$, to appear.

22. K. Uhlenbeck, Removable singularities in Yang-Mills fields, *Commun. Math. Phys.* 83, 11 (1982).

23. K. Uhlenbeck, Connections with L_p bounds on curvature, *Commun. Math. Phys.* 83, 81 (1981).

24. R. Ward, A Yang-Mills-Higgs monopole of charge 2, *Commun. Math. Phys.* 79, 317 (1981).

25. R. Ward, Two Yang-Mills-Higgs monopoles close together, *Physics Lett.* 102B, 136 (1981).

26. R. Ward, Ansätze for self-dual Yang-Mills fields, *Commun. Math. Phys.* 80, 563 (1981).

27. E. Weinberg, Parameter counting for multimonopole solutions, *Phys. Rev.* D20, 936 (1979).

28. P. Goddard, J. Nuyts and D. Olive, Gauge theories and magnetic charge, *Nuclear Physics* B125, 1 (1977).

29. A. Schwarz, Magnetic monopoles in gauge theories, *Nuclear Physics* B112, 358 (1976).

Supported in part by the National Science Foundation under Grant PHY 79-16812.

Harvard University
Cambridge, Mass. 02138 USA

GAUGE FIELDS
AND COHOMOLOGY OF SPACES OF NULL GEODESICS

Yu.I.Manin

Steklov Mathematical Institute, Moscow

1. Introduction

In this talk I discuss some recent work on the application of cohomology to the understanding of the classical equations of motion, derived from the most general Lagrangeans including Yang-Mills, Dirac, Higgs fields, possibly in the curved space-time. A general picture is gradually emerging showing striking relationships with the supergravity approach and displaying the similar unification tendencies. The essential tool of the whole program is the generalized Radon-Penrose transform along the null geodesics.

2. Space-time U

Our space-time is a four-dimensional complex manifold U with the complex conformal structure. We fix two holomorphic spin bundles $S_{\pm}$ of rank 2 and the isomorphism of the sheaf of holomorphic 1-forms $\Omega^1 U$ with $S_+ \otimes S_-$. The conformal metric takes values in $\Lambda^2 S_+ \otimes \Lambda^2 S_- \subset S^2(\Omega^1 U)$. A metric in this conformal class is a section $g \in \Gamma(\Lambda^2 S_+ \otimes \Lambda^2 S_-)$.In our setting it is more convenient to use its spinorial decomposition $g = \varepsilon_+ \otimes \varepsilon_-$ where $\varepsilon_{\pm} \in \Gamma(\Lambda^2 S_{\pm})$. To work in the real space-time we can introduce the charge conjugation operator i.e. the real structure C: $U \xrightarrow{\approx} U$, $C_{\pm}: S_{\pm} \xrightarrow{\approx} C^*(\overline{S}_{\pm})$ with the usual properties. To save space I shall not take C into account in what follows.

3. Space of null geodesics L

A point of L is a holomorphic null geodesic in U.Working locally, we can always suppose that U is Stein, every null geodesic in U is connected and Stein and L is a 5-dimensional complex manifold. Set $L(x)$ = {geodesics passing through the point $x \in U$}. This is a quadric whose conormal bundle $N^*(x)$ in L does not depend on U and x up to isomorphism. Denote by I the subsheaf of $\Omega^1 L$ whose sections vanish on all vectors tangent to some $L(x)$. It is locally free of rank one.

4. Yang-Mills in Heaven

A Yang-Mills field on U is a holomorphic vector bundle E endowed with the holomorphic connection

$$\nabla: \quad E \to E \otimes \Omega^1 U.$$

The category of such pairs (E,∇) is equivalent to the category of vector bundles E_L on L trivial on all $L(x)$. The curvature form F_∇ can be calculated point-by-point in the following way: $F_\nabla(x)$ = (the characteristic class of E_L on the second infinitesimal neighbourhood of $L(x)$ in L) $\in$ $\in H^1(L(x), \mathrm{End}\, E_{L|L(x)} \otimes S^2 N^*(x))$. Using this we can in principle transcribe the Einstein vacuum equations in terms of L if we know the sheaves $(S_\pm)_L$. In fact, the (traceless) Ricci tensor on U vanishes iff $S_\pm$ are (anti) selfdual with respect to the spinorial connections $\nabla_\pm$.

5. Spinors on L

To construct the spinor bundles in terms of L alone we choose a) the decomposition $I = I_+ \otimes I_-$, where I_+ (resp. I_-) on $L(x)$ is isomorphic to $O(-1,0)$ (resp. $O(0,-1)$); b) two cohomology classes $\varepsilon_{\pm L_1} \in H^1(L, I_\pm^2)$ nonvanishing on $L(x)$. We interprete then $\varepsilon_{\pm L} \in \mathrm{Ext}\,(I_\pm^{-1}, I_\pm)$ as extensions

$$0 \to I_\pm \to \Sigma_\pm \to I_\pm^{-1} \to 0 \quad \text{and prove that } \Sigma_\pm = (S_\pm)_L .$$

The implied spinorial metrics $\varepsilon_\pm$ correspond to the $\varepsilon_{\pm L}$ under the canonical isomorphisms $H^1(L, I_\pm^2) = \Gamma(\Lambda^2 S_\pm)$, where $S_\pm$ also are reconstructed from $I_\pm$.

6. Dirac operator

Fix a Yang-Mills field (E,∇) on U, represented by the bundle E_L on L. A calculation shows that

$$H^1(E_L \otimes \Sigma_+(-2,0)) = \Gamma(E \otimes S_+), \quad H^1(E_L(-1,0)) = \Gamma(E \otimes S_-)$$

where $E_L(a,b) = E_L \otimes I_+^{-a} \otimes I_-^{-b}$. Note that the spinor fields on U are represented on L not by the sections of spinor bundles $\Sigma_\pm$ but by certain 1-cohomology classes. The map $H^1(E_L \otimes \Sigma_+(-2, 0)) \to H^1(E_L(-1, 0))$ induced by the map. $(\Sigma_+ \to I_+^{-1}) \otimes I_+^2$ is just the (left) Dirac operator in the Yang-Mills and gravitational background. It follows that the space of massless Dirac fields on U is

$$H^1(E_L(-3, 0) \otimes E_L(0, -3)).$$

7. Conformally flat case

If U is conformally flat i.e. essentially a domain in the Grassmannian of 2-planes in the Penrose twistor space T, then L is endowed by additional structures. We will use here only the infinitesimal neighbourhoods $L^{(j)}$ of L in $P(T) \times P(T^*)$. Using the identifications of various cohomology groups of L with fields on U as indicated earlier we see the following pattern.

a. E_L extends uniquely to $E_L^{(2)}$ on $L^{(2)}$ and the obstruction to the third extension in $H^2(\text{End } E_L(-3,-3))$ is identified with the Yang-Mills axial current $\nabla * F_\nabla$.

b. The left Dirac operator in the Yang-Mills background is the obstruction to the extension of 1-cohomology classes δ: $H^1(E_L^{(1)}(-1, 0)) \to H^2(E_L(-3, -2))$.

c. The Laplace operator is the similar obstruction map: $\square \sim \delta$: $H^1(E_L(-2,0)) \to H^2(E_L(-3,-1))$.

d. The solutions of the Yang-Mills ϕ^4-system $\square\phi_\pm = \phi_\pm\phi_\mp\phi_\pm$, $\phi_\pm = \Gamma(\text{End } E \otimes \Lambda^2 S_\pm)$, are in (1-1) correspondence with the second extensions of the bundle $E_L(-1, 0) \otimes E_L(0, -1)$.

8. Problems

The first unsolved question concerns the definition of the neighbourhoods $L^{(j)}$ in the general curved case. Following the remarkable suggestion by Witten [2] one should bring in the picture the fermionic coordinates of the supergravity instead of commuting nilpotents. To do this effectively we lacked spinor bundles on L. The recent construction explained in n°5 hopefully remedies this. But the main appeal of the cohomological approach lies in its possible applications to the quantum fields. It is encouraging that the two-point classical Green functions also have a very natural interpretation in terms of certain local cohomology.

9. Bibliographical indications

The conformally flat case is treated in [1],following the initial breakthrough in the articles by Witten [2] and Isenberg,Yasskin,Green [3]. The cohomological calculations of [4] go over without changes to the curved case. The spinor bundles on L are constructed in [5].The spaces of null geodesics are studied in detail in [6] . Cf. also the articles [7], [8].

References

1. Henkin G.M., Manin Yu.I., Phys.Lett. 95B (1980),405-408
2. Witten E., Phys.Lett. 78B (1978), 394-398
3. Isenberg J. Yassking Ph.B., Green P.S., Phys.Lett., 78B (1978), 464-468
4. Henkin G.M., Manin Yu.I., (to appear in Compositio Math.)
5. Manin Yu.I., Penkov I.B. (to appear in Funkc. Analiz)
6. Le Brun C.J., Jr., Oxford thesis (1980)
7. Manin Yu.I., in: Sovremennye Problemy Matematiki N 17, Moscow, VINITI, 1981
8. Manin Yu.I., in: Proc. IV Bulgarian Summer School of Theor. Phys., Primorsko, 1980

III. Methods from Constructive Field Theory

CONSTRUCTIVE QUANTUM FIELD THEORY: SCALAR FIELDS

Konrad Osterwalder

I. Introduction

The great success of quantum electrodynamics and more recently of nonabelian gauge theories as models for elementary particle physics is in sharp contrast to the poor mathematical understanding one has of these theories. As of now, no nontrivial mathematically well defined model of relativistic quantum field theory is known to exist (in 4-dimensional spacetime) and to exhibit all the features desired by physics. It is therefore an interesting challenge to

1) try to understand the basic mathematical structures of relativistic quantum field theory

and to

2) construct models of quantum field theory and to analyse their (physics) content such as the mass spectrum, bound states, the scattering operator etc.

Program 1) is called axiomatic quantum field theory. Here one starts from a list of properties any quantum field theory should have, such as the existence of a Hilbert space of states, Lorentz covariance, spectral condition, causality (the so-called axioms). From these assumptions many deep, in part experimentally verifiable results could be derived: The PCT theorem, the connection between spin and statistics, dispersion relations, the existence of superselection rules, the Goldstone theorem, a scattering theory, etc.

Program 2) is called constructive quantum field theory. Starting from specific formal Lagrangians one tries to construct (non perturbatively) models and to show that they satisfy the postulates of axiomatic quantum field theory. In 2- and 3-dimensional space time this goal has been fully achieved. The reduction of dimension diminishes the problems of renormalization without eliminating them completely. In 4 dimensions these problems have so far prevented the completion of the program: only partial results are known in this case.

The most successful method of constructive quantum field theory has been the Euclidean approach. Here the basic objects to be studied are the Schwinger functions or Euclidean Green's functions. In particular for purely bosonic models, they are mathematically the same as the correlation functions of some models of classical statistical mechanics. Furthermore they turn out to be the moments of some probability measure on function space. Thus both the methods of statistical mechanics and the theory of stochastic processes and of functional integration can be used to study quantum field theory models. On the other hand the work of field theorists has also led to progress in those fields.

In these lectures we first explain that part of axiomatic quantum field theory which is relevant for the Euclidean program. Then we discuss free boson fields, Gaussian measures on function space and explain the so-called Feynman-Kac formula. In the following chapters we introduce interactions and establish the two major technical results: the stability bound and the existence of the thermodynamic limit. We conclude these lectures with a survey of some of the major results of constructive quantum field theory.

II. Connection between Euclidean and relativistic formulation of quantum field theory

One of the standard formulations of relativistic quantum field theory is given in terms of the Wightman axioms [J, SW, BLT, Si, GJ]. For a real scalar field they are as follows:

- The states of the theory are described by unit rays in a (separable) Hilbert space $\mathcal{H}$
- On $\mathcal{H}$ there is a unitary representation $U(a,\Lambda)$ of the Poincaré group (a denoting a translation, Λ a homogeneous Lorentz transformation) Since $U(a,\mathbb{1})$ is unitary it can be written as $e^{i\,a\cdot P}$; $a\cdot P = a_\mu P^\mu$, $P = (P^0, P^1, P^2, P^3)$ being the energy-momentum operator.

The spectrum of P is in the closed forward light cone V_+ and 0 is a simple eigenvalue with eigenvector $\Omega \in \mathcal{H}$, the vacuum. Ω is normalized so that $\|\Omega\| = 1$.

- There exist field operators $\varphi(x)$ which are operator valued, tempered distributions. More precisely, for any $f \in \mathcal{S}(\mathbb{R}^4)$, the space of rapidly decreasing, infinitely often differentiable functions, there is an operator $\varphi(f) = \int \varphi(x)\, f(x)\, d^4x$ defined on a common dense domain $D \subset \mathcal{H}$.

  Furthermore $\Omega \in D$, $U(a,\Lambda) D \subset D$

  $\varphi(f)\; D \subset D$

  and $\varphi(\bar{f}) \subset \varphi(f)^*$.
- $U(a,\Lambda)\, \varphi(x)\, U(a,\Lambda)^{-1} = \varphi(\Lambda x + a)$ (Relativistic invariance of the fields)
- $\varphi(x)\, \varphi(y) - \varphi(y)\, \varphi(x) = 0$, if $(x-y)^2 > 0$ i.e. x-y space like (Locality, microscopic causality)
- The set of linear combinations of vectors of the form $\Pi_{k=1}^{N}\, \varphi(f_k)\Omega$, $f_k \in \mathcal{S}$, is dense in $\mathcal{H}$.

The set $(\mathcal{H}, U, \Omega, \varphi)$ will simply be called a Wightman theory if it has the above properties.

It turns out that a Wightman theory can be uniquely described in terms of the infinite set of distributions

$$W_n(x_1, \ldots x_n) = \langle \Omega, \varphi(x_1) \ldots \varphi(x_n)\Omega \rangle ,$$

the Wightman distributions.

From the spectral assumptions on P and from locality it follows that each $W_n(x_1, \ldots x_n)$ is boundary value of an analytic function $W_n(z_1, \ldots z_n)$, the Wightman function, whose analyticity domain contains the whole Euclidean space

$$E^{4n} = \{ z_1, \ldots z_n \mid \quad z_i = (ix_i^0 , \vec{x}_i) , \quad x_i^0 \in \mathbb{R} , \quad \vec{x}_i \in \mathbb{R}^3 \}$$

except for the set of points

$$O^{4n} = \{ (z_1 \ldots z_n) \in E^{4n} , \quad z_i = z_j \quad \text{for some} \quad i \neq j \}$$

Points in $E^{4n} - O^{4n}$ are called noncoincident Euclidean points. The restriction of $W_n(z_1, \ldots z_n)$ to these points,

$$S_n(x_1, \ldots x_n) = W_n(z_1, \ldots z_n)$$

$$z_i = (ix_i^0 , \vec{x}_i) , (z_1, \ldots z_n) \in E^{4n} - 0^{4n} ,$$

is called <u>n-point Schwinger function</u>. The crucial point to be made in this chapter is that any Wightman theory can be described uniquely by the set $\{S_n\}_{n=0}^{\infty}$ of its Schwinger functions.

Leaving aside the mathematical details, the meaning of the Schwinger functions can be explained easily:
Due to the spectral property, the unitary group $e^{ia\cdot P}$ can be extended to a <u>holomorphic semigroup</u> $e^{iz\circ P}$ for $z \in \mathbb{C}^4$ and $\text{Im } z \in V_+$ (Notice that the spectrum of $iz\cdot P$ is in the left hand plane, hence e^{izP} is a contraction). Writing (formally) $\varphi(z)$ for $e^{iz\cdot P} \varphi(0) e^{-iz\cdot P}$ we see that vectors of the form

$$\Psi(z_1, \ldots z_n) = \varphi(z_1) \varphi(z_2) \ldots \varphi(z_n)\Omega$$

are analytic functions of $z_1, \ldots z_n$ if

$$\text{Im } z_1 \in V_+ , \text{Im } (z_{k+1} - z_k) \in V_+ , k = 1, \ldots n-1$$

Now we set

$$W_n (z_1, \ldots z_n) = \langle \Omega , \Psi (z_1, \ldots z_n)\rangle$$

To extend W_n to its full domain of analyticity one has to use Lorentz invariance and locality. This domain is closed under permutations $(z_1, \ldots z_n) \longrightarrow (z_{\pi 1}, \ldots z_{\pi n})$ and W_n is a symmetric function, i.e.

$$W_n(z_1, \ldots, z_n) = W_n (z_{\pi 1}, \ldots z_{\pi n})$$

for any permutation π.

If we compute the scalar product of two vectors Ψ we easily find:

$$\langle \Psi(z_1, \ldots z_n) , \Psi(z_1', \ldots z_m')\rangle$$
$$= W_{n+m} (\bar{z}_n, \ldots \bar{z}_1, z_1', \ldots z_m')$$

Notice that for $z_1, \ldots z_n, z_1', \ldots z_m'$ in the analyticity domains of Ψ we have $\text{Im } \bar{z}_{n-1} - \bar{z}_n \in V_+$

$$\text{Im } z_1' - \bar{z}_1 \in V_+$$

$$\text{Im } z_m' - z_{m-1}' \in V_+$$

Restricting the Ψ's to noncoincident Euclidean points $(z_1,\ldots z_n)$, $z_k = (ix_k^0 , \vec{x}_k)$, $0 < x_1^0 < x_2^0 < \ldots < x_n^0$ we set

$$\Psi_E(x_1,\ldots x_n) = \Psi(z_1,\ldots z_n)$$

Then $S(x_1,\ldots x_n) = \langle \Omega, \Psi_E(x_1,\ldots x_n)\rangle$

and

$$\langle \Psi_E(x_1,\ldots x_n) , \Psi_E(x_1',\ldots x_m')\rangle = S_{n+m} (\theta x_n,\ldots \theta x_1, x_1',\ldots x_m') \tag{1}$$

Here $x \longrightarrow \theta x \equiv (-x^0,\vec{x})$ corresponds to

$$z = (ix^0,\vec{x}) \longrightarrow \bar{z} = (-ix^0,\vec{x}) .$$

From (1) we derive a crucial property of the Schwinger functions. Let

$$X = \Sigma_{n=0}^N \int \Psi_E(x_1,\ldots x_n) f_n(x_1,\ldots x_n) dx$$

be a vector in $\mathcal{H}$. Here the test functions $f_n(x_1,\ldots x_n)$ have to be chosen to be zero unless $0 < x_1^0 < x_2^0 < \ldots < x_n^0$.

Then

$$0 \leq \|X\|^2 = \Sigma_{n,m} \int S_{n+m} (\theta x_n,\ldots \theta x_1, x_1',\ldots x_m') \cdot \bar{f}_n (x_1,\ldots,x_n) f_m(x_1',\ldots,x_m') dx\, dx'$$

$$\equiv \Sigma S_{n+m} (\Theta \bar{f}_n \times f_m)$$

We call this property of the Schwinger functions <u>reflection positivity</u>.

Continuing the list of properties of the Schwinger functions: Lorentz invariance of W_n implies <u>Euclidean invariance</u> of S_n:

$$S_n(x_1,\ldots x_n) = S_n(Rx_1+a,\ldots Rx_n+a)$$
$$R \in SO(4),\ a \in \mathbb{R}^4$$

The Wightman functions $W_n(z_1,\ldots z_n)$ are symmetric. Hence <u>symmetry</u> of the S_n:

$$S_n(x_1,\ldots x_n) = S_n(x_{\pi 1},\ldots x_{\pi n})$$

for any permutation π.

The nondegeneracy of the vacuum Ω implies the <u>cluster property:</u> for any $a \neq 0$, $a \in \mathbb{R}^4$

$$\lim_{\lambda\to\infty} S_{n+m}(x_1,\ldots x_n, y_1 + \lambda a,\ldots y_m + \lambda a) = S_n (x_1,\ldots x_n) \cdot S_m (y_1,\ldots y_m)$$

Finally the S_n have some regularity properties. From the Wightman axioms it follows that the Schwinger functions are tempered distributions in the sense that

$$S_n(f) = \int S_n(x_1,\ldots,x_n)\ f(x_1,\ldots,x_n)\ dx$$

is bounded in absolute value by some Schwartz norm of f, where $f \in \mathcal{S}(\mathbb{R}^{4n})$ and $f(x_1,\ldots x_n)$ together with all its partial derivatives is zero if $x_i = x_j$ for some $i \neq j$. However, the Schwartz norm may depend on n in an arbitrary way. Here we want to assume that for some fixed L, r and c the Wightman distributions satisfy

$$|W_n(f)| < c\cdot(n!)^L\ |f|_{n\cdot r} \qquad \text{(WO)}$$

$$\text{and} \quad |f|_k = \sup_{\substack{x_1,\ldots x_n \\ \Sigma\alpha_{i\mu} \leq k}} \left|(1+\Sigma\, x_i^2)^{k/2}\ \Pi\left(\frac{\partial}{\partial x_i^\mu}\right)^{\alpha_{i\mu}} f(x_1\ldots x_n)\right|$$

Then there are K, s and c_1 such that for all $f \in \mathcal{S}(\mathbb{R}^{4n})$, vanishing together with all the partial derivatives in points where $x_i = x_j$ for some $i \neq j$, the Schwinger functions satisfy

$$|S_n(f)| < c_1(n!)^K\ |f|_{n\cdot s} \qquad \text{(EO)}$$

The properties of $\{S_n\}$ discussed so far characterize them as the Schwinger functions of a unique Wightman theory.

Reconstruction Theorem [OS]

There is a one to one correspondence between Schwinger functions $\{S_n\}$ satisfying

Ⓔ Reflection positivity, Euclidean invariance, symmetry, cluster property, regularity (EO)

and Wightman theories $(\mathcal{H},U,\Omega,\varphi)$ with (WO).

The practical meaning of this theorem is that a Wightman theory can be obtained by constructing a set of Schwinger functions, veryfying properties Ⓔ and then reconstructing $(\mathcal{H},U,\Omega,\varphi)$.

Remarks:

1. With appropriate modifications the same result can be proven for arbitrary spinor fields [OS]. Generalizations to gauge field theories are also possible [FOS].

2. Strengthening of the regularity assumption leads to additional results, such as the existence of time ordered products [EE] and the existence of algebras of local observables [DF].
3. In the case of scalar bosons the reconstruction of $(\mathcal{H}, U, \Omega, \varphi)$ can be simplified if one lumps together all the $\{S_n\}$ in the generating functional

$$S(f) = \Sigma_{n=0}^{\infty} \frac{i^n}{n!} S_n(f \times f \times \ldots \times f), \quad f \in \mathcal{S}(\mathbb{R}^4)$$

and reformulates Ⓔ accordingly. (EO) has to be strengthened considerably [GJ, F1].
4. The whole scheme explained here is valid even if the number of dimensions of space time is different from 4.
5. The equivalence of (EO) and (WO) is a slight improvement of the results in [OS]. It is due to [L].

III. Free scalar fields

For a free scalar field of mass m_o one finds

$$\begin{aligned} S_2(x_1, x_2) &= C(x_1 - x_2) \\ &= \frac{1}{(2\pi)^4} \int \frac{e^{ip(x_1-x_2)}}{p^2 + m_o^2} d^4p \\ &= (-\Delta + m_o^2)^{-1} (x_1, x_2) \end{aligned}$$

and

$$S_n(x_1, \ldots x_n) = \begin{cases} 0 \quad \text{if } n \text{ is odd} \\ \sum\limits_{\text{pairings}} \prod\limits_{k=1}^{n/2} C(x_{i_k} - x_{j_k}), \text{ otherwise} \end{cases}$$

Here $p^2 = (p^0)^2 + (p^1)^2 + (p^2)^2 + (p^3)^2$ is the Euclidean scalar product (p, p), the same for px_1 etc. The sum over pairings runs over all partitions of $(1, \ldots n)$ into $n/2$ pairs $(i_1, j_1) \ldots (i_{n/2}, j_{n/2})$. We leave it as an exercise to the reader to verify that the $\{S_n\}$ thus defined satisfy all properties Ⓔ.

A nice feature of the Schwinger functions of the free field is that they are the vacuum expectation values of a <u>free Euclidean field</u> $\Phi(x)$, operating on a Fock space $\mathcal{F}_E$. More precisely let

$$\Phi(x) = \frac{1}{(2\pi)^2} \int \frac{d^4p}{(p^2+m_o^2)^{1/2}} e^{ipx} [A(p) + A^*(-p)]$$

$$\text{with} \quad [A(p), A^*(p')] = \delta^4(p-p')$$

$$[A(p), A(p')] = [A^*(p), A^*(p')] = 0$$

then

$$S_n(x_1, \dots x_n) = \langle \Omega_o, \Phi(x_1) \dots \Phi(x_n) \Omega_o \rangle$$

where Ω_o is the free vacuum vector, characterised by $A(p)\Omega = 0$, all p. Notice in particular that

$$[\Phi(x_1), \Phi(x_2)] = 0 \quad \text{for all } x_1, x_2,$$

in contrast to the commutation relations for free Minkowski fields.

Generalized free Euclidean fields Φ^G may be defined similarly by their 2-point function

$$\langle \Omega_o, \Phi^G(x_1) \Phi^G(x_2) \Omega_o \rangle = C^G(x_1, x_2)$$

where C^G has to be a positive semidefinite bilinear form on $\mathcal{S}(\mathbb{R}^4) \times \mathcal{S}(\mathbb{R}^4)$ in order to guarantee that

$$0 \leq \| \Phi^G(f)\Omega_o \|^2 = \int \bar{f}(x)\, C^G(x,y)\, f(y)\, dx\, dy$$

Generalized free Schwinger functions are then introduced

$$S_n^G(x_1, \dots x_n) = \langle \Omega_o, \Phi^G(x_1) \dots \Phi^G(x_n) \Omega_o \rangle$$

These $\{S_n^G\}$, however, have little to do with relativistic quantum field theory unless they satisfy reflection positivity.

<u>Theorem</u> [GJ]: Reflection positivity holds for $\{S_n^G\}$ if and only if it holds for $S_2^G = C^G$, i.e. if

$$\int \bar{f}(-x^0, \vec{x})\, S_2^G(x,y)\, f(y^0, \vec{y})\, dx\, dy \geq 0$$

for all $f \in \mathcal{S}(\mathbb{R}^4)$ with $f(x^0, \vec{x}) = 0$ unless $x^0 > 0$.

The introduction of interaction in chapter IV will make it necessary to <u>regularize</u> the free field, i.e. to suppress the contributions from large p-values to the integrand of

$$C(x,y) = \frac{1}{(2\pi)^4} \int \frac{e^{ip(x-y)}}{p^2+m_o^2} d^4p$$

A first possibility would be to replace

$$\tilde{C}(p) = \frac{1}{p^2+m_o^2} \quad \text{by} \quad \tilde{C}_{\mathcal{H}}(p) = \frac{\chi_{\mathcal{H}}(|p|)}{p^2+m_o^2}$$

where $\chi_{\mathcal{H}}(x)$ is the characteristic function of $[-x,x]$.

This regularization is not appropriate for many purposes, because the Schwinger functions obtained from $\Phi_{\mathcal{H}}$ violate reflection positivity. A better choice, at least from this point of view, would be

$$\tilde{C}_{\mathcal{H}}(p) = \frac{\chi_{\mathcal{H}}(|\vec{p}|)}{p^2+m_o^2}$$

If the momenta in the p^0-direction are not affected then reflection positivity remains valid.

Another possibility which preserves reflection positivity is the lattice regularization. Here the large p-contributions are eliminated by a drastic change of the small x-behaviour of C. Instead of allowing arbitrary values in $\mathbb{R}^4$ for the variables $x(y)$ they are now restricted to a lattice $\delta\cdot\mathbb{Z}^4$, i.e.

$$x^i = \delta\cdot n^i \quad , \quad n^i \text{ integer}$$

$$\delta = \text{"lattice constant"} \; .$$

The Laplacian Δ is replaced by the finite difference Laplacian Δ_δ. For a function f on $\delta\mathbb{Z}^4$ we set

$$(\partial^k f)(x) = \frac{1}{\delta} \; [f(x+\delta e_k) - f(x)]$$

$$(\partial^{k*} f)(x) = \frac{1}{\delta} \; [f(x-\delta e_k) - f(x)]$$

$$-\Delta_\delta = \sum_k (\partial^k)^* \partial^k = \sum_k \partial^k (\partial^k)^* = \nabla_\delta^* \cdot \nabla_\delta$$

Here e_k is a unit vector parallel to the k-axis.

The lattice free field Φ_δ is now defined by

$$\langle \Omega_o, \Phi_\delta(x) \; \Phi_\delta(y)\Omega_o \rangle$$

$$= C_\delta \; (x-y) = (-\Delta_\delta + m_o^2)^{-1} \; (x,y)$$

$$= \frac{1}{(2\pi)^4} \int_{|p_i| < \pi/\delta} d^4p \; e^{ip(x-y)} \; [2\delta^{-2}(4 - \Sigma\cos\delta p_k) + m_o^2]^{-1}$$

The introduction of the lattice approximation discretizes the problem: for each lattice point x in δZ^4 we have a field operator $\Phi_\delta(x)$, but we can even make the number of degrees of freedom <u>finite</u>, by restricting the lattice to a finite volume Λ. In terms of a finite volume lattice free field $\Phi_{\delta,\Lambda}$ we define regularized Schwinger functions by

$$S_n^{\delta,\Lambda}(x_1,\dots x_n) = \langle \Omega_o, \Phi_{\delta,\Lambda}(x_1)\dots\dots \Phi_{\delta,\Lambda}(x_n)\Omega_o\rangle$$

$$(2) \qquad = \begin{cases} 0 \text{ if } n \text{ is odd} \\ \sum\limits_{\text{pairings}} \prod\limits_{k=1}^{n/2} C_{\delta,\Lambda}(x_{i_k}, x_{j_k}) \text{ otherwise} \end{cases}$$

for $x_i \in \delta\mathbb{Z}^4 \cap \Lambda$. $C_{\delta,\Lambda}$ is the two-point function of $\Phi_{\delta,\Lambda}$; its precise form will be given later.

The crucial point is that the $S_n^{\delta,\Lambda}$ can be realized as <u>moments of a Gaussian measure on $\mathbb{R}^N$</u>, where N is the number of lattice points in Λ [GRS]. To see this we denote by $(q_1,\dots q_N)$ the coordinates of a point in $\mathbb{R}^N$. Then the Gaussian measure $d\mu_A$ with mean O and covariance A is given by

$$d\mu_A(q) = \frac{1}{Z_o} e^{-\frac{1}{2}\Sigma q_i A^{-1}_{ij} q_j} \Pi\, dq_k$$

where A is a positive definite N x N matrix (the <u>covariance</u>) with inverse A^{-1} and Z_o is a normalization factor such that $\int_{\mathbb{R}^N} d\mu_A(q) = 1$. The moments of $d\mu_A$ are

$$\int q_{r_1}\dots q_{r_n}\, d\mu_A(q) =$$

$$(3) \qquad \begin{cases} 0 \text{ if } n \text{ is odd} \\ \sum\limits_{\text{pairings}} \prod\limits_{k=1}^{n/2} A_{i_k j_k} \text{ if } n \text{ is even} \end{cases}$$

Here the sum runs over pairings $\{(i_1 j_1),\dots (i_{n/2}\, j_{n/2})\}$ of the numbers $\{r_1,\dots r_n\}$.

Comparing formulas (2) and (3) we see that they are the same, if we substitute x_k for r_k and $C_{\delta,\Lambda}(x_k,x_\ell)$ for $A_{r_k r_\ell}$. In other words

$$S_n^{\delta,\Lambda}(x_1,\dots x_n) = \int q_{x_1}\dots q_{x_n}\, d\mu_{C_{\delta,\Lambda}}(q)$$

$$d\mu_{C_{\delta,\Lambda}} = \frac{1}{Z_o} e^{-\frac{1}{2}\sum\limits_{x,y\in\Lambda} q_x C^{-1}_{\delta,\Lambda}(x,y)\, q_y} \prod_{x\in\Lambda\cap\delta\mathbf{Z}^4} dq_x$$

To define $C_{\delta,\Lambda}$ as a finite volume approximation of C_δ we set e.g.

$$\sum_{x,y\in\Lambda} q_x C^{-1}_{\delta,\Lambda}(x,y)\, q_y = \sum_{x\in\Lambda} \delta^4[(\nabla_\delta q)^2_x + m_o^2 q_x^2]$$

$$= \sum_{\substack{x,y:n.n.\\ \in\Lambda}} \delta^2 (q_x - q_y)^2 + \sum_{x\in\Lambda} \delta^4 m_o^2\, q_x^2$$

where the first sum runs over pairs of nearest neighbour lattice points x,y, both of them in $\Lambda\cap\delta\mathbf{Z}^4$.

With this choice $d\mu_{C_{\delta,\Lambda}}$ is the Gibbs measure of a classical lattice spin system with (unbounded) spin q_x at each site x and with nearest neighbour interaction. This establishes the connection between boson quantum field theory and classical statistical mechanics. Introducing interaction into the quantum field theory model will not change this picture: the interaction being local, the Gibbs measure will simply be changed by a factor of the form $\prod_{x\in\Lambda} e^{P(q_x)}$ with P typically a polynomial. In the language of statistical mechanics this means just a change of the a priori single spin distribution.

If we now take the limits $\Lambda\to\mathbb{R}^4$, $\delta\to 0$ the measures $d\mu_{C_{\delta,\Lambda}}$ converge to a Gaussian measure $d\mu_C$ on function space $\mathcal{S}'(\mathbb{R}^4)$,[GV] (Convergence of the characteristic function). The limiting measure may be thought of as

$$d\mu_C = \frac{1}{Z_o} e^{-\frac{1}{2}\Sigma\,(q_x,(-\Delta + m_o^2)q_x)} \prod_{x\in\mathbb{R}^4} dq_x$$

but this expression is just formal. The reason is that $d\mu_C$ cannot be split into a Gaussian exponential and a Lebesque measure on an infinite dimensional space, the later being ill defined.

The free field Schwinger functions can now be expressed as moments of the measure $d\mu_C$

$$\langle\Omega_o, \Phi(x_1)\ldots\Phi(x_n)\Omega_o\rangle = \int q_{x_1}\ldots q_{x_n}\, d\mu_C(q) \qquad (4)$$

The quantity q_x is a distribution in $\mathcal{S}'(\mathbb{R}^4)$ whose values are random variables, i.e. for

$$f\in\mathcal{S}(\mathbb{R}^4)$$
$$q(f) = \int q_x\, f(x)\, dx$$

are random variables whose joint distributions are given by (4). Furthermore q_x is a Markov process, a fact that has been exploited for field theory in [N] and in [GRS].

To be able to work with powers of the fields $\Phi(x)$ we have to introduce <u>Wick ordering</u>. In the Fock space formalism the definitions are straightforward: Set

$$\Phi^-(x) = \frac{1}{(2\pi)^2} \int \frac{d^4p}{(p^2+m_o^2)^{1/2}} e^{ipx} A(p) \quad , \ \Phi^+(x) = (\Phi^-(x))^*.$$

Then $\Phi(x) = \Phi^+(x) + \Phi^-(x)$. Now define

$$:\Phi(x_1)\dots \Phi(x_n): = :\prod_{i=1}^{n} (\Phi^+(x_i) + \Phi^-(x_i)):$$

$$= \sum_{\{\underline{i},\underline{j}\}} \Phi^+(x_{i_1})\dots \Phi^+(x_{i_k}) \ \Phi^-(x_{j_1})\dots \Phi^-(x_{j_\ell})$$

where the sum runs over all different partitions of $\{1,\dots n\}$ into two sets $\{i_1,\dots i_k\}$ and $\{j_1,\dots j_\ell\}, k + \ell = n$.

In the language of functional integration Wick ordering can be introduced, too, but not with the help of splitting Φ into Φ^+ and Φ^-; this would destroy the commutativity of the fields. We denote by $\mathcal{F}_{\leq n}$ the space of all polynomials $P_n(q)$ of order n or less in the random variables $q(f)$, $f \in \mathcal{S}(\mathbb{R}^4)$, equipped with the scalar product

$$(P_n, P'_n) = \int \bar{P}_n(q)\, P'_n(q)\, d\mu_C\, (q)$$

The orthogonal complement of $\mathcal{F}_{\leq n-1}$ in $\mathcal{F}_{\leq n}$ is denoted by $\mathcal{F}_n$ and is called the <u>n-particle</u> subspace. The completion of $\bigoplus_{n=0}^{\infty} \mathcal{F}_n$ is the space $L^2(\mathcal{S}',d\mu_C)$ of square integrable functions of the random fields q_x. It is just another representation of the Fock space $\mathcal{F}_E$. Now let E_n be the operator of orthogonal projection in $L^2(\mathcal{S}',d\mu_C)$ onto $\mathcal{F}_n$. Then we define:

$$:q(f_1)\dots q(f_n): = E_n\, q(f_1)\dots q(f_n)$$

It is an easy exercise to show that

$$:q(f)^n: = \alpha^n H_n\, (\alpha^{-1} q(f))$$

where $H_n(x)$ is the n'th order Hermite polynomial and
$\alpha^2 = \int q(f)^2\, d\mu_C = \langle \Omega_o, \Phi(f)\ \Phi(f)\ \Omega_o \rangle$.

We conclude this section with a discussion of the <u>Feynman-Kac</u> formula for free fields. It allows us to obtain the free Hamiltonian H_o in

terms of Euclidean quantities. The reconstruction theorem of section II relates a subspace of $\mathcal{F}_E$ to Minkowski Fock space $\mathcal{F}_M$. Let $\underline{f} = (f_0, f_1, \dots f_N)$ be a sequence of elements $f_k \in \mathcal{S}(\mathbb{R}^{4k})$ and $f_k(x_1, \dots x_k) = 0$ unless $0 \le x_1^0 \le x_2^0 \le \dots$.
Then the vector

$$X(\underline{f}) = \sum_{k=0}^{N} \int f_k(x_1, \dots x_k)\ \Phi(x_1) \dots \Phi(x_k)\ dx\ \Omega_0$$

is in $\mathcal{F}_E$, the set of all such vectors is denoted by $\mathcal{F}_E^+$. On $\mathcal{F}_E^+$ we can define a <u>new scalar product</u>

$$\langle\!\langle X(\underline{f}) \ , \ X(\underline{g}) \rangle\!\rangle = \sum_{n,m} S_{n+m}\ (\Theta\bar{f}_n \times g_m) \quad .$$

By reflection positivity, $(\mathcal{F}_E^+, \langle\!\langle , \rangle\!\rangle)$ modulo vectors with vanishing $\langle\!\langle , \rangle\!\rangle$ norm is a prehilbert space whose completion we denote by $\mathcal{F}_M$. Let W be the natural mapping of $\mathcal{F}_E^+$ into $\mathcal{F}_M$, and $(\ , \ )$ the scalar product in $\mathcal{F}_M$. Then

$$(WX(\underline{f}) \ , WX(\underline{g})) = \langle\!\langle X(\underline{f}) \ , \ X(\underline{g}) \rangle\!\rangle$$

Now let U^t, $t \ge 0$, be the semigroup on $\mathcal{F}_E^+$ defined by

$$U^t X(\underline{f}) = \sum_{k=0}^{N} \int f_k(x_1, \dots x_n) \prod_{i=1}^{n} \Phi(x_i^0 + t, \vec{x}_i)\ dx\ \Omega_0$$

Then it can be shown that

$$\langle\!\langle U^t X(\underline{f}) \ , \ U^t X(\underline{f}) \rangle\!\rangle \quad \le \quad \langle\!\langle X(\underline{f}) \ , \ X(\underline{f}) \rangle\!\rangle$$

$$\langle\!\langle U^t X(\underline{f}) \ , \ X(\underline{g}) \rangle\!\rangle \quad = \quad \langle\!\langle X(\underline{f}) \ , \ U^t\ X(\underline{g}) \rangle\!\rangle$$

This allows us to conclude that U^t can be lifted to a (strongly continuous) one parameter semigroup of selfadjoint contractions on $\mathcal{F}_M$ and

$$WU^t\ X(\underline{f}) = e^{-tH_0}\ WX(\underline{f}) \quad .$$

This is the Feynman-Kac formula for free fields.
Furthermore, if we choose for $k = 1, \dots n$, $x_k = (x_k^0, \vec{x}_k)$ in $\mathbb{R}^4$ such that

$$0 \le x_1^0 \le x_2^0 \le \dots \le x_n^0 \ , \text{ then}$$

$$W\ \Phi(x_1) \dots \Phi(x_n)\ \Omega_0 = e^{-x_1^0 H_0}\ \varphi(0,\vec{x}_1)\ e^{-(x_2^0 - x_1^0)H_0}\ \varphi(0,\vec{x}_2) \dots \varphi(0,x_n)\,\Omega_M$$

where $\varphi(0,\vec{x})$ is the free Minkowski field at time 0 and Ω_M the vacuum vector in F_M, i.e. $\Omega_M = W\Omega_0$.

IV. Interacting Fields

Schwinger functions which are moments of a Gaussian measure describe generalized free fields (if the covariance is chosen such that properties Ⓔ hold). To construct a theory describing interacting particles we want to find Schwinger functions satisfying Ⓔ which are the moments of a non-Gaussian measure.

Let us begin with a Lagrangian

$$\frac{1}{2}\,((\nabla\phi)^2 + m_o\phi^2) + \mathcal{L}_I(\phi)$$

where $\mathcal{L}_I(x)$ is a function bounded from below,

e.g. $\mathcal{L}_I(x) = \lambda\,(x^4 + \sum_{k=0}^{3} a_k\, x^k)$. Then a formal calculation leads to Schwinger functions of the form

$$\begin{aligned} &S_n(x_1,\dots x_n) \\ &= \frac{1}{Z_1}\int q_{x_1}\cdots q_{x_n}\; e^{-\int \mathcal{L}_I(q_x)dx - \frac{1}{2}\int (q_x,(-\Delta+m_o^2)q_x)dx}\;\Pi dq_x \\ &= \frac{Z_o}{Z_1}\int q_{x_1}\cdots q_{x_n}\; e^{-\int \mathcal{L}_I(q_x)dx}\; d\mu_C(q) \end{aligned} \tag{5}$$

i.e. the S_n are moments of a measure on $\mathcal{S}'(\mathbb{R}^4)$

$$d\mu(q) \sim e^{-\int \mathcal{L}_I(q_x)dx}\, d\mu_C(q)$$

where $d\mu_C$ is the Gaussian measure introduced in the last section. Unfortunately these expressions are mathematically ill defined. For most $q \in \mathcal{S}'(R^4)$ the exponent $\int \mathcal{L}_I(q_x)dx$ is divergent. The divergences are in general due both to the local singularities (ultraviolet or high momentum divergences) and to the large x behaviour (infinite volume divergences) of q.

The following procedure leads from the formal ansatz (5) to a mathematical construction of Schwinger functions:

1) The formal expressions are regularized until they are mathematically well defined quantities.
2) After a renormalization the regularizations are removed by limit procedures.
3) Properties Ⓔ are verified for the Schwinger functions obtained by this construction.

The <u>regularization</u> of $\int \mathcal{L}_I(q_x)\,dx$ has to eliminate both the ultraviolet and the volume divergences. An ultraviolet cutoff is introduced either by putting the whole system on a lattice or by replacing

$$q_x \equiv q(x) \longrightarrow q_{\varkappa}(x) = (\tilde{\chi}_{\varkappa} * q)\ (x)$$

where $\tilde{\chi}_{\varkappa}$ is the Fourier transform of the characteristic function of $\{p \mid |p| \leq \varkappa\}$ or of $\{p \mid |\vec{p}| \leq \varkappa\}$ or of some other fast decreasing function of p with $\lim_{\varkappa \to \infty} \chi_{\varkappa}(p) = 1$.

The infinite volume integration is replaced by $\int \ldots g(x)dx$, where g is some sufficiently smooth function with rapid decrease at large $|x|$. E.g. we may choose g to be the characteristic function of a finite volume Λ .

Regularized Schwinger functions are now defined by

$$S_n^{\varkappa,g}(x_1,\ldots x_n) = \frac{1}{Z}\int q(x_1)\ldots q(x_n)\ e^{-\int \mathcal{L}_I(q_{\varkappa}(x)g(x)dx}d\mu_C$$

Here and in the following $Z = Z(\varkappa,g)$ will always be a normalization constant chosen such that $S_0^{\varkappa,g} = 1$.

The <u>renormalization</u> of the Schwinger functions requires in general a ($\varkappa$-dependent) change of the coefficients in $\mathcal{L}_I(q)$ such that the limit $\varkappa \longrightarrow \infty$ can be taken safely. In four-dimensional space time this is a very difficult problem and so far it cannot be dealt with propererly in a non perturbative fashion. The problem is reduced in complexity without being eliminated completely, if we reduce the dimension of space time to two or to three (the ultraviolet divergencies are less serious in lower dimensions). In <u>two-dimensional space time</u> e.g. with

$$\mathcal{L}_I(q) = \lambda(q^{2n} + \sum_{k=0}^{2n-1} a_k q^k) = \lambda P(q)$$

renormalization is obtained by replacing

$$\mathcal{L}_I(q_{\varkappa}(x)) \longrightarrow \mathcal{L}_I^{ren}(q_{\varkappa}) = :\mathcal{L}_I(q_{\varkappa}):$$

i.e. by simply Wick-ordering $\mathcal{L}_I$. Notice that this operation amounts to a change of the coefficients a_k in $P(q)$,

$$:P(q): = q^{2n} + \sum_{k=0}^{2n-1} \hat{a}_k(\varkappa)\, q^k$$

with some $\hat{a}_k(\varkappa)$ divergent in $\varkappa$, at most $\sim (\ln \varkappa)^n$.

With the proper renormalization of $\mathcal{L}_I$ one shows that the limit

$$S_n(x_1,\ldots x_n) = \lim_{\substack{g \to 1 \\ \varkappa \to \infty}} S_{n,ren}^{\varkappa,g}$$

$$= \lim_{\substack{g\to 1 \\ \varkappa\to\infty}} \frac{1}{Z} \int q(x_1)\dots q(x_n)\, e^{-\int \mathcal{L}_I^{ren}(q_\varkappa(x))\, g(x)dx}\, d\mu_C$$

exists and is independent of the particular choice of the regularizations. (In four-dimensional space time a field strength renormalization will be necessary, too, i.e. $q_\varkappa(x)$ has to be replaced by $z^{-1/2}(\varkappa)\, q_\varkappa(x)$ with $z(\varkappa)$ chosen appropriately).

The last step in the construction will be the verification of properties (E). With a proper choice of the regularization, reflection positivity is true for $S^{\varkappa,g}_{n,ren}$, hence it remains true in the limit. Symmetry is valid for $S^{\varkappa,g}_{n,ren}$ (for all regularizations) hence also for S_n. The regularity and clustering properties have to be shown to hold for $S^{\varkappa,g}_{n,ren}$, uniformly in $\varkappa$ and g, then they hold for S_n. Finally Euclidean invariance holds for S_n if we show the independence of the limit from the choice of the cutoffs.

Instead of the functional integral formalism we could have used the free Euclidean field to describe the above construction. In that notation we would have found

$$S^{\varkappa,g}_{n,ren}(x_1,\dots x_n) = \frac{1}{Z} \left\langle \Omega_o, \Phi(x_1)\dots\Phi(x_n)\, e^{-\int :\mathcal{L}_I(\Phi(x)):\, g(x)dx}\, \Omega_o \right\rangle$$

It is instructive to generalize the Feynman-Kac formula. Using the notations of the end of section III we find that

$$W \int :\Phi^k_\varkappa(0,\vec{x}):\, h(\vec{x})d\vec{x}\; X = \int :\varphi^k_\varkappa(0,\vec{x}):\, h(\vec{x})d\vec{x}\; WX$$

where $\Phi_\varkappa(x) = (\tilde{\chi}_\varkappa * \Phi)(x)$ is the field ultraviolet regularized in the spatial components only, i.e. $\chi_\varkappa = \chi_\varkappa(\vec{p})$ is the characteristic function of $\{p \mid |\vec{p}| < \varkappa\}$. Correspondingly $\varphi_\varkappa(0,\vec{x})$ is the regularized time zero Minkowski field. Generalizing this formula, we get

$$W \int :\mathcal{L}_I(\Phi_\varkappa(t,\vec{x})):h(\vec{x})d\vec{x}\; U^t X = e^{-tH_o} \int :\mathcal{L}_I(\varphi_\varkappa(0,\vec{x})):h(\vec{x})d\vec{x}\; WX$$

This formula allows us to prove the

Theorem

$$W\, e^{-\int_0^t dx^0 \int d\vec{x}\; :\mathcal{L}_I(\Phi_\varkappa(x^0,\vec{x})):\; h(\vec{x})}\, U^t X$$
$$= e^{-t(H_o + \int :\mathcal{L}_I(\varphi_\varkappa(0,\vec{x})):\; h(\vec{x})d\vec{x})}\, WX$$

The exponent on the right hand side of this expression is of course just (-t) times the regularized Hamiltonian of the interacting theory. As a corollary we get

Corollary

$$\langle \Omega_o,\ e^{-\int_{-t}^{t} dx^0 \int d\vec{x}\ :\mathcal{L}_I(\Phi_{\varkappa}(x^0,\vec{x})):\ h(\vec{x})}\ \Omega_o\rangle$$

$$= \int e^{-\int_{-t}^{t} dx^0 \int d\vec{x}\ :\mathcal{L}_I(q_{\varkappa}(x)):\ h(\vec{x})}\ d\mu_C(q)$$

$$\| e^{-t(H_o + \int :\mathcal{L}_I(\phi_{\varkappa}(0,\vec{x})):\ h(\vec{x})d\vec{x})}\ \Omega_M \|^2$$

V. The Stability Bound

In the next two sections we discuss the major steps in the construction of the no-cutoff limit of the regularized, renormalized Schwinger functions. For simplicity we restrict ourselves to models in 2-dimensional space time with

$$\mathcal{L}_I^{ren}(q\ ) = \lambda : P(q\ ) : \qquad , \ P(q) = q^{2n} + \text{lower order terms.}$$

In this section we first analyze the limit $\varkappa \to \infty$ of $:P(q_\varkappa):$ and then of $e^{-\lambda :P(q_\varkappa):}$. The analysis will be carried out in the space $L^2(\mathcal{S}', d\mu_C)$, the space of functions on $\mathcal{S}'$ which are square integrable with respect to the measure $d\mu_C$ and more generally in the spaces $L^p(\mathcal{S}', d\mu_C)$, the spaces of functions F on $\mathcal{S}'$ with finite p-norm $\| F \|_p = (\int |F(q)|^p d\mu_C)^{1/p}$. See [N ,G] and [Si,GJ].

Theorem: Let g be a non-negative function in $L^2(\mathbb{R}^2)$ with compact support and

$$P_{\varkappa,g} = \int : P(q\ (x)) : g(x)\ dx$$

Then a) $\lim_{\varkappa\to\infty} P_{\varkappa,g} \equiv P_g$ exists in all $L^p(\mathcal{S}', d\mu_C)$

b) There are constants $\varepsilon > 0$ and $c = c(g)$ such that

$$\| P_g - P_{\varkappa,g} \| \leq c(p-1)^n \varkappa^{-\varepsilon}$$

To illustrate this theorem and its proof we take $p = 2$ and consider for $\varkappa < \varkappa'$

$$\| \int :q_\varkappa^k(x):\ g(x)dx - \int :q_{\varkappa'}^k(x):g(x)dx\|_2^2 =$$

$$\langle \Omega_o,\ (\int (:\Phi_\varkappa^k(x): - :\Phi_{\varkappa'}^k(x):)\ g(x)dx)^2 \Omega_o\rangle$$

Using the formula for moments of Gaussian measures (or equivalently: expanding the right hand side in terms of Feynman diagrams) we find that the expression equals

$$\text{constant} \int |\tilde{g}(\sum_{i=1}^{n} p_i)|^2 (\prod_{i=1}^{n} \chi_{\varkappa}(p_i) - \chi_{\varkappa'}(p_i))^2 \prod_{i=1}^{n} \frac{1}{p_i^2 + m_0^2} \, dp_i$$

$$\leq \text{constant} \int_{\text{some } |\vec{p}_i| > \varkappa} |\tilde{g}(\sum_{i=1}^{n} p_i)|^2 \prod \frac{1}{p_i^2 + m_0^2} \, dp_i \leq \text{constant } \varkappa^{-\varepsilon} .$$

For an analysis of the exponential of $P_{\varkappa,g}$ we need a lower bound.

<u>Theorem:</u> With the assumptions as before, there is a constant $c = c(g)$, such that

$$P_{\varkappa,g} \geq -c(\ln \varkappa)^n \qquad \text{"Wick bound"}$$

Here $P_{\varkappa,g}$ has to be considered as a function on $\mathcal{S}'(\mathbb{R}^2)$.

The polynomial P being of the form (q^{2n} + lower order) is of course bounded from below. The Wick ordering, however, introduces $\varkappa$-dependent coefficients in the lower order terms. E.g.

$$:q^4(x): = q_{\varkappa}^4(x) - 6\, c_{\varkappa}\, q_{\varkappa}^2(x) + 3\, c_{\varkappa}^2$$

$$= (q_{\varkappa}^2(x) - 3\, c_{\varkappa})^2 - 6\, c_{\varkappa}^2 \geq -6\, c_{\varkappa}^2$$

where $c_{\varkappa} = \int q_{\varkappa}^2(x)\, d\mu_C = \frac{1}{2\pi} \int \frac{\chi_{\varkappa}^2(p)}{p^2 + m_0^2} \, d^2p \leq \text{const.} \ln \varkappa$

Extending this argument to arbitrary polynomials P one proves the theorem.

This theorem shows that $e^{-\lambda P_{\varkappa,g}}$ is a <u>bounded</u> function on $\mathcal{S}'$ and is in $L^p(\mathcal{S}', d\mu_C)$ for all $p \geq 1$. For $\varkappa \to \infty$, however, it becomes unbounded.

<u>Theorem:</u> $e^{-\lambda P_g} \in L^p(\mathcal{S}', d\mu_C)$ for all $p \geq 1$

The main consequence of this result is that

$$d\nu_g(q) = Z_g^{-1} \, e^{-\lambda P_g(q)} \, d\mu_C(q)$$

with $Z_g = \int e^{-\lambda P_g(q)} d\mu_C(q)$ defines a measure of total weight 1 on $\mathcal{S}'(\mathbb{R}^2)$. Its moments define the finite volume Schwinger functions

$$S_n^g(x_1, \dots x_n) = \int q(x_1) \dots q(x_n) \, d\nu_g(q)$$

The idea of the proof is to show that the set of points q in $\mathcal{S}'$, where $e^{-\lambda P_g(q)}$ is large has small $d\mu_C$-measure. More precisely we set

$$\mathcal{A}_{\varkappa} = \{ q \in \mathcal{S}' \mid |P_g(q) - P_{\varkappa,g}(q)| \geq 1 \}$$

Then

$$\int_{\mathcal{A}_{\varkappa}} d\mu_C \;\leq\; \int |P_g(q) - P_{\varkappa,g}(q)|^p \, d\mu_C$$
$$= \; \|P_g - P_{\varkappa,g}\|_p^p$$
$$\leq \; c^p (p-1)^{np} \varkappa^{-\varepsilon p}$$

by the first theorem of this section. Choosing p such that the last expression is minimized we find that for sufficiently large $\varkappa$ and some $\delta > 0$

$$\int_{\mathcal{A}_{\varkappa}} d\mu_C \;\leq\; e^{-\varkappa^{\delta}}$$

This combined with the Wick bound proves the theorem.

We now improve this theorem by taking care of the g-dependence of the estiamtes.

Theorem: Let g be the characteristic function χ_Λ of a finite (sufficiently regular) region $\Lambda \subset \mathbb{R}^2$ whose area is $|\Lambda|$. Then with $P_g \equiv P_\Lambda$

$$\| e^{-\lambda P_\Lambda} \|_p \;\leq\; c_1 \, e^{c_2 |\Lambda|}$$

for some constants c_1, c_2.

Via the Feynman-Kac formula this bound is connected with the bound

$$H_o + \lambda \int_{-\ell}^{\ell} : P(\varphi(0,x)) : dx \;\geq\; - c\,\ell$$

For this reason it is called the linear lower bound or the stability bound. It has been a crucial step in any construction of a quantum field theory model.

VI. The Thermodynamic Limit

The limit $g \longrightarrow 1$ (or $\Lambda \longrightarrow \mathbb{R}^2$) is called the infinite volume limit or to stress the analogy with statistical mechanics the thermodynamic limit. There are two entirely different methods to deal with this limit.

First method: Using correlation inequalities one shows monotonicity in the volume Λ. Together with a uniform bound this yields existence of a unique limit. The method works for all values of λ but not for all choices of the polynomial P [N,GJ,Si].

Second method: Using a cluster expansion one proves an exponential cluster decay property for finite volume Schwinger functions. This leads immediately to a proof of unique convergence as $\Lambda \longrightarrow \mathbb{R}^2$. This method is in general more complicated than the first one but it leads to much more detailed information about the limit. It can be applied whenever λ and P are such that sufficiently strong exponential clustering holds (i.e. if one is sufficiently far away from a "critical point") [GJS,GJ].

In these lectures we only discuss the first method. To prove correlation inequalities we approximate the finite volume $d\nu_g$ of the last section by a lattice measure. For simplicity we take again $g = \chi_\Lambda$. Then

$$d\nu^N_{\Lambda,\delta} = \frac{1}{Z^N} \exp - \Big[\frac{1}{2} \sum_{\substack{x,y:n.n\\ x,y\in\Lambda}} (q_x - q_y)^2 + \frac{1}{2} \sum_{x\in\Lambda} \delta^2 m_0^2 q_x^2 + \lambda \sum_{x\in\lambda} \delta^2 :P(q_x):\Big] \prod_{x\in\Lambda\cap\delta\mathbb{Z}^2} dq_x$$

and

$$d\nu^D_{\Lambda,\delta} = \frac{1}{Z^D} \exp\Big[\sum_{\substack{x,y:n.n.\\ x,y\in\Lambda}} q_x q_y - \frac{1}{2} \sum_{x\in\Lambda} (m_0^2\delta + 4) q_x^2 - \lambda\Sigma\delta^2 :\, P(q_x):\Big] \cdot \prod_{x\in\Lambda\cap\delta\mathbb{Z}^2} dq_x$$

are lattice approximations to $d\nu_g$ with Neumann and Dirichlet boundary conditions, respectively.

Remarks: 1) The powers of δ are not the same here as they were in the corresponding formula in section III, where we treated the 4-dimensional case. 2) The difference between the exponents of the two lattice measures are just contributions from the boundary $\partial\Lambda$ of Λ. Other boundary conditions are possible. 3) For simplicity we define Wick ordering in: $P(q_x)$: with respect to the continuum free measure $d\mu_C$.

Notice that $d\nu^D_{\Lambda,\delta}$ can be written as

$$d\nu^D_{\Lambda,\delta} = \frac{1}{Z^D} \exp\Big[\sum_{\substack{x,y:n.n\\ x,y}} q_x q_y\Big] \prod_{x\in\Lambda\cap\delta Z^2} e^{-Q_\delta(q_x)} dq_x$$

i.e. it is the Gibbs measure of a nearest neighbour ferromagnetic classical lattice spin system (similarly for $d\nu^N_{\Lambda,\delta}$, but with modifications at $\partial\Lambda$) .

The simplest inequalities, given in the following theorem, are the Griffith inequalities, see e.g. [Ru].

Theorem: Let P = even polynomial + linear term. Then

$$\int q_{x_1} \cdots q_{x_n} \, d\nu^B_{\Lambda,\delta} \geq 0 \qquad \text{(Griffith I)}$$

$$\int q_{x_1}\cdots q_{x_n}\, q_{y_1}\cdots q_{y_m}\, d\nu^B_{\Lambda,\delta} \geq \qquad \text{(Griffith II)}$$
$$\int q_{x_1}\cdots q_{x_n}\, d\nu^B_{\Lambda,\delta}\cdot\int q_{y_1}\cdots q_{y_m}\, d\nu^B_{\Lambda,\delta}$$

for both boundary conditions $B = N$ or D.

These inequalities remain true as we pass to the limit $\delta \longrightarrow 0$. We set $C_{\Lambda,B} = (-\Delta_{\Lambda,B} + m_o^2)^{-1}$, where $\Delta_{\Lambda,B}$ is the Laplacian on $L_2(\mathbb{R}^2)$ with Dirichlet ($B = D$) or Neumann ($B = N$) boundary conditions on $\partial\Lambda$. Then finite volume Schwinger functions are defined as in the previous section by

$$S_n^{\Lambda,B}(x_1,\ldots x_n) = \frac{1}{Z_{\Lambda,B}}\int q(x_1)\ldots q(x_n)\; e^{-\lambda P_\Lambda(q)}\, d\mu_{C_{\Lambda,B}}(q)$$

They are the limits as $\delta \longrightarrow 0$ of the moments of $d\nu^B_{\Lambda,\delta}$ and thus they satisfy

$$S_n^{\Lambda,B}(x_1,\ldots x_n) \geq 0 \qquad \text{(Griffith I)}$$

$$S_{n+m}^{\Lambda,B}(x_1,\ldots x_n,\, y_1,\ldots y_m) \geq S_n^{\Lambda,B}(x_1,\ldots x_n) S_m^{\Lambda,B}(x_1,\ldots x_m) \qquad \text{(Griffith II)}$$

<u>Remark:</u> Here and in the following statements about the Schwinger functions should be understood in a distributional sense, e.g. for all $f \in \mathcal{S}(\mathbb{R}^{2n})$ and $f(x_1,\ldots x_n)$ non-negative,

$$S_n^{\Lambda,B}(f) = \int S_n^{\Lambda,B}(x_1,\ldots x_n)\, f(x_1,\ldots x_n)\, dx \geq 0$$

is the precise formulation of (Griffith I).

<u>Main Theorem:</u> For all $f \in \mathcal{S}(\mathbb{R}^{2n})$, f non-negative

a) $S_n^{\Lambda,D}(f) \geq 0$

b) $S_n^{\Lambda,D}(f)$ is monotonically increasing in Λ

c) $S_n^{\Lambda,D}(f)$ is uniformly bounded in Λ

<u>Corollary:</u> $\lim_{\Lambda \nearrow \mathbb{R}^2} S_n^{\Lambda,D}(f) = S_n(f)$ exists

Statement a) of the Main Theorem is just (Griffith I)

To verify b) it suffices to show monotonicity for the lattice approximations of $S_n^{\Lambda,D}(x_1,\ldots x_n)$, i.e. for

$$\int q_{x_1}\cdots q_{x_n}\, d\nu^D_{\Lambda,\delta} .$$

We take two regions $\Lambda_1 \subset \Lambda_2$ and interpolate $d\nu^D_{\Lambda_1,\delta}$ and $d\nu^D_{\Lambda_2,\delta}$ by setting

$$d\nu(\underline{J}) = Z(\underline{J})^{-1} \exp\Big[\sum_{\substack{x,y:nn\\ x,y\in\Lambda_2}} J_{xy}\, q_x\, q_y\Big] \prod_x e^{-Q_\delta(q_x)}\, dq_x$$

where $\underline{J} = \{J_{xy} \,|\, x,y \in \Lambda_2\ ,\ 0 \leq J_{xy}\}$ and $Z(\underline{J})$ normalizes $d\nu(\underline{J})$ to total weight 1.

Introducing $\underline{J}^{(1)} : J^{(1)}_{xy} = \begin{cases} 1 & \text{for } x,y \in \Lambda_1 \\ 0 & \text{otherwise} \end{cases}$

and $\underline{J}^{(2)} : J^{(2)}_{xy} = 1$ for all $x,y \in \Lambda_2$

we see that $d\nu^D_{\Lambda_i,\delta} = d\nu(\underline{J}^{(i)})$, $i = 1,2$.

Therefore the desired monotonicity property follows, if we prove that $\int q_{x_1}\cdots q_{x_n}\, d\nu(\underline{J})$ is monotonically increasing in all the J_{xy} or that for all $x,y \in \Lambda_2$: $\frac{\partial}{\partial J_{xy}} \int q_{x_1}\cdots q_{x_n}\, d\nu(\underline{J}) \geq 0$.

$$\begin{aligned} &\frac{\partial}{\partial J_{xy}} \int q_{x_1}\cdots q_{x_n}\, d\nu(\underline{J}) \\ &= \int q_{x_1}\cdots q_{x_n}(q_x q_y)\, d\nu(\underline{J}) - \frac{\partial Z(\underline{J})}{\partial J_{xy}} \cdot Z(\underline{J})^{-1} \cdot \int q_{x_1}\cdots q_{x_n} d\nu(\underline{J}) \\ &= \int q_{x_1}\cdots q_{x_n}(q_x q_y)\, d\nu(\underline{J}) - \int q_x q_y\, d\nu(\underline{J}) \cdot \int q_{x_1}\cdots q_{x_n}\, d\nu(\underline{J}) \end{aligned}$$

This last expression is non-negative by (Griffith II) which holds for $d\nu(\underline{J})$, too.

The uniform bound claimed in c) does not follow from correlation inequalities. It can be proven by applying reflection positivity (multiple reflection method) and the linear lower bound of section V. See e.g. [GJ]. Notice that this proof of the existence of the thermodynamic limit works only if P is an even polynomial, up to a possible linear term.

Besides Griffith I/II many other correlation inequalities can be proven for boson quantum field theory models. They are useful in the study of phase transitions and the approach to the critical point and other problems.

VII. Further Results

We conclude these lectures with a summary of major results proven for a large class of quantum field theory models, by various methods.

Let the dimension d of space time be 2 and $\mathcal{L}_I(q) = \lambda(q^{2n} + \sum_{k=0}^{2n-1} a_k q^k) = \lambda P(q)$. Then

1) For all $\mathcal{L}_I$ the infinite volume Schwinger functions exist. They satisfy all the properties Ⓔ with a possible exception of the cluster

property. Thus they define a unique Wightman theory, possibly with a non-unique vacuum vector [GRS,GJ]

2) For λ/m_0^2 sufficiently small

a) the cluster property holds, the vacuum is unique [GJS].

b) the joint spectrum of the energy-momentum operator (H,P) is as indicated in Fig. 1 or as in Fig. 2 [B,DE].

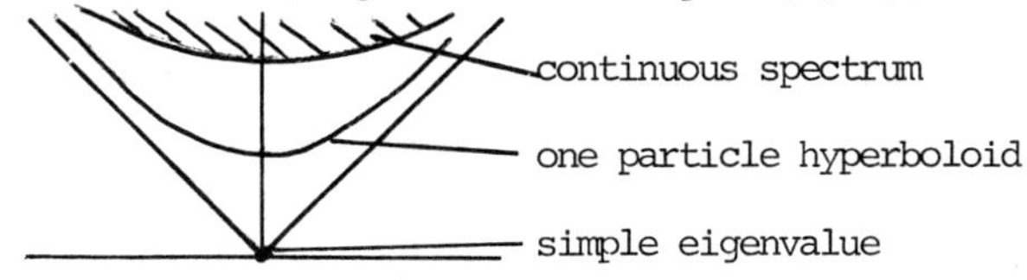

Fig. 1 (e.g. for $\mathcal{L}_I(q) = \lambda(q^4 + a_2 q^2)$

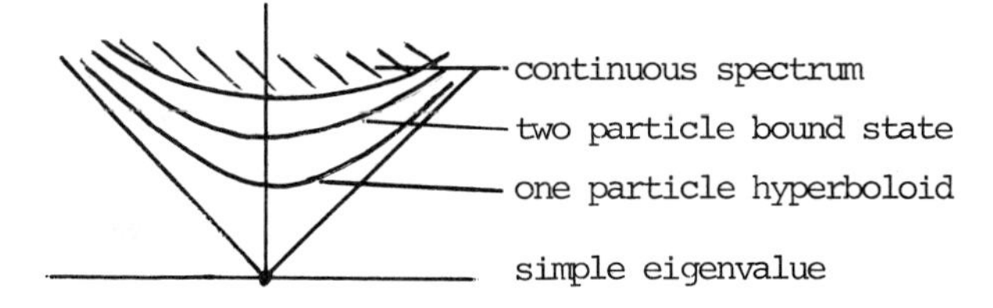

Fig. 2 (e.g. for $\mathcal{L}_I(q) = \lambda(q^6 - q^4)$; the mass of the bound state is:

$$m_b = 2\, m_0 \left(1 - \frac{9}{8}\left(\frac{\lambda}{m_0^2}\right)^2 + O(\lambda^3)\right)$$

c) There exist time-ordered products and the scattering matrix is not $\mathbb{1}$, i.e. the model describes nontrivial scattering [OSe,EEF]

d) The Schwinger functions, the Wightman distributions, the S-matrix elements and the masses of the particles described by the models have <u>asymptotic</u> power series expansions in powers of the parameter λ[D]. In particular for $\mathcal{L}_I(q) = \lambda(q^4 + \text{lower order})$ the power series for the Schwinger functions are <u>Borel summable</u>. This means that the Schwinger functions are uniquely determined by their Taylor coefficients (= Euclidean Feynman integrals). The same holds for the physical mass [EMS].

e) The relativistic field operator $\varphi(x)$, reconstructed from the Schwinger functions, satisfies a field equation

$$(\partial_t^2 - \partial_x^2 + m^2)\, \varphi(t,x) = -\lambda \vdots Q'(\varphi) \vdots$$

Here $: :$ means Wick ordering with respect to the physical vacuum and Q is a polynomial that agrees with P in leading order. The relation between P and Q is one to one [S].

3) Let $\mathcal{L}_I(q) = \sigma^{-2}(q^2 - \sigma^2)^2 + \mu q$, $m_0 = 1$. Then
   a) for $\mu \neq 0$ the vacuum is unique ($\mathcal{L}_I$ has a unique minimum).
   b) for $\mu = 0$ and σ sufficiently large there are two different vacua (phases). They correspond to the two minima of $\mathcal{L}_I(q)$ and they are interchanged under the transformation $q \longrightarrow -q$. This is called symmetry breaking and will be explained in A. Jaffe's contribution to this volume [GJS 2].

   There are choices of $\mathcal{L}_I(q)$ such that two different vacua coexist without a symmetry being broken [F2] or such that more than two vacua coexist [Ga,Su] .

Other models in d = 2 dimensions for which most of the above results have been established are the sine-Gordon model with $\mathcal{L}_I(q) = \lambda \sin(\varepsilon q)$, the Hoegh-Krohn model with $\mathcal{L}_I(q) = \lambda \exp \alpha q$ and the Abelian Higgs model.

In d = 3 dimensions the model $\mathcal{L}_I(q) = \lambda q^4$ has been studied extensively and most of the above results have been shown to hold.

In this model renormalization requires more than just Wick ordering. This is the reason for the greatly increased complexity of all the methods and constructions.

For references for these models we refer the reader to [GJ] where the most updated and most complete bibliography on constructive quantum field theory can be found.

References

a) Textbooks

[BLT] Bogoliubov, N.N., Logunov, A.A. and Todorov, R.T. (1975). Introduction to Axiomatic Quantum Field Theory (Benjamin)

[GJ] Glimm, J. and Jaffe, A. (1981). Quantum Physics (Springer-Verlag)

[J] Jost, R. (1965). The General Theory of Quantized Fields (American Math. Soc.)

[Ru] Ruelle, D. (1969). Statistical Mechanics (Benjamin)

[Si] Simon, B. (1974). The $P(\phi)_2$ Euclidean (Quantum) Field Theory (Princeton University Press)

[SW] Streater, R.F. and Wightman, A. (1964). PCT, Spin and Statistics, and all That (Benjamin)

[Erice] Velo, G. and Wightman, A.S. (1973). Constructive Quantum Field Theory, Proc. of the 1973 International School of Math. Physics in Erice, Springer Lecture Notes in Physics, Vol. 25

[GV] Gelfand, I. and Vilenkin, N. (1964). Generalized Functions, Vol. IV (engl. transl.), Academic Press

b) Research articles

[B] Burnap, C., Ann.Phys. 104, 184 (1977) and Harvard University Thesis (1966)

[DE] Dimock, J. and Eckmann, J.-P., Commun.Math.Phys. 51, 41 (1976)

[DF] Driessler, W. and Fröhlich, J., Ann.Inst. Henri Poincaré, 27, 221 (1977)

[EE] Eckmann, J.-P. and Epstein, H., Commun.Math.Phys. 64, 95 (1979)

[EEF] Eckmann, J.-P., Epstein, H. and Fröhlich, J., Ann.Inst. Henri Poincaré 25, 1 (1976)

[EMS] Eckmann, J.-P., Magnen, J. and Sénéor, R., Commun.Math.Phys. 39, 251 (1975)

[F1] Fröhlich, J., Helv.Phys.Acta 47, 265 (1974) and Adv.Math. 23, 119 (1977)

[F2] Fröhlich, J., Acta Phys. Austr. Suppl. 15, 133 (1976)

[FOS] Fröhlich, J., Osterwalder, K. and Seiler, E., Gauge theories without ghosts, to appear

[Ga] Gawedzki, K., Commun.Math.Phys. 59, 117 (1978)

[G] Glimm, J., Commun. Math.Phys. 8, 12 (1968)

[GJ1] Glimm, J. and Jaffe, A., Phys.Rev. 176, 1945 (1968), Ann.Math. 91, 362 (1970), Acta Math., 125, 204 (1970), J.Math.Phys. 13, 1568 (1972)

[GJS1] Glimm, J., Jaffe, A. and Spencer, T., in [Erice] and Ann.Math. 100, 585 (1974)

[GJS2] Glimm, J., Jaffe, A. and Spencer, T., Commun.Math.Phys. 45, 203 (1975), Ann.Phys. 101, 610 and 631 (1976)

[GRS] Guerra, F., Rosen, L. and Simon,B., Commun.Math.Phys. 27, 10 and 29, 233 (1973), 41, 19 (1975), Ann.Math. 101, 111 (1975)

[L] Lang, G., unpublished manuscript

[N] Nelson, E., contribution to [Erice] and J.Funct.Anal. 11, 211 (1972), 12, 97 and 211 (1973)

[OSe] Osterwalder, K. and Sénéor, R., Helv.Phys.Acta 49, 525 (1976)

[OS] Osterwalder, K. and Schrader, R., Commun.Math.Phys. 31, 83 (1973) and 42, 281 (1975). See also contribution to [Erice]

[S] Schrader, R., Fortschr. Physik 22, 611 (1974)

[Su] Summers, S., Ann.Inst. Henri Poincaré 34, 173 (1981) and Harvard University Press

K. OSTERWALDER
MATHEMATIK, ETH
CH-8092 ZÜRICH, SWITZERLAND

CONSTRUCTIVE QUANTUM FIELD THEORY : FERMIONS.

by

E. Seiler*
Max-Planck-Institut für Physik und Astrophysik
- München -

Lectures given at the International Summer School "Gauge Theories : Fundamental Interactions and Rigorous Results". Poiana Brasov, Romania, August/September 1981.

Contents :

*Present address : IHES., 35, route de Chartres, 91440 Bures-sur-Yvette (France)

1. Introduction.

In the standard euclidean approach to constructive Bose quantum field theory one constructs a probability measure for these fields ; its moments are the Schwinger functions (=euclidean Green's functions) of the theory. Obviously these moments are symmetric under the interchange of any pair of fields and thus conform to Bose statistics.

This makes already clear that fermions cannot be treated this way since their Schwinger functions have to be antisymmetric.

Matthews and Salam [1] proposed a formal way around this problem long before constructive field theory existed : They showed how one could formally "integrate out" the fermions and thereby produce an effective measure for the bosonic fields of the theory. The effective measure always contains a Fredholm determinant of an operator that does not really have such a determinant ; for this reason a large part of these lectures will be concerned with giving a precise meaning to these objects and discussing their properties.

A typical Matthews-Salam formula is the following one for the electron propagator in Quantum Electrodynamics (QED) :

$$\langle\psi(x)\overline{\psi}(y)\rangle \equiv G_F'(x,y) = \frac{1}{N}\int e^{-1/2\int F_{\mu\nu}^2}\, G_F'(x,y;A)\det\frac{\not\partial+iA+M}{\not\partial+M} \qquad (1.1)$$

where

$$G_F'(x,y;A) = (\not\partial+i\not A+M)^{-1}(x,y)$$

is the kernel of the resolvent of the covariant Dirac operator and N is a normalization constant.

A constructive approach could consist in giving a precise meaning to (1.1) and similar expressions and verifying the Osterwalder-Schrader axioms (cf. [2]) without worrying about a possible justification of (1.1). It turns out, however, that the easiest way to obtain some of these properties, in particular Osterwalder-Schrader (or reflection) positivity is to derive (1.1) (or a more precise version of it) from a more basic structure, either a lattice approximation which allows to give a meaning to the "integration" over fermions, or a field theoretic Hamiltonian.

I will discuss both approaches a little later.

Once the Matthews-Salam formulae are established, one more or less proceeds as in bosonic theories : Cutoffs that are still present, like ultraviolet cutoffs on the Bose fields and volume cutoffs have to be removed to obtain a euclidean quantum field theory. The program has been completed, as in the case of purely bosonic models, only in low dimension (two or at most three). This is again due to our insufficient understanding of nonperturbative renormalization. Nevertheless I will offer some glimpses of four dimensional theories that are supposed to describe aspects of the real world, such as quantum electrodynamics (QED_4) and quantum chromodynamics (QCD_4) . The approach taken here might be able to shed some light on certain intriguing aspects of these theories connected with the key words "instantons, CP violation, θ-vacua, U(1) problem".

2. Lattice Fermions.

Euclidean lattice fermions at first just seem to be some game invented by someone who loves to play around with formalisms. But amazingly they really do encode some genuine physics as should become clearer in the following lectures.

The basic idea is due to F.A. Berezin [3]. He considers a complex Grassmann algebra A generated by $\psi_1,\dots,\psi_N$ (i.e. the ψ_i are "anti-commuting variables") and defines on it a linear operation, frivolously called "integration", by the rules

$$\int 1 d\psi_i = 0 \ , \ \int \psi_i d\psi_k = \delta_{ik} \tag{2.1a}$$

$(i,k = 1,\dots,N)$

$$d\psi_i \, d\psi_k = -d\psi_k \, d\psi_i \tag{2.1b}$$

So if $f \in A$, $f = \Sigma a_{i_1 \dots i_k} \psi_{i_1} \dots \psi_{i_k}$ (we think of f as a "function of $\psi_1,\dots,\psi_N$"), forming $\int f \, d\psi_i$ simply eliminates ψ_i from all the terms that contain it, after multiplying them by -1 for each step required to move ψ_i to the right end ; terms not containing ψ_i are annihilated.

It is straightforward to see that

$$\int f \, d\psi_1 \ldots d\psi_N = a_{12\ldots N} \tag{2.2}$$

If A is an antisymmetric $N \times N$ matrix, $A = (\alpha_{ik})_{i,k=1}^{N}$, we denote

$$\sum_{i,k=1}^{N} \psi_i \, \psi_k \, \alpha_{ik} \quad \text{by} \quad (\psi, A\psi)$$

and compute

$$\int e^{(\psi,A\psi)} d\psi_1 \ldots d\psi_N = \Sigma \frac{1}{N!} 2^{-N} \varepsilon_{i_1 \ldots i_N} \alpha_{i_1 i_2} \ldots \alpha_{i_{N-1} i_N} \equiv \mathrm{Pf}(A) \tag{2.3}$$

($\mathrm{Pf}(A)$ is called the Pfaffian of A ; $\varepsilon_{i_1 \ldots i_N}$ is the sign of the permutation $i_1, \ldots, i_N$ of $1, \ldots, N$). Also

$$\int \psi_\ell \, \psi_m \, e^{(\psi,A\psi)} d\psi_1 \ldots d\psi_N = \mathrm{Pf}(A)(A^{-1})_{m\ell} \tag{2.4}$$

etcetera.

In most applications the generators come in two groups $\psi_1, \ldots, \psi_N$ and $\bar\psi_1, \ldots, \bar\psi_N$. If now $A = (\alpha_{ik})_{i,k=1}^{N}$ is any $N \times N$ matrix we abbreviate $\sum_{i,k=1}^{N} \bar\psi_i \, \alpha_{ik} \, \psi_k$ by $(\bar\psi, A\psi)$ and compute

$$\int e^{(\bar\psi,A\psi)} d\psi_1 \ldots d\psi_N \, d\bar\psi_1 \ldots d\bar\psi_N = \det A \tag{2.5}$$

$$\int \bar\psi_\ell \, \psi_m \, e^{(\bar\psi,A\psi)} d\psi_1 \ldots d\psi_N \, d\bar\psi_1 \ldots d\bar\psi_N = \det A \quad A^{-1}_{m\ell} \tag{2.6}$$

etc.

Note also the "change of variable formula" :

If

$$\psi_i = \sum_{k=1}^{N} B_{ik} \, \psi'_k$$

then

$$\int f(\psi_1, \ldots, \psi_N) \, d\psi_1 \ldots d\psi_N = \int \tilde f(\psi'_1, \ldots, \psi'_N) \det B^{-1} d\psi'_1 \ldots d\psi'_1$$

where (2.7)

$$\tilde f(\psi'_1, \ldots, \psi'_N) = f(\sum_k B_{1k}\psi_k, \ldots, \sum_k B_{Nk}\psi_k)$$

In short :

$$d\psi_1 \ldots d\psi_N = \det B^{-1} \, d\psi'_1 \ldots d\psi'_N \tag{2.8}$$

(2.5) and (2.6) suggest to define a "Gaussian" Grassmann integration as follows : Let

$$d\mu_C(\psi_1,\dots,\overline{\psi}_N) \equiv \det C \; e^{-(\overline{\psi},C^{-1}\psi)} d\psi_1\dots d\psi_N \;\; d\overline{\psi}_1\dots d\overline{\psi}_N \tag{2.9}$$

Then

$$\int \overline{\psi}_\ell \, \psi_m \, d\mu_C = C_{m\ell} \tag{2.10}$$

and similar expressions hold for higher moments, for instance

$$\int \overline{\psi}_{\ell_1} \, \psi_{m_1} \, \overline{\psi}_{\ell_2} \, \psi_{m_2} \, d\mu_C = C_{m_1\ell_1} \, C_{m_2\ell_2} - C_{m_2\ell_1} \, C_{m_1\ell_2} \tag{2.11}$$

The formulas (2.10),(2.11) have obvious generalizations to infinite dimensional Grassmann algebras : For instance, if $\mathcal{H}$ is a Hilbert space, C a bounded operator on $\mathcal{H}$,$f \to \psi(f)$ ($f \to \overline{\psi}(f)$) a linear (antilinear) map from $\mathcal{H}$ into the set of generators of a Grassmann algebra $\mathcal{A}$, we may define $d\mu_C$ by

$$\int e^{\psi(g)\,\overline{\psi}(f)} \, d\mu_C = \det(1+(g \otimes f)C) \tag{2.12}$$

($f,g \in \mathcal{H}$, $g \otimes f$ is the rank 1 operator $\varphi \to g(f,\varphi)$). (2.12) determines a linear map from $\mathcal{A}$ into $\mathbb{C}$ by expansion in f and g :

$$\int \prod_{i=1}^{N} (\psi(g_i) \, \overline{\psi}(f_i)) d\mu_C = \det((f_i,Cg_k))_{i,k=1}^{N} \tag{2.13}$$

It is not hard to obtain the infinite dimensional "integration" prescription (2.12), (2.13) as a limit from the finite dimensional one given before ; I leave the details as an exercise.

C is sometimes called the covariance operator in analogy with true Gaussian measures. There is a useful "integration by parts" formula:

Define a fermionic derivative by

$$\frac{\partial}{\partial\psi_i} \psi_1 \dots \psi_N = (-1)^{i+1} \psi_1 \dots \psi_{i-1} \, \psi_{i+1} \dots \psi_N$$

etc. by linearity.

If then u is an odd element of the Grassmann algebra we have

$$\int \psi_i \, u \, d\mu_C = \sum_k \int C_{ik} \frac{\partial u}{\partial \overline{\psi}_K} d\mu_C$$

We can now approach the definition of a free fermion field on the lattice. The naïve choice is the following :

Let Λ be a finite piece of the simple cubic lattice $a\mathbb{Z}^d$,

$$S_\Lambda \equiv \frac{1}{2} \sum_{<xy>} \overline{\psi}(x)\, \gamma_{xy}\, \psi(y)\, a^{d-1} + M \sum_{x\in} \overline{\psi}(x)\, \psi(x)\, a^d \tag{2.14}$$

where the first sum is over nearest neighbor pairs ("links") <xy> ;

$\gamma_{xy} = \gamma_\mu$ if <xy> points in the positive μ direction

$(\mu = 0,1,\ldots,d-1)$

$\gamma_{xy} = -\gamma_{yx}$

and $\{\gamma_\mu | \mu = 0,1,\ldots,d\}$ are the usual euclidean Dirac matrices that are hermitean and obey

$$\{\gamma_\mu, \gamma_\nu\} = 2\delta_{\mu\nu} \quad (\mu,\nu = 0,1,\ldots,d) \tag{2.15}$$

For traditional reasons we write γ_5 for the "extra" matrix γ_d. $\overline{\psi}(x)\gamma_{xy}\,\psi(y)$ is short for $\sum_{\alpha,\beta} \overline{\psi}_\alpha(x)(\gamma_{xy})_{\alpha\beta}\,\psi_\beta(y)$. 2.14 leads to the following "Gaussian measure"

$$d\mu_\Lambda \equiv \frac{1}{Z_\Lambda}\, e^{-S_\Lambda} \prod_{x\in\Lambda} d\psi_\alpha(x)\, d\overline{\psi}_\alpha(x) \tag{2.16}$$

with

$$Z_\Lambda = \int e^{-S_\Lambda} \prod_{\substack{x\in\Lambda \\ \alpha}} d\psi\,(x)\, d\overline{\psi}_\alpha(x)$$

It is easy to take the limit $\Lambda \nearrow a\mathbb{Z}^d$. We obtain the Gaussian measure

$d\mu_C$

characterized by the covariance operator

$$G_F = (M + \frac{1}{a} \sum_\mu \frac{1}{2}\, \gamma_\mu (T_\mu - T_\mu^*))^{-1} \quad \text{on } \ell^2(a\mathbb{Z}^d) \tag{2.17}$$

T_μ stands for the unitary translation by 1 lattice unit in the $+\mu$ direction. We often write e^{iaP_μ} for T_μ .

The problem with (2.16) is the following : T_μ has continuous spectrum in the unit circle. Therefore for $a \to 0$ C receives contributions both from $e^{iaP_\mu} \approx 1$ and $e^{iaP_\mu} \approx -1$: It describes 2^d fernionic particles instead of one. A more precise way of seeing this is the following : For $x_o - y_o > 0$

$$G_F(x,y) = \int (M+\frac{i}{a}\gamma_\mu \sin P_\mu a)^{-1} e^{iaP_\mu(x-y)_\mu} dP \tag{2.18}$$

can be written as a Laplace transform by closing the path of integration for P_o as follows :

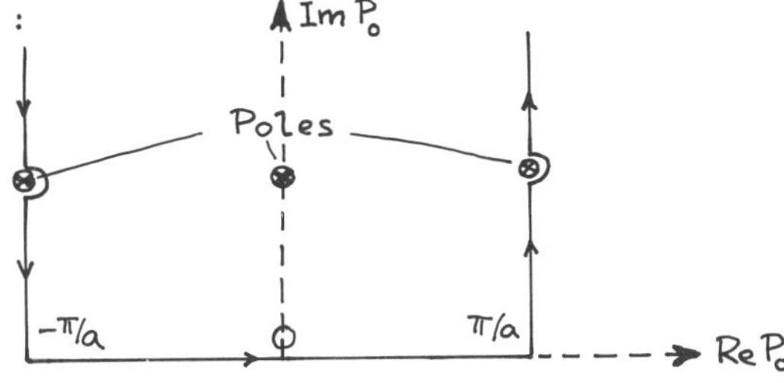

This picks up a residue at the poles in P_o determined by

$$\frac{1}{a^2}\sin^2(P_o a) = -\frac{1}{a^2}\sum_{i=1}^{d-1}\sin^2(aP_i) - M^2 \quad \text{or with } P_o = iE$$

$$\frac{1}{a^2}\text{sh}^2(aE) = \frac{1}{a^2}\sum_{i=1}^{d-1}\sin^2(aP_i) + M^2 \tag{2.19}$$

$$\text{ch } 2aE = 1 + a^2M^2 + 2\sum_{i=1}^{d-1}\sin^2(aP_i)$$

This means that for $a \to 0$ E stays finite for $P_i = O(a)$ or $P_i = \pm\frac{\pi}{a} \pm O(a)$.

In computing Feynman graphs with closed fermion loops this proliferation will show up by producing extra combinatorial factors.

Wilson [4] proposed the following device to avoid these problems : Replace the "action" S_Λ by

$$S_\Lambda^W \equiv \frac{1}{2}\sum_{\langle xy\rangle} \overline{\psi}(x)(1+\gamma_{xy})\,\psi(y)\,a^{d-1} + (M-\frac{d}{a})\sum \overline{\psi}(x)\psi(x)\,a^d \tag{2.20}$$

or equivalently G_F by

$$G_F^W \equiv (M + \frac{i}{a}\sum_\mu \gamma_\mu \sin aP_\mu - \frac{1}{a}\sum(1-\cos aP_\mu))^{-1} \tag{2.21}$$

Now even for $M = 0$ the action is no longer invariant under the "chiral transformations" $\psi \to e^{i\alpha\gamma_5}\psi$, $\overline{\psi} \to \overline{\psi}\, e^{i\alpha\gamma_5}$.

In [5] we proposed a slightly different modification ; but it is quite natural to consider a whole family of actions parametrized by a "chiral angle θ" :

$$S_\Lambda \equiv \frac{1}{2} \sum_{<xy>} \overline{\psi}(x)(e^{i\theta\gamma_5} + \gamma_{xy})\psi(y)a^{d-1} + \sum_x \overline{\psi}(x)(M - e^{i\theta\gamma_5}\frac{d}{a})\psi(x)a^d \qquad (2.22)$$

This corresponds to

$$G_F^\theta \equiv (M + \frac{i}{a}\Sigma\gamma_\mu \sin aP_\mu - \frac{1}{a}\sum_\mu (1-\cos aP_\mu)e^{i\theta\gamma_5})^{-1} \qquad (2.23)$$

Wilson's choice corresponds to $\theta = 0$, the one of ref. [5] to $\theta = \frac{\pi}{2}$; (2.20) was proposed in [6].

It is now easy to see that (2.20), (2.21) does not suffer from the proliferation problem and has the correct formal continuum limit, but at the prize of breaking chiral invariance. This will lead to interesting phenomena in the continuum limit such as anomalies and θ-vacua (cf. Chapter 10).

Let us now discuss our lattice free fermion fields in more details. Their most important property is physical (Osterwalder-Schrader) positivity. This means the following :

Assume that our finite lattice Λ splits into two sublattices Λ_+ and Λ_- corresponding to $x^o > 0$ or $x^o < 0$, respectively, and that Λ is symmetric with respect to reflection r in the hyperplane $t = 0$.

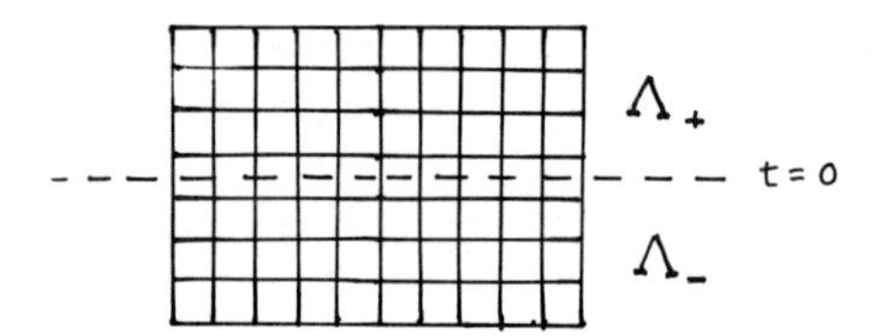

Accordingly we define subalgebras A_+ and A_- of our Grassmann algebra generated by $\psi(x), \overline{\psi}(x)$ with $x \in \Lambda_+$ (Λ_-) and a map ϑ (time reflection) from A_+ to A_- by

$$\vartheta\psi(x) = \overline{\psi}(rx)\gamma_o \qquad (2.24a)$$

$$\vartheta\overline{\psi}(x) = \gamma_o \psi(rx) \qquad (2.24b)$$

and the requirement that ϑ be an antilinear anti-isomorphism from A_+ to A_- .

We then have

Theorem 2.1 : If $f \in A_+$,

$$\int f \vartheta f \, d\mu_{C^0} \geq 0 \tag{2.25}$$

Proof : Let P be the convex cone generated by elements of the form $f \vartheta f$ $(f \in A_+)$ with f either even or odd. Clearly for $F \in P$

$$\int F \, d\psi_1 \ldots d\bar{\psi}_N \geq 0 \tag{2.26}$$

Note that P is closed under multiplication, so if $F \in P$ e^F is also $\in P$.

Now split $S^\theta_\Lambda = S^\theta_+ + S^\theta_- + S^\theta_c$ where $S^\theta_+ \in A_+$, $S^\theta_- \in A_-$ and S_c contains the terms coupling across the $t = 0$ hyperplane. Clearly $e^{-S^\theta_+ - S^\theta_-} = e^{-S^\theta_+} \vartheta e^{-S^\theta_+} \in P$.

We are almost done if we can show that $-S^\theta_c \in P$ and this will be implied by

$$-\bar{\psi}(rx) \frac{1}{2} (e^{i\theta\gamma_5} + \gamma_o)\psi(x) \in P$$

This is true because it can be written as

$$- \psi(x) \, e^{-i\frac{\theta}{2}\gamma_5} \frac{1+\gamma_o}{2} e^{i\frac{\theta}{2}\gamma_5} \psi(x) = - (P_\theta\psi(x))(P_\theta\psi(x))$$

$$= (P_\theta\psi(x)) \vartheta P_\theta\psi(x)$$

where $P_\theta = e^{-i\frac{\theta}{2}\gamma_5} \frac{1+\gamma_o}{2} e^{i\frac{\theta}{2}\gamma_5}$ is a projection.

Therefore (2.25) is proven provided f is either even or odd. But this restriction is obviously irrelevant since there are no even-odd cross terms. □

This remarkable positivity shows that there is actually already a "physical" Hilbert space behind this simple structure : Clearly we have a positive semidefinite scalar product on A_+ . If we denote by N the set of elements $f \in A_+$ obeying $\int f \vartheta f = 0$ then $H \equiv \overline{A_{+}/N}$ is a Hilbert space.

If we take the infinite volume limit (at least in time direction), it is easy to see [5,6] that there is a "transfer matrix" T that shifts by two lattice units in the positive time direction and obeys

$$0 \leq T \leq 1 \tag{2.27}$$

We often write

$$T = e^{-2aH} \tag{2.28}$$

and interprete H as the Hamiltonian (this requires, strictly speaking, that $T > 0$ which has been shown by Lüscher [7] in a slightly different setting ; otherwise we might just banish the null space of T from our Hilbert space). The next step is to couple Bose fields to our lattice fermions and derive lattice versions of the Matthews-Salam formulae.

Let us first consider a Yukawa coupling by adding a term

$$S_\Lambda^Y \equiv \sum_x \bar{\psi}(x)\, \Gamma\psi(x)\, \phi(x) \tag{2.29}$$

($\Gamma = 1$ or $i\gamma_5$) to the action ; ϕ is either an external field or a lattice random field with a probability distribution determined by its own action.

We then obtain

$$\int e^{-S_\Lambda^Y - S_\Lambda^\theta} d\psi_1 \ldots d\bar{\psi}_N / \int e^{-S_\Lambda^\theta} d\psi_1 \ldots d\bar{\psi}_N = \int e^{-S_{Y,\Lambda}} d\mu_C$$

$$= \det(1+K(\phi)) \tag{2.30}$$

where

$$K(\phi) = G_F\, \Gamma\, \phi \tag{2.31}$$

and similarly

$$\int \psi(x)\, \bar{\psi}(y)\, e^{-S_{Y,\Lambda}} d\mu_C = G_F^1(x,y;\phi) \det(1+K(\phi)) \tag{2.32}$$

where (2.33)

$$G_F'(x,y;\phi) = (G_F^{-1} + \Gamma\phi)^{-1}(x,y) = ((1+K(\phi))^{-1} G_F)(x,y)$$

(we suppressed the label θ).

Coupling to gauge fields proceeds in a similar way : A lattice gauge field is a map from links (=nearest neighbor pairs) <xy> into a compact group g :

$$<xy> \to g_{xy}$$

obeying

$$g_{xy} = g_{yx}^{-1}$$

Furthermore, we have to pick a unitary representation U of G and assume that the Grassmann algebra A is now generated by

$$\{\psi_{\alpha a}(x) \, , \, \bar{\psi}_{\alpha a}(x)\}_{x,\alpha,a}$$

where x runs through the lattice, α runs through a basis in "spin space" and $\psi_{\alpha a}(x)$ for fixed α, x runs through a basis of the representation space of U ; similarly $\bar{\psi}_{\alpha a}$ runs through a basis of the representation space of $\bar{U}$, the conjugate representation to U. Denoting $\sum_{\substack{\alpha,a\\ \beta,b}} \bar{\psi}_{\alpha a}(x) \, U(g_{xy})_{ab} \, \Gamma_{\alpha\beta} \, \psi_{\beta b}(y)$ by $\bar{\psi}(x) \, U(g_{xy}) \, \Gamma \, \psi(y)$

we define now the minimally coupled action for the fermions by

$$\begin{aligned} S_\Lambda^\theta(g) = {} & \frac{1}{2} \sum_{<xy>} \bar{\psi}(x)(e^{i\theta\gamma_5} + \gamma_{xy}) U(g_{xy}) \psi(y) a^{d-1} \\ & + M \sum a^d \, \bar{\psi}(x)\psi(x) \; - \frac{d}{a} \sum a^d \, \bar{\psi}(x) e^{i\theta\gamma_5} \psi(x) \end{aligned} \tag{2.34}$$

and compute

$$\int e^{-S_\Lambda^\theta(g)} d\psi_1 \ldots d\bar{\psi}_N / \int e^{-S_\Lambda^\theta(\mathbb{1})} d\psi_1 \ldots d\bar{\psi}_N = \det(1+K(U)) \tag{2.35}$$

where

$$K(U)(x,y) = \sum_{x'} G_F(x,x')(e^{i\theta\gamma_5} + \gamma_{xy})(U(g_{x'y}) - 1)\frac{1}{a} \tag{2.36}$$

Similarly

$$\frac{\int e^{-S_\Lambda^\theta(g)} \psi(x) \, \bar{\psi}(y) \; d\psi_1 \ldots d\bar{\psi}_N}{e^{-S_\Lambda^\theta(g)} \; d\psi_1 \ldots d\bar{\psi}_N} = (1+K)^{-1} G_F \tag{2.37}$$

These are examples of lattice Matthews-Salam formulae.

It should be kept in mind that the lattice gauge field $\{g_{xy}\}$

often is thought of as arising from a continuum gauge field A_μ by the prescription

$$g_{xy} = P \exp i \int_{<xy>} A_\mu \, dx^\mu$$

where the path ordered integral is over the link from x to y . A_μ is in the end to be considered as a random gauge field ("random connection") to be integrated with its own measure, or sometimes as an external field.

We note an important inequality :

Theorem 2.2. ("Diamagnetic bound") : $\det (1+K(U))| \leq 1$

The proof exploits physical positivity in the form of iterated Schwarz inequalities [8] .

The name "diamagnetic bound" is used in quotation marks because it does not express diamagnetism, but rather a paramagnetic property of the fermions ; it looks, however, much like the diamagnetic bound for bosons [8].

3. Digression : Lattice Gauge Fields.

We touched upon lattice gauge fields already in the previous section. Here I just want to give a general idea what they are and what they are good for. F. Guerra will have a lot more to say about them in his lectures [9] . The concept of lattice gauge fields is due to Wegner [10] and Wilson [11] .

A lattice gauge field is a map from the links $<xy>$ of the lattice into the gauge group G (assumed to be compact)

$$<xy> \rightarrow g_{xy} \tag{3.1}$$

such that

$$g_{xy} = g_{yx}^{-1} \tag{3.2}$$

As a lattice replacement of the Yang-Mills action Wilson [11] proposed

$$- S = \frac{1}{g^2} \sum_P \mathrm{tr} \prod_{<xy> \in \partial P} U(g_{xy}) \tag{3.3}$$

where the sum is over the elementary loops ("plaquettes") P of the lattice and U is a locally faithful representation of G . (3.3) generalizes the prescription given by Wegner [10] for $G = Z_2$. $-S$ is obviously biggest when $\prod_{<xy> \in \partial P} g_{xy} \equiv g_{\partial P} = \mathbb{1}$ for all P ; it is not hard to see (and well known) that S has the standard continuum limit, at least for lattice gauge fields induced by smooth continuum fields (where G has to be a Lie group, of course) : If a is the lattice constant, we have

$$\mathrm{tr}\, U(g_{\partial P}) = \mathrm{tr}\, U(e^X) \tag{3.4}$$

with

$$X = a^2 F_P + O(a^3) \in \mathcal{G} \tag{3.5}$$

where $\mathcal{G}$ is the Lie algebra of G and F_P is the Yang-Mills fields strength (curvature) : If P is a plaquette in the $\mu\nu$ directions with positive orientation ,

$$F_P = F_{\mu\nu} = \partial_\mu A_\nu - \partial_\nu A_\mu + \frac{1}{2} [A_\mu, A_\nu].$$

From (3.4) and (3.5) we obtain

$$\mathrm{Re}\, \mathrm{tr}\, U(g_{\partial P}) = \mathrm{tr}\, \mathbb{1} - \frac{a^4}{2} \mathrm{tr}\, U(F_P)^2 + O(a^5) \tag{3.6}$$

where $U(F_P)$ is the matrix representing $F_P \in \mathcal{G}$ in the representation of $\mathcal{G}$ induced by U .

(3.1) leads to the measure $e^{-S} \prod_{<xy>} dg_{xy}$ where dg is the Haar measure on G .

(3.3) is by no means the only possible lattice gauge action ; many others have been proposed [12] that all have the property that the fields take values in the compact space G and the measure favors $g_{\partial P} = \mathbb{1}$.

The good thing about lattice gauge theories is that they allow to analyse and even compute physical properties in a nonperturbative way. For instance all of these models have convergent strong coupling expansions that allow to prove a property related to the "confinement" of quarks [4,5] . If G is discrete, in dimension 3 and higher there is always a low temperature phase without "confinement" (see [6] , for instance). The most interesting is of course a compact non-abelian Lie group G (for instance SU(2) or SU(3) . If the dimension is four (or less), it is believed that confinement in this sense holds for all couplings (for the continuum limit it is necessary to go to weak coupling) ; nobody, however, has even come close to finding a proof of this.

But there are the remarkable and encouraging numerical studies of Creutz [13] which indicate that this is likely to be true.

What is not so clear, however, is to what extent this effect is produced by making the field space compact when going on the lattice. This is certainly a drastic step, even though it has a fairly natural look. It should be noted, however, that the existence of continuum limits of lattice theories in which the fields are compact so far could only be shown in the case of the two-dimensional Ising model [14]. It is possible to set up a lattice approximation that leaves the fields in the Lie algebra just like their continuum counterparts : this approximation seems to be closer to the continuum theory even though it has a somewhat ugly look. A worrisome fact is that such models do not show confinement in Wilson's sense [15] in four dimensions. The same difference between compact and noncompact lattice versions can be seen more rigorously in the three-dimensional U(1) model (QED_3) [16] .

This is not the place to enter into a detailed discussion of these matters that are not sufficiently well understood at present ; I just want to point out that there are interesting and important problems to be solved.

So let us return to our main theme : Fermions and determinants.

4. Modified Determinants and Renormalized Determinants.

In this section we have to prepare some functional analysis that is needed for the continuum limit of the Matthews-Salam formulae see [17,18]).

We are dealing with operators on a separable Hilbert space $\mathcal{H}$.

If C is a trace class operator , $z \in \mathbb{C}$

$$\det(1 - zC)$$

is the well known Fredholm determinant that may be defined for instance by

$$\det(1 - zC) = \sum_{n=o}^{\infty} (-z)^n \operatorname{tr} \Lambda^n(C) \tag{4.1}$$

Here $\Lambda^n(C)$ is the operator induced by C on the n-fold antisymmetric tensor product H ,i.e.

$$C(\varphi_1 \wedge \ldots \wedge \varphi_n) \equiv C\varphi_1 \wedge \ldots \wedge C\varphi_n$$

It is easy to see that

$$\operatorname{tr} \Lambda^n(C) \,| \leq \frac{1}{n!} (\|C\|_1)^n$$

and hence

$$|\det(1 - zC)| \leq \exp |z| \, \|C\|_1 \tag{4.2}$$

($\|C\|_1$ is the trace norm).

The operators occurring in the Matthews-Salam formulae will typically not be trace class but a suitable power will be. This motivates the consideration of the following spaces of compact operators : For $q \geq 1$

$$I_q \equiv \{\text{compact operators } C \mid \; |C|^q \equiv (C^*C)^{q/2} \text{ trace class}\}$$

with the norms

$$\|C\|_q \equiv (\operatorname{tr} |C|^q)^{\frac{1}{q}}$$

There is a natural definition of modified determinants adapted to these spaces :

<u>Definition</u> 4.1 : Let $C \in I_q$. Then

$$\det_q(1 - C) \equiv \det(1 - A_q(C))$$

where

$$A_q(C) = 1 - (1-C) \exp \sum_{k=1}^{q-1} \frac{1}{k} C^k$$

Remark : It is easy to see that $A_q(C)$ is in I_1 and hence this definition makes sense.

Theorem 4.2 : For $C \in I_q$ $\det_q(1 - zC)$ is an entire function of z ; its zeros are exactly the inverse eigenvalues of C .

Proof : See [17] .

To see what this definition means in the field theoretic context, let us look at QED_4 . There we would like to define $\det(1 + K(A))$ where $K(A) = G_F \gamma_\mu A_\mu$ (cf. eq.(2.36)). It will be easier if we consider instead

$$K(A) = (-\Delta + m^2)^{-3/4} (i\not\partial + m)\not A(-\Delta+m^2)^{-1/4} \tag{4.3}$$

which formally would give the same determinant. As long as A_μ is a decent field (for instance in $\bigcap_{p\geq 1} L^p$) it is readily seen that $K(A) \in I_q$ for any $q > 4$, so $\det_5(1 + K(A))$ makes sense. To see what we have done, look at the expansion (convergent for "small" A)

$$\log \det_5(1 - K(A)) = - \sum_{n\geq 5} \frac{1}{n} \operatorname{Tr} K^n$$

$\operatorname{Tr} K(A)^n$ corresponds to the Feynman graph

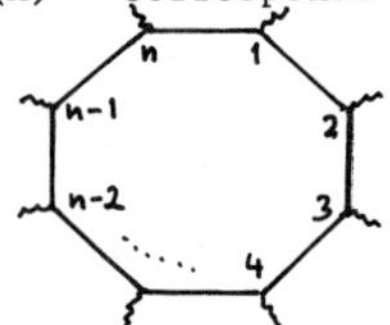

So by replacing det with $\det_5$ we have deleted the vacuum polarization graphs with 4 or less corners. This is of course not legitimate; they have to be restored in properly renormalized form.

Let us denote by $\operatorname{Tr}_{ren} K(A)^n$ the renormalized version of $\operatorname{Tr} K(A)^n$. We then define

Definition 4.3 : $\det_{ren}(1 + K) = \det_q(1 + K) \exp \sum_{n=1}^{q-1} \operatorname{Tr}_{ren}(-K)^n$

(for $K \in I_q$).

Remark : Of course $\det_{ren}$ is not unique ; it contains the usual nonuniqueness of renormalization. In QED_4 and QCD_4 the nonuniqueness consists only of a scale parameter related to field strength renormalization (see Section 7).

For the discussion of the continuum limit of lattice determinants we need

Theorem 4.4 : $\det_q(1-C)$ is a Lipschitz continuous function of $C \in \mathcal{I}_q$; similarly

$$\Lambda^k(C)\,\det_q(1-C)$$

is Lipschitz continuous on $\mathcal{I}_q$ if we use the uniform topology (given by the operator norm) in the image space.

For a proof we refer to [18].

5. The Continuum Limit.

We study the continuum limits of the Matthews-Salam formulae for given Bose fields with sufficient regularity. This means that the Bose fields are either external or are random fields with a cutoff. We state the results for gauge fields ; the Yukawa case is completely analogous. For simplicity we will also set $\theta = 0$.

Then the lattice Dirac operator is of the form (cf. Section 2)

$$\not{\partial}^a + R^a$$

where

$$\not{\partial}^a = \frac{i}{a}\sum_\mu \gamma_\mu \sin a P_\mu \tag{5.1}$$

$$R^a = \frac{1}{a}\sum_\mu (\cos a P_\mu - 1) \tag{5.2}$$

and P_μ is related to the lattice translation T_μ by $e^{iaP_\mu} = T_\mu$.

If we define

$$A^a_\mu(x) = \frac{1}{ia}(U(g_{x,x+ae_\mu}) - \mathbb{1}) \tag{5.3}$$

(e_μ the unit vector in the $+\mu$ direction)

we can introduce a covariant lattice Dirac operator $\not{D}^a_A + R^a_A$ where

$$D^a_A = \not{D}^a + i\gamma_\mu A^a_\mu \cos aP_\mu \qquad (5.4)$$

$$R^a_A = R^a + iA^a_\mu \sin aP_\mu \qquad (5.5)$$

A simple convergence result is

Theorem 5.1 : If A^a converges to the continuum gauge field A (with compact support) in L^∞ then $K(A)$ converges to $K(A)$ in any I_p, $p > d$. Here $K(A)$ is given by (4.3) and $K^a(A^a)$ is its lattice analogue that was defined in eq.(2.36). With our present notation and for $\theta = 0$

$$K^a(A^a) = (C^a_{\mathbb{1}})^{3/4}(-\not{D}^a+R^a+M)(\not{D}^a_A-\not{D}^a+R^a_A-R^a)(C^a_{\mathbb{1}})^{1/4} \qquad (5.6)$$

with

$$C^a_{\mathbb{1}} = (\not{D}^a+R^a+M)(-\not{D}^a+R^a+M))^{-1}$$

$$\equiv (-\Delta^a+(R^a+M)^2)^{-1} \qquad (5.7)$$

$$\Delta^a = (\not{D}^a)^2$$

Remark : In order to make sense of the statement one has to imbed the lattice Hilbert space (essentially $\ell^2(a\mathbb{Z}^d)$) into the continuum Hilbert space ; this is done in the obvious way of spreading the lattice values over a lattice cell (square, cube or hypercube) to obtain a piecewise constant L^2 function. This imbedding is clearly isometric.

Proof of Theorem 5.1 : First one shows that $K^a(A^a)$ is uniformly bounded in I_p $(p>d)$. This is seen easily by using the fact that an operator A of the form $A = f(p)g(x)$ (i.e. multiplication in x-space by g followed by multiplication in p-space by f) obeys

$$\|A\|_q \le \|f\|_q \, \|g\|_q \, C \quad (q \ge 2)$$

(see [19]).

Similarly I_q convergence of a sequence $A_n = f_n(p)g_n(x)$ is implied by L^q convergence of f_n and $g_n (q \ge 2)$.

Because we assumed only L^∞ convergence of A^a_μ , this has to be supplemented by

<u>Lemma</u> 5.2 : Let A_n converge to A in I_p, (B_n) be a uniformly bounded sequence of operators such that $B_n \to B$, $B_n^* \to B^*$ strongly. Then $A_n B_n \to AB$ in I_p. (See [20] for a proof).

For convergence of the low order terms one has to resort to the estimation of lattice Feynman graphs which is a pain but does not pose any fundamental problems ; this has only been carried out completely for simple cases like QED_2.

So the following two statements are not yet theorems (but more obvious truths than conjectures) :

<u>Statement</u> 5.3 : For QCD_4, QED_4, there is a function $g(a)$ such that

$$\det_2(1 + K^a(A^a))\, e^{-\frac{1}{g(a)^2}\int \|F\|^2}$$

converges to a nonzero limit as $a \to 0$. Here $\|F\|^2 = \|dA + \frac{1}{2}[A,A]\|^2$ is the classical action of the gauge field.

The limit is

$$\det_{ren}(1 + K(A))$$

<u>Statement</u> 5.4 : For QCD_3, QED_3

$$\lim_{a\to 0} \det(1 + K^a(A^a)) = \det_{ren}(1 + K(A))$$

A^a is derived from a smooth continuum gauge field A.

The following is, however, proven [6,20,21] :

<u>Theorem</u> 5.5 : Let $d=2$. Then $\lim_{a\to 0} \det(1 + K^a(A^a))$ exists and is equal to $\det_{ren}(1 + K^a(A^a)) \neq 0$, provided $A^a \to A$ in the norm

$$\|A\|_{\infty,\alpha} \equiv \|A\|_\infty + \int_{\Lambda\times\Lambda} dxdy\, \frac{\|A(x)-A(y)\|^2}{|x-y|^{2+\alpha}}$$

for some $\alpha > 0$. Λ is the support of A^a, A.

Proof : The stronger norm is needed only for the 2nd order vacuum polarization. Note that

$$\|A\|_{\infty,\alpha} = \|A\|_{\infty} + c_\alpha \int |\hat{A}(k)|^2 |k|^\alpha d^2k$$

$$\mathrm{Tr}\ K^a(A^a)^2 = \int d^2k\ \hat{A}^a_\mu(k)\hat{A}^a_\nu(k)\Pi^a_{\mu\nu}(k)$$

where $\Pi^a_{\mu\nu}$ is given by the second order vacuum polarization graphs [graph] + [graph]. The stronger norm is needed because in the continuum limit $\Pi_{\mu\nu}(k) = O(\log |k|)$ for large k .

Finally let us discuss convergence of Schwinger functions in external gauge fields. The relevant result is

Theorem 5.6 : Let $A^a \to A$ in such a way that the renormalized determinant converges to a nonzero limit. Then

$$G'^a_F(x,y;A) \equiv [(C^a_{1\!\!1})^{\frac{1}{4}}(1 + K^a(A^a))^{-1}(C^a_{1\!\!1})^{\frac{3}{4}}(-\not{D}^a+M)]\ (x,y) \qquad (5.8)$$

converges as $a \to 0$ in the sense of operator norms (interpreting (5.8) as kernels of integral operators).

Proof : By Lipschitz continuity of $(1 + K)^{-1}\ \det_p(1 + K)$ in $K \in \mathcal{I}_p$ and Theorem 5.1 together with the assumed convergence of the low order terms (properly renormalized) the proof becomes trivial □

6. Hamiltonian Approach

The Matthews-Salam formulae can also be derived using relativistic fermion fields, along the lines given in [22] (which, however, contains an error).

Let $\psi(\vec{x})$, $\bar{\psi}(\vec{x})$ be relativistic time 0 Dirac fields ; $\psi_\kappa(\vec{x}) = (h_\kappa * \psi)(\vec{x})$ a cutoff version, where for instance h_κ may be chosen to be

$$h_\kappa(\vec{x}) = \frac{1}{\pi^{\frac{d-1}{2}}}\ \frac{1}{\kappa^{\frac{d-1}{2}}}\ e^{-\frac{\kappa\vec{x}^2}{2}}$$

H_o is the free Dirac Hamiltonian in second quantized form.

At first we will couple these fields to an external time dependent field ; for simplicity we discuss the case of a scalar Yukawa coupling:

Let $\phi(\vec{x},t)$ be a continuous real valued function of compact support in $\vec{x}$ for all t. Consider the time dependent Hamiltonian

$$H_\kappa(t) \equiv H_o - V_\kappa(t) \tag{6.1}$$

$$V_\kappa(t) \equiv \int : \overline{\psi}_\kappa(\vec{x})\ \psi_\kappa(\vec{x}) : \phi(\vec{x},t) d\vec{x} \tag{6.2}$$

The differential equation

$$\dot{U}(t) = - H(t)\ U(t) \tag{6.3}$$

is solved by the product integral

$$U(t) \equiv T \exp(- \int_o^t H(t)dt) \tag{6.4}$$

which may be rigorously defined by its power series expansion

$$U(t) = \sum_{n=o}^{\infty} (-1)^n \int_{0 < s_1 < \ldots < s_n < t} ds_1 \ldots ds_n\ e^{-(t-s_n)H_o}\ V(s_n)\ e^{-(s_n - s_{n-1})H_o}\ V(s_{n-1}) \ldots V(s_1)\ e^{-s_1 H_o} \tag{6.5}$$

Norm convergence of (6.5) is assured since V is bounded.

Matthews-Salam formulae are now obtained by taking expectation values in states of the Dirac field. For instance, if we take the vacuum Ω_o, we obtain

$$(\Omega_o, U(t)\Omega_o) = \det{}_2(1-K_\kappa(\phi)) \tag{6.6}$$

where

$$K_\kappa(\phi) = G_\kappa(\vec{x}-\vec{y},t-t')\ \phi(y,t') \quad \text{and} \quad G_\kappa(\vec{x},t) = h_\kappa * G * h_\kappa$$

where G is the euclidean Green's function of the free Dirac field.

Similarly, if we take the matrix elements of 1-fermion states $\psi(f_i)\Omega_o$ we obtain

$$(\psi(f_1)\Omega_o, U(t)\psi(f_2)\Omega_o) = \det{}_2(1-K_\kappa(\phi)) \int G_\kappa(\vec{x},\vec{y};t,t';\phi)\ f_1(\vec{x})\ f_2(\vec{y})\ d\vec{x}\ d\vec{y} \tag{6.7}$$

where $G_\kappa(\vec{x},\vec{y};t,t';\phi)$ is the cutoff euclidean Green's function in the presence of the external field ϕ.

We now promote ϕ to a random field with continuous sample paths and a time translation invariant probability distribution that obeys a Markov property with respect to t=const hyperplanes ; for instance ϕ may be taken to be the free field with a cutoff in space and in the spatial moments. We denote the corresponding probability measure by $d\mu(\phi)$. By the Feynman-Kac-Nelson formula for ϕ [23] we see that for instance

$$(\Omega_o, e^{-H_\kappa t} \Omega_o) = \int \det_2(1+K_\kappa(\phi))\, d\mu(\phi) \tag{6.8}$$

where Ω_o is now the bare vacuum for bosons and fermions,

$$H_\kappa = H_{o,\mathrm{fermion}} + H_{o,\mathrm{Boson}} + V_\kappa .$$

(The existence of the integral on the right hand side of eq. (6.8) requires a little argument ; one can for instance introduce a cutoff on the size of ϕ and use the inequality

$$\det_2(1+K_\kappa(\phi)) \leq \exp \frac{1}{2} \|K_\kappa(\phi)\|_2 \leq \exp(\mathrm{const} \;\; \|\phi\|_\infty^2) .$$

Removal of this cutoff can be considered part of the general stability problem - see below.

Alternatively one can split $K_\kappa = L_\kappa + M_\kappa$ with $L_\kappa \in I_1$, $\|M_\kappa\|_2 < \varepsilon$. Using some determinant inequalities (cf.[18] one obtains $|\det_2(1+K_\kappa(\phi))| \leq \exp 2\, \|L_\kappa\|_1 \;\; \exp \frac{1}{2} \|M_\kappa\|_2^2$ which implies integrability. See also [25]).

Removal of cutoffs can be discussed as in other approaches such as the lattice (see Chapters 8,9); it is in fact a little easier. For gauge fields the most convenient way to prove analogous formulas uses the "temporal gauge" $A_o = 0$.

The advantage of this "semi-euclidean" approach (cf.[24]) is that physical positivity is always manifest.

7. Relation to other definitions of determinants.

In this section I will try to clarify the relation of the renormalized determinants defined above to currently popular other determinant defintions that are based on heat kernels and ζ-functions of elliptic operators.

These alternative definitions are useful for the discussion of certain properties.

The first point is to realize that for most purposes it suffices to consider only positive operators since (formally)

$$|\det(1+K(A))|^2 = \frac{\det(\not{D}+\not{A}+M)^*(\not{D}+\not{A}+M)}{\det(\not{D}+M)^*(\not{D}+M)} \tag{7.1}$$

The first simple fact is then

Theorem 7.1 : Let $A > 0$, $A+B > 0, \varepsilon > 0$, $A^{-\frac{1}{2}}BA^{-\frac{1}{2}} \in I_1$. Then

$$\det(1+(A+\varepsilon)^{-\frac{1}{2}}B(A+\varepsilon)^{-\frac{1}{2}}) = \exp\{-\int_0^\infty \frac{dt}{t} \operatorname{Tr}(e^{-t(A+B)}-e^{-tA})e^{-t\varepsilon}\} \tag{7.2}$$

Proof : First note that both sides make sense : For the left hand side this is obvious by our assumption; for the right hand side note that

$$e^{-t(A+B)}-e^{-tA} = -\int_0^t e^{-(t-s)(A+B)}Be^{-sA}ds \tag{7.3}$$

(This is the so-called Duhamel formula). Iterating (7.3) we obtain the expansion

$$e^{-t(A+B)}-e^{-tA} = \sum_{n=1}^{\infty} (-1)^n \int_{\substack{0\le s_i \\ \Sigma s_i \le t}} ds_1\dots ds_n$$

$$e^{-(t-s_1-\dots-s_n)A}Be^{-s_1A}\dots Be^{-s_nA} \tag{7.4}$$

which is easily seen to be norm convergent [22].

(6.4) also converges in the trace norm :

$$\text{Let}\quad K \equiv \|A^{-\frac{1}{2}} B A^{-\frac{1}{2}}\|_1$$

$$C(t) \equiv \sup_{t \geq t' \geq 0} \sqrt{t'}\, e^{-t'} \leq \sqrt{t}$$

Then the n-th term on the right hand side of (7.4) is bounded by

$$C^n K^n \int_{\substack{s_i \geq 0 \\ \Sigma s_i \leq t}} ds_1 \ldots ds_n (s_1 \ldots s_n (t-s_1 \ldots -s_n))^{-\frac{1}{2}} \tag{7.5}$$

Now note that

$$\int_0^t ds(s(t-s))^{-\frac{1}{2}} = \pi$$

Carrying out the integration over s_n in (7.5) produces therefore

$$C^n K^n \pi \int_{\substack{s_i \geq 0 \\ \Sigma s_i \leq t}} ds_1 \ldots ds_{n-1} = \pi (CK)^n \frac{t^{n-1}}{(n-1)!}$$

and hence

$$\begin{aligned} \|e^{-t(A+B)} - e^{-tA}\|_1 &\leq \pi CK\, e^{tCK} \\ &\leq \pi t^{\frac{1}{2}} K e^{t^{\frac{3}{2}} K} \end{aligned} \tag{7.6}$$

This is a poor bound for large t. But for $t \geq 1$ we can telescope as follows :

$$\begin{aligned} e^{-t(A+B)} - e^{-tA} &= e^{-(t-1)(A+B)} (e^{-A-B} - e^{-A}) \\ &+ (e^{-(t-1)(A+B)} - e^{-(t-1)A}) e^{-A} \end{aligned} \tag{7.7}$$

Iterating this we find easily

$$\|e^{-t(A+B)} - e^{-tA}\|_1 \leq \text{const } t \quad \text{for} \quad t \geq 1 \tag{7.8}$$

These estimates show that the right hand side of (7.2) is an absolutely convergent integral.

To see equality of both sides of (7.2) we take the logarithmic derivative with respect to ε of both sides. The left hand side yields

$$-\frac{1}{2}\operatorname{Tr}(1+C_\varepsilon)^{-1}((A+\varepsilon)^{-1}C_\varepsilon+C_\varepsilon(A+\varepsilon)^{-1}) = -\operatorname{Tr}(A+\varepsilon)^{-\frac{1}{2}}(1+C_\varepsilon)^{-1}C_\varepsilon(A+\varepsilon)^{-\frac{1}{2}}) \tag{7.9}$$

where

$$C_\varepsilon \equiv (A+\varepsilon)^{-\frac{1}{2}}B(A+\varepsilon)^{-\frac{1}{2}} \tag{7.10}$$

(here we used the rule

$$\frac{d}{ds}\log\det(1+C(s)) = \operatorname{Tr}[(1+C(s))^{-1}C'(s)]$$

A little computation shows that

$$(A+\varepsilon)^{-\frac{1}{2}}(1+C_\varepsilon)^{-1}C_\varepsilon(A+\varepsilon)^{-\frac{1}{2}} = (A+\varepsilon)^{-1}-(A+B+\varepsilon)^{-1} \tag{7.11}$$

so the logarithmic derivatives of both sides of (7.2) are equal to

$$\operatorname{Tr}((A+B+\varepsilon)^{-1}-(A+\varepsilon)^{-1})$$

Since both sides of (7.2) have the same limit as $\varepsilon \to \infty$ (namely 1), (7.2) is proven. □

An easy generalization of Theorem 7.1 is

<u>Theorem 7.2</u>. Let $A > 0$, $A+B > 0$, $\varepsilon > 0$, $A^{-\frac{1}{2}}BA^{-\frac{1}{2}} \in I_p$. Then

$$\log\det_p(1+(A+\varepsilon)^{-\frac{1}{2}}B(A+\varepsilon)^{-\frac{1}{2}}) = (-1)^p\int_0^\infty \frac{dt}{t}\int_{\substack{0\le s_i\\ \Sigma s_i\le t}} ds_1\ldots ds_p\, e^{-t\varepsilon}$$

$$\operatorname{Tr} e^{-(t-s_1-\ldots-s_p)(A+B)}Be^{-s_1A}\ldots.Be^{-s_pA} \tag{7.11}$$

<u>Remark</u> : Of course one can take the limit $\varepsilon \to 0$ in (7.2) and (7.11).

Proof of the Theorem 7.2. Iterating the Duhamel formula 7.3 p times and use of the definition of $\det_p$ proves (7.11) for $A^{-1/2}BA^{-1/2} \in I_1$

The left hand side is continuous in I_p by Theorem 4.4 so the right hand side is also. By arguments similar to the ones used in the proof of Theorem 7.1 it can be seen that I_p limits of the right hand side or (7.11) are still of the same form, so (7.11) follows by approximating $A^{-1/2}BA^{-1/2}$ by I_1 operators. □

Remark : A formula similar to (7.2) can be found already in Schwinger's work [20]; (7.2) is often used as a definition in the literature, but I think it is important to have equality with the standard Fredholm definition.

The next goal is to obtain closed expressions for renormalized determinants. We first give the definition of the analytically renormalized determinant which is identical to the so-called ζ-function definition of Seeley [27]. Afterwards we show that under certain conditions this coincides with the definition given earlier.

We now specialize to the kind of operators occurring in field theory. So $A+B > 0$ and $A > 0$ will now be 2nd order elliptic operators on $\mathbb{R}^d$ (or sometimes another d-dimensional manifold); we assume that $A+B$ and A have the same leading symbol and that $A^{-1/2}BA^{-1/2} \in I_q$ for some q. Then it is well known from the De-Witt-Seeley expansion [27,28,29,30] or from perturbation theory using Feynman graphs that we have for the pth order term in (7.4).

Property 7.3 :

$$a_p(t) \equiv (-1)^p \int_{\substack{0 \le s_i \\ \Sigma s_i \le t}} ds_1 \ldots ds_p \operatorname{Tr} e^{-(t-s_1-\ldots-s_p)A} Be^{-s_1A} \ldots Be^{-s_pA} =$$

$$= c(B)t^{-r(p)} + O(t^{-r(p)+1/2})$$

where $r(p) < \frac{d}{2}$, $p \geq q$, $r \in \mathbb{Z}$, and $r(p)$ decreases with p.

Examples :

$$(1) \quad A = -\Delta \ , \ B = V : a_p(t) = (-1)^p c_p(V) t^{p-\frac{d}{2}}) \quad \text{(see below)} \tag{7.12}$$

(2) $d = 4$:

$$A \equiv H_F(0) = \not{D}^* \not{D} = -\Delta$$

$$B+A = H_F(A) = (\not{D}+i\not{A})^*(\not{D}+i\not{A}) = -\Delta_A + \frac{1}{2}\Sigma_{\mu\nu}F_{\mu\nu} \qquad (\Sigma_{\mu\nu} = \frac{i}{2}[\gamma^\mu,\gamma^\nu])$$

Here

$$a_p(t) = t^{p-\frac{d}{2}} c_p(A) + O(t^{p-\frac{d}{2}+1}) \tag{7.13}$$

$$c_1(A) = 0 \tag{7.14}$$

$$c_2(A) = \frac{1}{3}||F||^2 \tag{7.15}$$

(see for instance [30] ; actually it can be seen that $c_p(A) = 0$ for all odd p) .

Property 7.3 allows to prove the following

Lemma 7.4 : Assume that Property 7.3 holds and $A^{-1/2}BA^{-1/2} \in I_q$. Then

$$\log \det{}^{(\alpha)}(1+(A+\varepsilon)^{-1/2}B(A+\varepsilon)^{-1/2}) \equiv$$

$$-\int_0^\infty \frac{dt}{t^{1-\alpha}} \operatorname{Tr}(e^{-t(A+B)}-e^{-tA})e^{-t\varepsilon} = -\operatorname{Tr}((A+B+\varepsilon)^{-\alpha}-(A+\varepsilon)^{-\alpha})\Gamma(\alpha)$$

is a meromorphic function of α in the whole complex plane; it has simple poles at $\alpha = -r(p)$.

Proof : Clearly $\log \det^{(\alpha)}$ is holomorphic in α for $\operatorname{Re}\alpha > q-1$ (this follows from the expansion (7.4)). By extracting the first p terms of the expansion (7.4) we obtain holomorphy of the remainder for $\operatorname{Re}\alpha > q-p$. The extracted terms produce simple poles because

$$\int_0^\infty dt\, t^{\alpha-1} t^{-r(p)} e^{-\varepsilon t} = \Gamma(\alpha-r(p))\varepsilon^{-\alpha+r(p)} \quad .$$

Remark : With the usual definition $\zeta(\alpha;A) = \operatorname{Tr} A^{-\alpha}$ we can write

$$\log \det{}^{(\alpha)}(1+(A+\varepsilon)^{-1/2} B(A+\varepsilon)^{-1/2} =$$

$$= \Gamma(\alpha)(\zeta(\alpha;A+\varepsilon) - \zeta(\alpha;A+B+\varepsilon))$$

whenever the right hand side makes sense (for instance if we are on a compact manifold).

Definition 7.5 : The analytically renormalized determinant is defined as

$$\log \det_{AR}(1+(A+\varepsilon)^{-1/2}B(A+\varepsilon)^{-1/2}) \equiv$$

$$\lim_{\alpha\to o} (\frac{d}{d\alpha} + \log \mu^2)\alpha \log \det^{(\alpha)}(1+(A+\varepsilon)^{-1/2}B(A+\varepsilon)^{-1/2})$$

Remark : Whenever the ζ function exist, this may also be written as

$$(\zeta'(\alpha;A+\varepsilon)-\zeta'(\alpha;A+B+\varepsilon))\Big|_{\alpha=o} +$$

$$\log \mu^2 \lim_{\alpha\to o} (\zeta(\alpha;A+\varepsilon)-\zeta(\alpha;A+B+\varepsilon)) .$$

Def. 7.5 makes sense provided Property 7.3 holds and $A^{-1/2}BA^{-1/2} \in I_q$ vor some $q > 1$. The indeterminacy contained in the $\log \mu^2$ corresponds to a choice of scale (or field strength normalization).

Lemma 7.6 :

$$\log \det_{AR}(1+(A+\varepsilon)^{-1/2}B(A+\varepsilon)^{-1/2}) =$$

$$= \frac{1}{r(1)!} \int_o^\infty dt \log(t\mu^2)(\frac{d}{dt})^{r(1)+1} \mathrm{Tr}(e^{-(A+B)t}-e^{-At})t^{r(1)}e^{-\varepsilon t}$$

Proof : We integrate by parts $r(1)+1$ times the formula for $\log \det^{(\alpha)}$ assuming $\mathrm{Re}\,\alpha > r(1))$ and obtain

$$\log \det^{(\alpha)}(1+(A+\varepsilon)^{-1/2}B(A+\varepsilon)^{-1/2})$$

$$= - \int_o^\infty dt\ t^{\alpha-1-r(1)}(t^{r(1)}\mathrm{Tr}(e^{-t(A+B+\varepsilon)}-e^{-t(A+\varepsilon)}))$$

$$= (-1)^{r(1)} \frac{1}{r(1)-\alpha} \frac{1}{r(1)-\alpha-1} \cdots \frac{1}{-\alpha}$$

$$\int_o^\infty dt\ t^\alpha (\frac{d}{dt})^{r(1)+1}(t^{r(1)}\mathrm{Tr}\ e^{-t(A+B+\varepsilon)}-e^{-t(A+\varepsilon)}))$$

which makes sense for $\mathrm{Re}\,\alpha > -1$.

Applying Def. 7.5 we see that

$$\log \det_{AR}(1+(A+\varepsilon)^{-1/2}B(A+\varepsilon)^{-1/2}) =$$

$$= \frac{1}{r(1)!}(-1)^{r(1)}\int_o^\infty dt \log t \left(\frac{d}{dt}\right)^{r(1)+1}$$

$$\{t^{r(1)}\mathrm{Tr}(e^{-(A+B+\varepsilon)t}-e^{-(A+\varepsilon)t})\}$$

$$+ \log \mu^2 \frac{1}{(r(1))!}(-1)^{r(1)} \int_o^\infty dt\left(\frac{d}{dt}\right)^{r(1)+1}$$

$$\{t^{r(1)}\mathrm{Tr}(e^{-(A+B+\varepsilon)t}-e^{-(A+\varepsilon)t})\}$$

which proves the lemma. □

To make contact with our earlier definition we have to assume again that the operators $A+B$ and A are elliptic differential operators of the kind encountered in field theory. It then turns out that Def. 7.5 just amounts to using analytic dimensional renormalization for the divergent low order terms

$$\mathrm{Tr}\ K^p \quad (K = (A+\varepsilon)^{-1/2}B(A+\varepsilon)^{-1/2}) \ , \quad p \leq \sup_{r(q)\geq o} q$$

This is best seen in the examples introduced earlier :

(1) $A = -\Delta \ , \ B = V$

(2) $A = -\Delta \ , \ A+B = -\Delta_A + 1/2\Sigma_{\mu\nu}F_{\mu\nu}$.

For example (1) we obtain for instance

$\mathrm{Tr}_{AR}K = 0$

$\mathrm{Tr}_{AR}K^2$ is obtained by taking the regular part at $\alpha = 0$ of

$\int_o^\infty a_2(t)t^{\alpha-1}dt$ where

$$a_2(t) = \int_o^t ds\ \mathrm{Tr}\ e^{(t-s)\Delta}Ve^{s\Delta}V \tag{7.16}$$

A computation gives

$$a_2(t) = \int_o^t ds\ e^{-p^2 t - \frac{k^2}{4} t + (p\cdot k)(t-2s)} e^{-\varepsilon t} |\hat{V}(k)|^2 d^4k$$

$$= \pi^2 \int_o^1 dx\ e^{-tk^2 x(1-x) - \varepsilon t} |\hat{V}(k)|^2 d^4k \tag{7.17}$$

and

$$\int_o^\infty a_2(t) t^{\alpha-1} dt = \pi^2 \frac{1}{\Gamma(\alpha)} \iint_o^1 dx |\hat{V}(k)|^2 \ (k^2 x(1-x) - \varepsilon)^{-\alpha} d^4k \tag{7.18}$$

which gives the well known analytically renormalized value of the graph

V ⋯ V

if we apply the evaluator $\lim\limits_{\alpha\to o} (\frac{d}{d\alpha} + \log \mu^2)\alpha$ to it.

For example (2) it is convenient to use directly the De Witt-Seeley expansion of the integrand

$e^{-t(A+B)}$ which gives [30]

$$\mathrm{Tr}(e^{t(\Delta_A + 1/2\Sigma\cdot F)} - e^{t\Delta}) = \int 1/3 \sum_{\mu,\nu} F_{\mu\nu}^2 + O(t) \tag{7.19}$$

which shows that

$$\log \det_{ren}(1+K) = \lim_{\delta\to o}\{-\int_\delta^\infty \frac{dt}{t} \mathrm{Tr}(e^{-t(A+B)} - e^{-tA}) e^{-\varepsilon t}$$

$$- 1/3 \log(\delta\mu^2) \int F^2\} \tag{7.20}$$

so that we are dealing with a definition that is equivalent to the usual one using counterterms.

This connection between Def. 7.5 and the usual definition involving counterterms is more general. One can see that the counterterms have the form of the residues of the poles of $\det^{(\alpha)}$ to the right of 0, or, equivalently, of the coefficients of the terms in an asymptotic expansion (de Witt-Seeley expansion) of

$$\mathrm{Tr}(e^{-t(A+B)} - e^{-tA})$$

that have a power t^{-r} $(r \geq 0)$:

The reason is quite simple : Consider the typical term

$$\int_o^\infty dt\ t^{\alpha-1-r} e^{-\varepsilon t} = \varepsilon^{r-\alpha}\Gamma(\alpha-r)$$

It has poles at $\alpha = r-q$, $q = 0,1,\ldots,r$ with residues $\frac{\varepsilon^q}{q!}(-1)^q$.

On the other hand

$$\int_\delta^\infty dt\ t^{-1-r} e^{-\varepsilon t}$$

has an expansion in powers of δ near $\delta = 0$ of the form

$$\sum_{q=o}^{r-1} \delta^{-r}(\delta\varepsilon)^q c_q + \log(\delta\mu^2)\varepsilon^r c_r$$
$$+ \text{ regular terms.} \tag{7.21}$$

The singular terms can be used as "counterterms" and we have

$$\lim_{\delta\to o}\Big(-\sum_{q=o}^{r-1} \delta^{-r}(\delta\varepsilon)^q c_q - \varepsilon^r c_r \log \delta\mu^2 + \int_\delta^\infty dt\ t^{\alpha-1-r} e^{-\varepsilon t}\Big) =$$

$$= \lim_{\alpha\to o}\Big(\varepsilon^{r-\alpha}\Gamma(\alpha-r) - \sum_{q=o}^{r} \frac{\varepsilon^q}{q!}(-1)^q \frac{1}{\alpha-r+q}\Big)$$

It is instructive to compute the counterterm in 2nd order for example (1) : It is determined by the residue at $\alpha = 0$ of

$$\int_o^\infty a_2(t)\, t^{\alpha-1} dt$$

and a simple computation gives for it

$$\pi^2 \int V(x)^2 dx \tag{7.22}$$

So the counterterm has the usual local structure.

<u>Remark</u> : Our Def. 7.5 depends only on one parameter μ^2 . In dimension $d > 4$ it is legitimate, however, to add finite counterterms of the form of the residues of $\det^{(\alpha)}$ at $\alpha = 0,1,\ldots,r(1)$.

So we have seen that the various definitions of renormalized determinants coincide, wherever they are simultaneously applicable. What could not be clarified so far is the status of renormalized determinants of massless fermions in external long range fields such as instantons (cf. [31]). The ζ-function definition gives an answer, but it is not clear how it ties in with more conventional renormalization. The problem there is that ε has to be sent to zero, but $A^{-1/2}BA^{-1/2}$ is not even a compact operator [32], and it is not known whether the t-integral converges at large t. If it doesn't, the ζ-function definition involves both ultraviolet and infrared counterterms, and the latter are hard to interprete physically.

Because the question has been raised, I want to conclude this chapter with a remark about the "diamagnetic" inequality in the continuum (cf. Theorem 2.2 : The fact is that it <u>fails</u> in dimension $d \geq 4$.

This can be seen by a simple scaling argument. Let us consider $d = 4$ for definiteness and let

$$A_\mu^{(\lambda)}(x) \equiv \lambda A_\mu(\lambda x) .$$

We consider

$$G(M,\mu,\lambda) \equiv \log|\det_{AR}(1+K(A))|^2$$

$$= \int_0^\infty dt \log(t\mu^2) \frac{d}{dt} \mathrm{Tr}(e^{-tH_F(A^{(\lambda)})} e^{-tH_F(0)}) \tag{7.23}$$

By unitary scaling or just dimensional analysis it is easy to see that

$$G(M,\mu,\lambda) = G(\frac{M}{\lambda}, \frac{\mu}{\lambda}, 1) \tag{7.24}$$

On the other hand

$$G(M,\mu,\lambda) = G(M,\mu_o,\lambda) - 1/3 \, ||F||_2^2 \log(\frac{\mu}{\mu_o})^2 \tag{7.25}$$

Combining (7.24) and (7.25) we get

$$G(M,\mu,\lambda) = G(\frac{M}{\lambda},\mu,1) + 1/3 \, ||F||_2^2 \log \lambda^2 \tag{7.26}$$

Sending $\lambda \to \infty$ we see that $G(M,\mu,\lambda) \to \infty$ if we note that for an external field A_μ of sufficiently fast decay at ∞ (for instance $A_\mu \in \bigcap_{p\geq 1} L^p$) we can send the fermion mass to zero.

The "diamagnetic" inequality, on the other hand, would require $G \leq 0$. This failure of the lattice inequality in the continuum is clearly due to the infinite counterterm that was required. The same effect leads also to the failure of the diamagnetic bound of Schrader and R. Seiler [33] in dimension $d \geq 4$ (cf. also [54]).

8. Stability in Finite Volume.

A crucial and usually the hardest part in the construction of a (euclidean) quantum field theory is always the proof that the partition function Z_Λ with the proper counterterms for a finite volume Λ has a limit when the ultraviolet cutoff is removed and that $|\log Z|$ is bounded linearly in the volume. Similar bounds for the "unnormalized" Schwinger functions $Z_\Lambda S_\Lambda$ can normally be obtained without much extra work. These bounds correspond to showing a lower bound for the Hamiltonian that is linear in the volume.

I first sketch a proof that for a fixed volume $0 < Z < \infty$ for two-dimensional models like QED_2 or Y_2 .

The partition functions are defined as follows : We introduce an ultraviolet cutoff κ into the Bose field (A or ϕ) . A space time cutoff may also be introduced by modifying this measure (for instance by imposing 0-Dirichlet boundary conditions); other possibilities are the use of antiperiodic boundary conditions for the fermions, or multiplication of the Bose fields by a characteristic function χ_Λ of a region Λ . We also introduce a coupling constant g multiplying K .

As a vacuum energy counterterm we may simply use

$$E_\kappa \equiv \frac{g^2}{2} \int \operatorname{Tr} K_\kappa^2 \, d\mu_\Lambda$$

where $d\mu_\Lambda$ is the Gaussian measure for the free Bose field with appropriate boundary conditions.

Stability in a finite volume then means

$$0 < \int \det_{ren}(1+gK_\kappa) e^{E_\kappa} d\mu_\Lambda \leq C < \infty \tag{8.1}$$

where C is independent of κ .

We note the following facts :

(1) $K_\kappa \in \mathcal{I}_2$ a.s.

(2) $E_\kappa = O(\log \kappa)$

(3) $\int e^{pE_\kappa} |\det_{ren}(1+gK_\kappa)|^p d\mu_\Lambda \leq \exp O(\log \kappa)$ for all $p \geq 1$

(1) and (2) are elementary (using Feynman graphs); for QED_2 (3) follows from (2) and the "diamagnetic inequality" (Theorem 2.2)

$$|\det_{ren}(1+gK_\kappa(A))| \leq 1 \tag{8.2}$$

For Y_2 it is easiest to use the bound

$$|\det_2(1+gK_\kappa(\phi))| \leq \exp \frac{g^2}{2} \|K_\kappa\|_2^2 \tag{8.3}$$

which implies

$$|\det_{ren}(1+gK_\kappa)e^{E_\kappa}| \leq \exp \frac{g^2}{2} (\|K_\kappa\|_2^2 + \mathrm{Tr}_{ren} K_\kappa^2) \tag{8.4}$$

For large fermion mass (8.4) implies (3) directly; otherwise one has to throw in some determinant inequalities (see [18]).

The crucial estimate is now contained in

<u>Theorem 8.1</u> : Let $\kappa' \geq \kappa$. Then there is a $p \geq 1$ such that

$$|\log(e^{E_\kappa}\det_{ren}(1+gK_\kappa)) - \log(e^{E_{\kappa'}}\det_{ren}(1+gK_{\kappa'}))|$$

is bounded by a "polynomial" in the Bose field of degree p that has an L^2 norm which is $O(\kappa^{-1/4})$.

An immediate consequence is

<u>Cor. 8.2</u> : (8.1) holds for large enough fermion mass.

<u>Proof</u> of Cor. 8.2 :

Let $u_\ell = \log \det_{ren}(1+gK_{\kappa_\ell})+E_{\kappa_\ell}$

$\kappa_\ell = \ell\mu \quad (\ell = 0,1,2,3,\ldots)$

$u_o = -\infty$

Then

$$\int e^{u_N} d\mu_\Lambda = \sum_{k=1}^{N} \int (e^{u_k} - e^{u_{k-1}}) d\mu_\Lambda$$

$$\leq \sum_{k=1}^{N} \int |u_k - u_{k-1}| e^{u_k + u_{k-1}} d\mu_\Lambda$$

$$\leq \sum_{k=1}^{N} ||u_k - u_{k-1}||_2 \, e^{c \frac{g^2}{M^2} \log k\mu}$$

$$\leq \sum_{k=1}^{N} (k\mu)^{-\frac{1}{4}} (k\mu)^{c \frac{g^2}{M^2}} < C < \infty$$

provided $c \frac{g^2}{M^2} < \frac{1}{4}$

Proof of the theorem (for QED_2) : Consider

$\log \det_4(1+K_\kappa)$. Let

$K(s) \equiv sK_\kappa + (1-s)K_{\kappa'}$. Then

$$|\log \det_4(1+K_\kappa) - \log \det_4(1+K_{\kappa'})| =$$

$$= |\int_o^1 ds \operatorname{Tr} K(s)^3 (1+K(s))^{-1} K'(s)|$$

$$= |\int_o^1 ds \operatorname{Tr} K(s)^2 \not{A}_s G_F' K'(s)|$$

$$\leq \frac{1}{M} ||K(s)||_3 \, ||K(s)\not{A}_s||_3 \, ||K'(s)||_3$$

$$\leq \frac{1}{3M} (||K(s)||_3^3 + ||K(s)\not{A}||_3^3 + ||K'(s)||_3^3)$$

Now

$$\int ||K(s)||_3^6 d\mu_\Lambda = O(1)$$

$$\int ||K(s)\not{A}_s||_3^6 \, d\mu_\Lambda = O(\log \kappa)$$

$$||K'(s)||_3^6 \, d \quad = O(\kappa^{-1}) \ .$$

This suffices to prove the theorem, if we also note that

$$\int |\mathrm{Tr} : K_{\kappa}^{2} : - \mathrm{Tr} : K_{\kappa'}^{2} : |^{2} \, d\mu_{\Lambda} = O(\kappa^{-1})$$

where $\mathrm{Tr} : K_{\kappa}^{2} : = \mathrm{Tr} K_{\kappa}^{2} - \frac{2}{g^{2}} E_{\kappa}$. □

This very simple proof which combines elements of [34,35] works only for large fermion mass. But this restriction can be removed without too much trouble [36,18].

In higher dimensions a proof of this simple structure cannot work. However a more elaborate cutting up procedure (phase space cell expansion) that was invented for ϕ_3^4 [37] can be used to prove stability in Y_3 [38].

9. Thermodynamic Limit.

The thermodynamic limit for fermion theories can be analyzed with methods that are completely analogous to the ones known from bosonic theories : Correlation inequalities (see for instance [39]) and cluster expansions [40].

Since this school is dedicated to gauge theories we discuss a little bit how the thermodynamic limit for QED_2 can be obtained by the method of the cluster expansion, based on [6].

First we state

Theorem 9.1 [6,42] : $|\log Z_{\Lambda}| \leq \mathrm{const}|\Lambda|$.

Proof (Idea) : In a box of sides L and T with antiperiodic b.c. we have

$$Z_{\Lambda}^{ap} = \frac{\mathrm{Tr} e^{-LH_{T}}}{\mathrm{Tr} e^{-LH_{o,T}}} = \frac{\mathrm{Tr} e^{-TH_{L}}}{\mathrm{Tr} e^{-TH_{o,L}}} \tag{9.1}$$

To fully justify these formulas requires some work [41], but they are very plausible from the lattice approximation.

(9.1) implies

$$|Z_\Lambda^{ap}| \leq e^{\text{const } LT}$$

by the convexity properties of $\log \operatorname{Tr} e^{-LH_T}$ in L etc. and the known properties of $\operatorname{Tr} e^{-LH_{o,T}}$ etc. On the other hand, using free b.c. it can be seen, using O.S. positivity that

$$Z_\Lambda^f \geq e^{\text{const } LT}$$

Finally we have

$$Z_\Lambda^f \leq c_1 e^{c_2(L+T)} Z_{LT}^{ap}$$

essentially because the trace of a positive operator is bigger than any expectation value (see [41] for more details). □

The next step are volume independent bounds on generating functionals ("ϕröhlich bounds") :

Theorem 9.2 : Let B be a finite rank operator on the fermion Hilbert space

$$(\bar{\psi}, B\psi) = \sum_{i=1}^{N} \bar{\psi}(g_i)\psi(f_i) \text{ . Then}$$

$$| < e^{(\bar{\psi},B\psi)+F(h)} >_{LT} | \leq c_o \exp \tfrac{1}{2} ||h||_2^2$$

$$\times \exp a \sum_{i=1}^{N} ||g_i||_{2;1} ||f_i||_{2;1}$$

where $||g||_{2;1} = \sum_\Delta ||g\chi_\Delta||_2$

(the sum is over a paving of $\mathbb{R}^2$ by unit squares Δ) .

The proof of this theorem [6] relies on chessboard bounds, Theorem 9.1 and the determinant inequality

$$|\det_{ren}(1+K+G_F B)| = |\det_{ren}(1+K)\det(1+G_F' B)| =$$

$$= |\det_{ren}(1+K)\det(1+ \sum_{i=1}^{N} g_i G_F' f_i)| \leq |\det_{ren}(1+K)| \exp(\tfrac{1}{M}\Sigma ||f_i||_2 ||g_i||_2)$$

To arrive at the cluster expansion for QED_2 it is convenient to include the second order vacuum polarization in the Gaussian measure for the electromagnetic field :

$$\det_{ren}(1+gK)\,d\mu_W(F) = \det_4(1+gK)\,d\mu_G(F)\,Z_o$$

where $d\mu_W$ is the white noise measure (free measure for the electromagnetic field F in two dimensions)

$$d\mu_G(F) = \frac{1}{Z_o}\, e^{-\frac{g^2}{2}\,\mathrm{Tr}:K:^2}\, d\mu_W(F)$$

which means that $d\mu_G(F)$ has a covariance given by

$$\hat{G}(k) = \frac{k^2}{k^2 + \frac{g^2}{\pi}\, T(k^2)}$$

in momentum space where

$$T(k^2) = 1 - \frac{4M^2}{|k|\sqrt{4M^2+k^2}}\ \mathrm{ar\,th}\ \frac{|k|}{\sqrt{4M^2+k^2}}$$

G is a massive covariance with mass gap $2M$.

The point is now that this Gaussian measure will only be weakly perturbed by the rest of the interaction contained in $\det_4$, provided $g^2/M^2 \ll 1$.

The cluster expansion [43] is designed for such a situation : It first decouples different squares, for instance by the use of Dirichlet b.c. and then reinserts these couplings step by step. The procedure converges if the mass of the underlying Gaussian measure is large and the perturbation of it is weak.

Since I cannot explain this fundamental method in more detail here, I refer to the literature (for instance [43] or [6]).

Other interesting results have been obtained for the Y_2 model : I mention the proof of Borel summability due to Renouard [44] and the proof of the occurrence of a phase transition based on a low temperature cluster expansion due to Bałaban and Gawędzki [45].

10. Anomalies, θ-Vacua and Other Interesting Subjects Related to Determinants.

As the heading indicates in this chapter we discuss various facts and speculations that are not directly connected to the constructive program but are closer to current physical thought and refer mostly to four dimensions; their common feature is that they have to do with fermion determinants.

a) Anomalies.

We saw in the beginning that lattice fermions required some strange modification in their kinetic term that affected their chiral properties. W. Kerler [46] argued that these extra terms directly produce the so-called anomalies in the divergence of the axial current that are associated with the names Adler, Bell and Jackiw [47] (but actually can be traced back to Schwinger [26]). Kerler's arguments are not completely correct, but his result is and his proof can be corrected using the careful study of lattice fermion Feynman graphs of Karsten and Smit [48]. As a consequence one obtains a connection between the parameter θ of the "θ-vacua" and the θ occurring in the lattice fermion action eq. (2.22) (cf. [49]).

Kerler's argument proceeds as follows : The linear map

$$\psi(x) \to e^{i\frac{\alpha_x}{2}\gamma_5}\psi(x)\ ;\ \bar{\psi}(x) \to e^{i\frac{\alpha_x}{2}\gamma_5}\bar{\psi}(x) \tag{10.1}$$

has determinant one. So if we make a change of variables according to this prescription in the fermionic "integral" (or just use the multiplication law for determinants) we obtain

$$\begin{aligned} &\int e^{-S^\theta_\Lambda(\{\alpha_x\})}\, d\psi_1 \ldots d\bar{\psi}_N \\ &= \int e^{-S^\theta_\Lambda(\{0\})}\, d\psi_1 \ldots d\bar{\psi}_N \end{aligned} \tag{10.2}$$

(assuming for the moment that we are on a finite lattice). Here $S^\theta_\Lambda(\{\alpha_x\})$ is obtained from S^θ_Λ (2.22) applying the substitution (10.1).

Taking the logarithm of (10.2), differentiating with respect to α_x and setting $\{\alpha_x\} = \{0\}$ produces the following Ward identity :

$$<\partial^a_\mu J^a_{5\mu}(x)> + 2M<J^a_5(x)> + <X^a> = 0 \qquad (10.3)$$

Here $\partial^a J^a_{5\mu}(x) = \frac{1}{a}\sum_\mu (J_{5\mu}(x+ae_\mu) - J_{5\mu}(x))$

$$J^a_{5\mu} = \frac{1}{2} : \overline{\psi}(x) i\gamma_5\gamma_\mu U(g_{x,x+ae_\mu})\psi(x+ae_\mu) :$$

$$+ \frac{1}{2} : \overline{\psi}(x+ae_\mu) i\gamma_5\gamma_\mu U(g_{x+ae_\mu,x})\psi(x) : \qquad (10.4)$$

$$J^a_5 = : \overline{\psi}(x) i\gamma_5 \psi(x) : \qquad (10.5)$$

$$X^a(x) = \sum_{\varepsilon=\pm1} \sum_\mu \frac{1}{a} [:\overline{\psi}(x) e^{i\theta\gamma_5} i\gamma_5 U(g_{x,x+a\varepsilon e_\mu})$$

$$+ :\overline{\psi}(x+\varepsilon ae_\mu) e^{i\theta\gamma_5} i\gamma_5 U(g_{x+\varepsilon ae_\mu,x})\psi(x):] - \frac{2d}{a} :\overline{\psi}(x) i\gamma_5 e^{i\theta\gamma_5}\psi(x) :$$

The Wick dots simply stand for subtraction of the free field $(U(g) \equiv 1)$ expectation value. In (10.3) we may take the infinite volume limit without problems provided g_{xy} approaches $\mathbb{1}$ suffiently fast for $|x| \to \infty$.

The crucial fact is now

<u>Theorem 10.1</u> : If the lattice gauge field $\{g_{xy}\}$ arises from a smooth continuum gauge field A_μ with topological charge density $q(x)$, then

$$\lim_{a\to o} <X^a(x)> = 2iq(x) \qquad (10.7)$$

(The topological charge density is given by the first Chern class $\frac{1}{2\pi}$ tr $U(F_{o1}(x))$ in $d = 2$, and by the second Chern class

$$\frac{1}{32\pi^2} \varepsilon_{\mu\nu\rho\lambda} \text{ tr } U(F_{\mu\nu})U(F_{\rho\lambda}) \text{ in } d = 4) .$$

The proof of this theorem uses a power counting argument to conclude that only finite low orders in perturbation theory contribute. In the end one is left with an integral

$$I_2 = \int d^2p \frac{\sum_\mu(1-\cos p_\mu)(\prod_\nu \cos p_\nu \sum_\mu(1-\cos p_\mu)-2 \sin^2 p_o \cos p_1)}{(\sum_\mu \sin^2 p +(\sum_\mu (1-\cos p_\mu))^2)^2} \qquad (10.8$$

in $d = 2$ and

$$I_4 = \int d^4p \frac{\sum_\mu (1-\cos p_\mu)(\prod_\nu \cos p_\nu \sum_\mu (1-\cos p_\mu) - 4\sin^2 p_o \prod_{\nu\neq o} \cos p_\nu)}{(\sum_\mu \sin^2 p_\mu + (\sum_\mu (1-\cos p_\mu))^2)^3} \tag{10.9}$$

in $d = 4$.

These integrals occur also in the perturbative discussion of ref. [48]; these authors have a clever identity that allows to conclude that

$$I_1 = -\pi \quad , \quad I_4 = -\frac{\pi^2}{2} \tag{10.10}$$

(in ref. [46] the second terms in I_1 and I_2 are missing and it is incorrectly claimed that the remaining parts give the correct answer (10.10)). A detailed version of this proof can be found in [49].

□

b) θ-Vacua.

θ-vacua are conventionally defined by adding a term

$$i\theta q(x) \tag{10.13}$$

to the euclidean Lagrangean. Notice that this, if it makes sense, will not produce a positive measure any more.

Here we want to argue that the same effect can be obtained by introducing the angle θ in the fermion action. The reason for this is

Lemma 10.2 : If $Q \equiv \int q(x)dx$ is an integer,

$$\lim_{a\to o} \frac{d}{d\theta} \log \det(1+K_\theta^a(A^a))$$
$$= iQ$$

provided the field strength obeys

$$\sup_{x,\mu,\nu} |F_{\mu\nu}(x)|(1+|x|^4) < \infty .$$

Proof : If we introduce a x dependent angle θ_x in the action and form

$$\frac{\partial}{\partial \theta_x} \log \det(1+K^a(A^a))\Big|_{\{\theta_x\} = \{\theta\}}$$

we see as before that this equals

$$\frac{1}{2} \langle \partial^a_\mu J^a_{5\mu}(x)\rangle + \frac{1}{2} \langle X^a(x)\rangle$$

Summing over all x and letting $a \to 0$ proves the lemma provided

(i) $\sum_x \partial^a_\mu \langle J^a_{5\mu}(x)\rangle = 0$

(ii) $\lim_{a \to o} \sum_x \langle X^a(x)\rangle a^d = iQ$

(i) follows from sufficient fall-off of $J^a_{5\mu}$ (which can be inferred from the fall-off assumption on $F_{\mu\nu}$ and the integrality of the topological charge) by the divergence theorem. (ii) is obtained by the dominated convergence theorem and Theorem 10.1. For more details see [49].

□

Note that Lemma 10.2 cannot hold for noninteger Q ; what breaks down is the argument that the divergence term does not contribute. We formulate the obvious consequence of Lemma 10.2.

Cor. 10.3 : Under the assumptions of Lemma 10.2

$$\lim_{a \to o} \det(1+K^a_\theta(A^a) = e^{iQ} \lim_{a \to o} \det(1+K^a_o(A^a))$$

It can be shown that the determinant is real for $\theta = 0,\pi$ and positive for one of these values (we don't know for sure which one).

Cor. 10.3 allows to introduce the θ-vacua through the fermion lattice approximation which has the advantage that periodicity in θ becomes manifest and one does not have to worry about the definition of topological charge for lattice fields or "rough" continuum fields.

There is the interesting possibility of spontaneous breakdown of CP symmetry at $\theta = 0$ or π , something that has been observed in other models [50,51,8] but not proven for QED_2 or QCD_4 .

c) Instantons.

Our methods fail for $M = 0$ and an external instanton field : $K(A)$ ceases to be a compact operator; in fact it will have spectrum in a whole disc [32].

The ζ function method, however, as well as other methods (for instance 't Hooft's [31]) give a result. What is the meaning of it ? It would become clearer if we knew the answer to the following Question : Does

$$\lim_{M^2 \to o} \int_o^\infty dt \log(t\mu^2) \frac{d}{dt} \mathrm{Tr}(e^{-tH_F(A)} - e^{-tH_F(o)}) e^{-M^2 t}$$

exist for long range fields A or does it diverge (presumably to $-\infty$)? (Possibly the zero modes of $H_F(A)$ have to be subtracted out).

If the limit exists it should of course be equal to 't Hooft's result [31].

The answer to the question is important for the interpretation of the computations based on the so-called "instanton gas" and in general for semiclassical computations in gauge theories, as well as ideas on chiral symmetry breaking in QCD_4 .

d) (In)Stability of QED_4 ?

Schwinger [26] in a famous paper computed the full vacuum polarization induced by a constant electromagnetic field (and thereby rederived a result obtained by Euler and Heisenberg [52]).

In euclidean terms his computation amounts to the following [53]: Let the field strength tensor $F_{\mu\nu}$ have eigenvalues $\pm E$, $\pm B$ $(E,B > 0)$. Then

$$\mathrm{Tr}\, e^{-tH_F(A)}(x,x) = \mathrm{tr}\, e^{1/2\Sigma\cdot Ft}\, e^{t\Delta_A}(x,x)$$

$$= 2(\cosh(\surd 2t(E+B)) + \cosh(\surd 2t(E-B)))$$

$$\frac{EB}{(4\pi)^2 \sinh Et \sinh Bt}$$

$$= \frac{1}{4\pi^2} EB \coth Et \coth Bt \tag{10.14}$$

(the trace referred to the Dirac algebra).

Fudging the thermodynamic limit a little bit this gives (cf. Chapter 7) :

$$\lim_{|\Lambda|\to\infty} \frac{1}{|\Lambda|} \log \det{}_{AR}(1+eK(A))$$
$$= \frac{1}{4\pi^2}\int_o^\infty dt \log(t\mu^2) \frac{d}{dt} (EB \coth Et \coth Bt - t^{-2})e^{-M^2 t} \equiv G(E,B) \tag{10.15}$$

Introducing $F \equiv (E^2+B^2)^{1/2}$ and changing t to t/F we obtain

$$G(E,B) = \frac{F^2}{4\pi^2} \int_o^\infty dt \log \frac{t\mu^2}{F}$$
$$\times \frac{d}{dt} (\frac{EB}{F^2} \coth \frac{ET}{F} \coth \frac{Bt}{F} - t^{-2})e^{-\frac{M^2}{F}t} = \frac{F^2}{12\pi^2} \log \frac{F}{\mu^2}$$
$$+ \frac{F^2}{4\pi^2}\int_o^\infty dt \log t \frac{d}{dt} (\frac{EB}{F^2} \coth \frac{Et}{F} \coth \frac{Bt}{F} - t^{-2})e^{-\frac{M^2}{F}t} \tag{10.16}$$

We split the last integration into $\int_o^1$ and $\int_1^\infty$. The first term contributes $O(F^2)$; the second part gives after integration by parts

$$- \frac{F^2}{4\pi^2}\int_1^\infty \frac{dt}{t} (\frac{EB}{F^2} \coth \frac{Et}{F} \coth \frac{Bt}{F} - t^{-2})e^{-\frac{M^2}{F}t}$$
$$= - \frac{EB}{4\pi^2}\int_1^\infty \frac{dt}{t} e^{-\frac{M^2}{F}t} + O(F^2)$$
$$= - \frac{EB}{4\pi^2}\int_{M^2/F}^\infty \frac{ds}{s} e^{-s} + O(F^2)$$
$$= \frac{EB}{4\pi^2} \log \frac{M^2}{F} + O(F^2) \tag{10.17}$$

So we have

$$G(E,B) = \frac{F^2}{12\pi^2} \log F(1 - \frac{3EB}{F^2}) + O(F^2) \tag{10.18}$$

Now $G(E,B)$ can be interpreted as a 1-loop effective potential; (10.18) says that it grows more than quadratically whenever $\frac{EB}{F^2} < \frac{1}{3}$ which raises doubts whether $\det_{AR}(1+eK(A))$ can be integrable with any Gaussian measure.

Of course the argument is not very stringent; it is certainly necessary to study the behavior for nonconstant fields (such an investigation has been initiated for the scalar determinant by Z. Haba [54]; but there the results are inconclusive because the ϕ^4 interaction that has to be present may change the situation).

It is also conceivable that the additional divergencies coming from the integration over the photon field will somehow rectify the situation, but this would be pretty miraculous.

It might be of interest to note that the one-loop effective potential for a pure Yang-Mills theory shows a similar behavior, but with a good sign in front ($-F^2 \log F$) which is caused by the ghost loops and directly related to asymptotic freedom [55]. Maybe asymptotic freedom is necessary for stability ?

It is too early to make any definite statement of that kind. All I want to say here is that there are important questions that might be answered by more detailed study of determinants.

* * *

My hope is that these lectures have convinced you that the Matthews-Salam formalism is a powerful tool in constructive field theory and that it suggests questions that are worth asking and have a chance of being answered.

References.

[1] M.T. Mathews, A. Salam, Nuovo Cim. 12, 367 (1948); Phys. Rev. 80, 440 (1950).

[2] K. Osterwalder, these proceedings.

[3] F.A. Berezin, "The Method of Second Quantization", Academic Press, New York 1966.

[4] K. Wilson, in "New Developments in Quantum Field Theory and Statistical Mechanics", M. Lévy and P.K. Mitter eds., Plenum Press, New York and London 1977 (Cargèse Summer School 1976).

[5] K. Osterwalder, E. Seiler, Ann. Phys. (N.Y.) 110, 440 (1978).

[6] E. Seiler, "Gauge Theories as a Problem of Constructive Quantum Field Theory and Statistical Mechanics", lecture notes Geneva 1981 (to appear in Springer lecture notes in physics).

[7] M. Lüscher, Comm. Math. Phys. 54, 283 (1977).

[8] D.C. Brydges, J. Fröhlich, E. Seiler, Ann. Phys. (N.Y.) 121, 227 (1979).

[9] F. Guerra, these proceedings.

[10] F. Wegner, J. Math. Phys. 12, 2259 (1971).

[11] K. Wilson, Phys. Rev. D10, 2445 (1974).

[12] J.M. Drouffe, Phys. Rev. D18, 1174 (1978).
N.S. Manton, Phys. Lett. 96B, 328 (1980).
A. Patrascioiu, J.B. Bronzan, "Does QCD Confine Color", Los Alamos preprint 1980.

[13] M. Creutz, Phys. Rev. D21, 2308 (1979); Phys. Lett. 45, 313 (1980).

[14] M. Jimbo, T. Miwa and M. Sato in "Mathematical Problems in Theoretical Physics", K. Osterwalder, ed. (Proceedings of the 1979 $M \cap \phi$ Conference Lausanne), Springer-Verlag, New York 1980.

[15] A. Patrascioiu, E. Seiler, I.O. Stamatescu, "Monte Carlo Study of Non-compact Lattice QCD", Max-Planck preprint MPI-PAE/PTH 43/81, to appear in Phys. Lett.

[16] M. Göpfert, G. Mack, "Proof of Confinement of Static Quarks in 3-Dimensional U(1) Lattice Gauge Theory for All Values of the Coupling Constant", DESY preprint 81/036 (1981), Comm. Math. Phys. (in press).

[17] B. Simon, Adv. Math. 24, 244 (1977); M. Reed, B. Simon, "Methods of Modern Mathematical Physics IV : Analysis of Operators", Academic Press, New York, San Francisco and London 1978; I.C. Goh'berg and M.G. Krein, "Introduction to the Theory of Nonselfadjoint Operators", Transl. AMS 18, Providence 1969; N. Dunford and J. Schwartz, "Linear Operators", Interscience, New York 1963.

[18] E. Seiler, B. Simon, J. Math. Phys. 16, 2289 (1975).

[19] E. Seiler, B. Simon, Comm. Math. Phys. 45, 99 (1976); B. Simon "Trace Ideals and their Applications", Cambridge University Press 1979.

[20] D.C. Brydges, J. Fröhlich and E. Seiler, Comm. Math. Phys. 71, 159 (1980).

[21] D. Weingarten, J. Challifour, Ann. Phys. (N.Y.) 123, 61 (1979).

[22] E. Seiler, B. Simon, Ann. Phys. (N.Y.) 97, 470 (1976).

[23] F. Guerra, L. Rosen, B. Simon, Ann. Math. 101, 111 (1975).

[24] D.C. Brydges, P. Federbush, J. Math. Phys. 15, 730 (1974).

[25] L. Rosen, book in preparation.

[26] J. Schwinger, Phys. Rev. 82, 664 (1951).

[27] R.T. Seeley, Symp. Pure Math. 10 (Chicago 1966), 288, AMS Proceedings, Providence 1967; Am. J. Math. 91, 963 (1969).

[28] B.S. de Witt "Dynamical Theory of Groups and Fields" (Gordon and Breach New York 1965); Phys. Rep. 19, 295 (1975).

[29] P.B. Gilkey, "The Index Theorem and the Heat Equation" (Publish or Perish, Boston 1974).

[30] E. Corrigan, P. Goddard, H. Osborn, S. Templeton, Nucl. Phys. B159, 469 (1979); V.N. Romanov and A.S. Schwarz, Theor. Math. Phys. 41, 190 (1979); C. Callias and C.H. Taubes, Comm. Math. Phys. 77, 229 (1980).

[31] G. 't Hooft, Phys. Rev. D14, 3432 (1976).

[32] E. Seiler, Phys. Rev. D22, 2412 (1980).

[33] R. Schrader, R. Seiler, Comm. Math. Phys. 61, 169 (1978).

[34] O.A. Mc Bryan, private communication.

[35] E. Seiler, Comm. Math. Phys. 42, 163 (1975).

[36] O.A. Mc Bryan, Comm. Math. Phys. 44, 237 (1975).

[37] J. Glimm, A. Jaffe, Fortschr. Phys. 21, 327 (1973).

[38] J. Magnen, R. Sénéor in : Third International Conference on Collective Phenomena, J. Lebowitz, J. Langer and W. Glaberson, eds., New York : The New York Academy of Sciences (1980).

[39] G. Battle, L. Rosen, J. Stat. Phys. 22, 123 (1980).

[40] A. Cooper, L. Rosen, Trans. AMS 234, 1 (1977); J. Magnen, R. Sénéor, Comm. Math. Phys. 51, 297 (1976).

[41] D.C. Brydges, J. Fröhlich, E. Seiler, Comm. Math. Phys. 79, 353 (1981)

[42] K.R. Itô, "Construction of Euclidean QED_2 via Lattice Gauge Theory, Bedford College preprint 1981.

[43] J. Glimm, A. Jaffe, T. Spencer in "Constructive Quantum Field Theory", G. Velo and A.S. Wightman, eds. (Erice 1973), Springer Verlag New York 1973.

[44] P. Renouard, Ann. Inst. H. Poincaré 27, 237 (1977); 31, 235 (1979).

[45] T. Bałaban, K. Gawędzki, "A low temperature expansion for the pseudoscalar Yukawa model of quantum fields in two space-time dimensions, Ann. Inst. H. Poincaré, to appear.

[46] W. Kerler, Phys. Rev. D23, 2384 (1981).

[47] S. Adler, Phys. Rev. 177, 2426 (1969); J.S. Bell, R. Jackiw, Nuovo Cim. 60A, 47 (1969).

[48] E. Seiler, I.O. Stamatescu, Lattice Fermions and θ-Vacua, Max-Planck preprint 1981.

[49] S. Coleman, R. Jackiw, L. Susskind, Ann. Phys. (N.Y.) 93, 267 (1975); J. Fröhlich, E. Seiler, Helv. Phys. Acta 49, 889 (1976).

[50] D.C. Brydges, J. Fröhlich, E. Seiler, Nucl. Phys. B152, 521 (1979).

[51] W. Heisenberg, H. Euler, Z. Phys. 98, 714 (1936).

[52] H. Hogreve, R. Schrader, R. Seiler, Nucl. Phys. B142, 525 (1978).

[53] Z. Haba, Wroclaw preprint, 1981.

[54] S.G. Matinyan, G.K. Savvidy, Nucl. Phys. B134, 539 (1978).

FIELD THEORIES AND SYMANZIK'S POLYMER REPRESENTATION

David Brydges
Department of Mathematics
University of Virginia

1. INTRODUCTION

About 13 years ago Symanzik [1] showed that the ϕ^4 euclidean field theory is isomorphic to the equilibrium statistical mechanics of interacting polymers. His observations have been a useful source of intuition on the problem of whether the ϕ^4 field theory is trivial or not in high dimensions, i.e., is it, despite its appearance, a free (or generalized free) field?

The ideas of the renormalization group provide another approach to this problem and from this heuristic basis [2] has arisen the conventional wisdom that in five or more dimensions ϕ^4 is a free field, in four dimensions it is a generalized free field. The last statement is a marginal conclusion from the renormalization group and more controversial. (In three and less dimensions non-trivial ϕ^4 field theories have been constructed [3].)

Recently Michael Aizenman showed [4] that ϕ^4_d in dimensions $d \geq 5$ is a generalized free field, i.e., a gaussian random field, at least when constructed as a limit of ϕ^4 lattice field theories. His methods have more to do with Symanzik's ideas than the renormalization group. He first proves that scaling limits of Ising models in more than 5 dimensions are gaussian by relating them to random walks and then approximates ϕ^4 with Ising models.

In these lectures, I am going to discuss a rather direct implementation of Symanzik's ideas which has been developed in [5]. The program is similar to Aizenman's (but sufficiently different to be a useful complement to his ideas). It also leads to a proof of a theorem on the triviality of ϕ^4 (and Ising models and non-linear σ models) in dimensions $d \geq 5$, along with partial results in $d \geq 4$, Frohlich [6]. Along the way it provides new proofs of many old and new correlation inequalities. One such set of inequalities says that renormalized perturbation theory provides alternating lower and upper bounds on correlation functions. This is potentially very useful because it means that perturbation theory can be used more directly in rigorous investigations than was previously possible.

The reader will see that these methods constitute a very elementary and attractive approach to certain field theoretic and statistical mechanical problems. Unfortunately, the price to be paid is that we seem to be limited to the ϕ^4 or $(\vec{\phi}^2)^2$ vector models and their limits such as the Ising model and non-linear σ models. (Some progress on gauge theories is made in [7].) Aizenman's methods will, in my opinion, be at least equally limited in scope. Thus, eventually, we are going to have to understand the renormalization group [8] in order to make progress on gauge theories.

My collaborators in this work are P. Federbush, J. Fröhlich, A. Sokal and T. Spencer. It gives me great pleasure to represent them at this summer school. I also am indebted to M. Aizenman for helpful comments and suggestions, clarifying my understanding of some of the inequalities discussed below. I also thank Jürg Fröhlich for permission to include a discussion of his theorem [6] before its publication.

2. Lattice Field Theories

Lattice field theories are finite difference approximations to (euclidean) field theories. The theories I shall discuss are also statistical mechanical models of ferromagnets.

Let L be a finite subset of a simple cubic lattice embedded in $\mathbf{R}^\nu$:

$$L \subset (\varepsilon\mathbf{Z})^\nu \subset \mathbf{R}^\nu .$$

ε is the lattice spacing. To each point x in L is assigned a random variable ϕ_x. I shall refer to the collection

$$\phi \equiv (\phi_x)_{x \in L}$$

as a "field configuration". The joint probability for the random variables $(\phi_x) \equiv \phi$ is of the form

$$e^{\frac{1}{2}\phi \mathcal{J} \phi} \prod_{x \in L} g(\phi_x^2)d\phi_x/\text{Normalization} \tag{2.1}$$

where g is any function $g: \mathbb{R}^+ \to \mathbb{R}^+$ which decays fast enough to ensure convergence and

$$\phi \mathcal{J} \phi \equiv \sum_{x,y \in L} \phi_x \mathcal{J}_{xy} \phi_y .$$

The matrix $\mathcal{J}$ will be assumed to be defined for all x,y in the infinite lattice $(\varepsilon\mathbb{Z})^\nu$ and

$$\mathcal{J}_{xy} = \mathcal{J}_{yx} \geq 0 \text{ if } x \neq y \tag{2.2}$$

(This means that the joint distribution favors field configurations for which the random variables are positive (or negative) in unison.) The coupling J is said to be ferromagnetic. We have required that $\mathcal{J}$ be defined for an infinite lattice because we will shortly be taking an infinite volume limit.

2.1 Notation

Given a function $F = F(\phi)$ we set

$$\langle F \rangle_L \equiv \frac{\int F \exp(\phi \mathcal{J} \phi/2) \,\Pi_{x \in L}\, g(\phi_x^2) d\phi_x}{\int \exp(\phi \mathcal{J} \phi/2) \Pi_{x \in L}\, g(\phi_x^2) d\phi_x} \tag{2.3}$$

This is called the "expectation" of the "observable" F.

2.2 Example (ϕ_ν^4 lattice field theory)

Let $\phi(x)$ be a function on $\mathbb{R}^\nu$. Define the functional $S(\phi)$ by

$$S(\phi) \equiv \tfrac{1}{2}\int (\nabla\phi)^2 d^\nu x + \tfrac{1}{2}\int \phi^2 d^\nu x + \int P(\phi) d^\nu x.$$

P is an even quartic polynomial

$$P(\phi) \equiv \frac{\lambda}{4}\phi^4 - \frac{1}{2} a \phi^2 \tag{2.4}$$

with $\lambda > 0$, a arbitrary real. The problem of constructing the ϕ_ν^4 euclidean lattice field theory can be stated in the following preliminary way: make sense of the formal object

"$\exp(-S(\phi))\mathcal{D}\phi$/Normalization"

where $\mathcal{D}\phi$ is the "flat measure"

$$"\mathcal{D}\phi \equiv \Pi_{x \in \mathbb{R}^\nu} d\phi(x)"$$

(This uncountable product of Lebesgue measures cannot be understood within the context of measure theory.) To obtain a well-defined object we first consider the finite difference approximation obtained by choosing in (2.3)

$$\begin{aligned} \mathcal{J}_{xy} &= \tilde{Z}\varepsilon^{\nu-2} \quad \text{if } x \neq y \text{ are nearest neighbors} \\ &= -Z.2\nu.\varepsilon^{\nu-2} - \varepsilon^{\nu} \text{ if } x = y \\ g(\phi_x^2) &= \exp \varepsilon^{\nu}(-\frac{\lambda}{4}\ \phi_x^4 + \frac{1}{2}\, a\ \phi_x^2). \end{aligned} \tag{2.5}$$

For the moment set the parameter $\tilde{Z} = 1$, then

$$\tfrac{1}{2}\ \phi \mathcal{J} \phi = \tfrac{1}{2}\textstyle\sum_{xy}\ \varepsilon^{\nu}(\phi_x - \phi_y)^2/\varepsilon^2 + \tfrac{1}{2}\textstyle\sum_x\ \varepsilon^{\nu}\ \phi_x^2$$

$$(\simeq \tfrac{1}{2}\textstyle\int(\nabla\phi)^2 + \tfrac{1}{2}\textstyle\int\phi^2\ )$$

The sum over xy is over nearest neighbors. Similarly the g's approximate the quartic terms due to (2.4).

The parameter $\tilde{Z}$ is known as the field strength renormalization. It will become relevant when we investigate the continuum limit more closely in section 7.

3. The Griffiths Correlation Inequality and the Infinite Volume Limit

We should expect the assumption that $\mathcal{J}_{xy} \geq 0$ if $x \neq y$ (ferromagnetism) to cause the variables ϕ_x to correlate in the sense that if one is held positive the others should have a tendency to be positive. A quantitative version of this intuition is the following theorem due to Griffiths:

3.1 Theorem

Suppose P,Q are polynomials in (ϕ_x), $x \in L$, with positive coefficients then

$$\langle PQ \rangle_L - \langle P \rangle_L \langle Q \rangle_L \geq 0$$

For a proof of this see [3].

We illustrate how this important inequality is used in the sequel by proving the well known fact that if F is a polynomial with positive coefficients and $\mathcal{J}$ vanishes on the diagonal then the expectation of F is increasing in the size of the lattice L:

$$L \subset L' \Longrightarrow \langle F\rangle_L \leq \langle F\rangle_{L'} \tag{3.1}$$

where of course we are assuming that F depends only on ϕ_x for $x \in L$. We first observe that if we modify the L' expectation,

$$\langle F\rangle_{L'} \to \langle F\rangle_{\underset{\sim}{t}}$$

by making the replacement

$$\Pi_{x \in L'}\, g(\phi_x^2) \to \Pi_{x \in L}\, g(\phi_x^2)\, \Pi_{y \in L' \sim L}\, g(\phi_y^2)\, e^{-t_y \phi_y^2}$$

in both numerator and denominator in the definition of $\langle F\rangle_{L'}$, then

$$\langle F\rangle_{\underset{\sim}{t} = 0} = \langle F\rangle_{L'}$$

$$\langle F\rangle_{\underset{\sim}{t} \to \infty} = \langle F\rangle_L$$

so it is sufficient to prove that

$$\frac{\partial}{\partial t_y} \langle F\rangle_{\underset{\sim}{t}} \leq 0$$

for all $y \in L' \sim L$. A simple calculation shows

$$\frac{\partial}{\partial t_y} \langle F\rangle_{\underset{\sim}{t}} = -\langle F\phi_y^2\rangle + \langle F\rangle\langle\phi_y^2\rangle$$

which is less than zero by the Griffiths inequality so the monotonicity of the expectation in the volume is proved.

By combining this fact with uniform bounds on the expectation (which will hold under additional hypotheses on g and $\mathcal{J}$) one can in many cases establish the existence of the <u>infinite volume limit</u>:

$$\lim_{L \nearrow (\varepsilon\mathbb{Z})^\nu} \langle F\rangle_L \equiv \langle F\rangle_\infty \tag{3.2}$$

3.2 Notation

Set, for any functions F,G of $\underset{\sim}{\phi}$

$$\langle F;G\rangle_L \equiv \langle FG\rangle_L - \langle F\rangle\langle G\rangle_L \tag{3.3}$$

3.3 Vector Models

If at each site we have a vector

$$\vec{\phi}_x = \phi_x^{(\alpha)};\ \alpha \geq 1,\dots,N$$

with N components and $g(\phi^2)$ is changed to $g(\vec{\phi}^2)$ along with

$$\underset{\sim}{\phi}\tilde{J}\underset{\sim}{\phi} \to \vec{\underset{\sim}{\phi}}.\tilde{J}\vec{\underset{\sim}{\phi}} \equiv \sum_{\alpha,x,y} \phi_x^{(\alpha)}\tilde{J}_{xy}\phi_y^{(\alpha)} \tag{3.4}$$

then we have the N component vector model. For such models is it reasonable to ask if, given P,Q are polynomials with positive coefficients in dot products $\vec{\phi}_x.\vec{\phi}_y$, $x,y \in L$,

$$\langle P;Q\rangle \geq 0? \tag{3.5}$$

This is known for N = 2, Ginibre [9], but not for $N \geq 3$.

4. Perturbation Theory

The following identity for a one dimensional gaussian integral is a trivial exercise in integrating by parts:

$$\int \phi F(\phi) e^{\tilde{J}\phi^2/2}\, d\phi = -\frac{1}{J}\int F'(\phi) e^{\tilde{J}\phi^2/2}\, d\phi$$

(where $\tilde{J} < 0$). It generalizes to the multidimensional version:

$$\int \phi_x F(\underset{\sim}{\phi}) e^{\frac{1}{2}\underset{\sim}{\phi}\tilde{J}\underset{\sim}{\phi}} = -\sum_z \tilde{J}^{-1}_{xz} \int \frac{\partial F}{\partial \phi_z}\, e^{\frac{1}{2}\underset{\sim}{\phi}\tilde{J}\underset{\sim}{\phi}} \tag{4.1}$$

(provided $\tilde{J}$ is negative definite) where $\tilde{J}^{-1}$ denotes the matrix inverse of $(\tilde{J}_{xy})$, $x,y \in L$. If we define

$$[F]^G \equiv \int F(\phi) e^{\frac{1}{2}\underset{\sim}{\phi}\tilde{J}\underset{\sim}{\phi}} \tag{4.2}$$

which is an unnormalized gaussian expectation, then we can rewrite this integration by parts formula as

$$[\phi_x F]^G = -\sum_z \tilde{J}^{-1}_{xz} \left[\frac{\partial F}{\partial \phi_z} \right]^G \tag{4.3}$$

This formula can be used repeatedly to evaluate expectations of any polynomial. Here are four examples to serve as exercises for the reader:

(1) $\langle \phi_x \phi_y \rangle^G \equiv [\phi_x \phi_y]^G / [1]^G = -\tilde{J}^{-1}_{xy}$

$\equiv \overline{\quad\quad}$ x y

We have just introduced a graphical notation, representing the matrix element $-\tilde{J}^{-1}_{xy}$ by a line with end-points labeled by x and y.

(2) $\langle \phi_{x_1} \phi_{x_2} \phi_{x_3} \phi_{x_4} \rangle^G = (-\tilde{J}^{-1}_{x_1 x_2}).(-\tilde{J}^{-1}_{x_3 x_4}) +$

$(-\tilde{J}^{-1}_{x_1 x_3}).(-\tilde{J}^{-1}_{x_2 x_4}) + (-\tilde{J}^{-1}_{x_1 x_4}).(-\tilde{J}^{-1}_{x_2 x_3})$

$\equiv$ $\overline{x_1 \quad x_2}$ / $\overline{x_3 \quad x_4}$ + $\overline{x_1 \quad x_3}$ / $\overline{x_2 \quad x_4}$ + $\overline{x_1 \quad x_4}$ / $\overline{x_2 \quad x_3}$ (4.4)

From now on we use the graphical notation exclusively.

(3) $\langle \phi_x^4 \rangle^G = 3$  x

This is the previous example disguised by taking $x_1 = x_2 = x_3 = x_4$. Since we identify end-points of lines we get loops

(4) $\langle \phi_x^4 \phi_y^4 \rangle^G = 9$  x y  + 72  + 24 x  y.

There exist rules for determining the combinatoric factors 9, 72, 24 from the topology of the graphs but the reader need not concern himself with this to understand the rest of these lectures. To convince yourself that this notation might be useful, try writing out example 4 without it.

The famous Feynman-Dyson-Schwinger series comes about as follows: consider the ϕ^4 lattice field theory. Let < > temporarily denote its expectation, (2.3), (2.5). Suppose we calculate the "two point function" $\langle\phi_x\phi_y\rangle$ by writing $\langle\phi_x\phi_y\rangle = [\phi_x\phi_y \exp{-V}]^G/[\exp{-V}]^G$ where

$$V = \frac{\lambda}{4}\sum_{x \in L} \varepsilon^{\nu}\phi_x^4 - \frac{a}{2}\sum_{x \in L} \varepsilon^{\nu}\phi_x^2 \quad .$$

Apply formula (4.3) to the numerator, choosing F to be $\phi_y \exp{-V}$. The derivative $\partial F/\partial\phi_z$ gives a term x——y (if z = y) and another one involving V'. On this second term use (4.3) again taking F to be ϕ_y.-V'.exp-V etc. You get an infinite series:

$$\langle\phi_x\phi_y\rangle = x\text{——}y - 2\lambda\ x\text{—○—}y + \ldots$$

4.1 Convention

Unlabeled vertices such as the one occurring in the second graph are (Riemann) summed over L. Here is a final example:

$$\langle\phi_{x_1}\phi_{x_2}\phi_{x_3}\phi_{x_4}\rangle = \begin{matrix} x_1\text{——}x_2 \\ x_3\text{——}x_4 \end{matrix} + \begin{matrix} x_1\text{——}x_3 \\ x_2\text{——}x_4 \end{matrix} + \begin{matrix} x_1\text{——}x_4 \\ x_2\text{——}x_3 \end{matrix}$$

$$- 6\lambda \begin{matrix} x_1 & & x_2 \\ & \times & \\ x_4 & & x_3 \end{matrix} + \ldots \qquad (4.5)$$

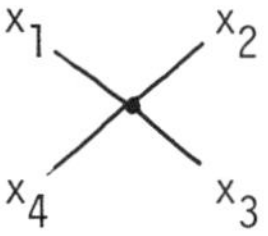

This perturbation series is known to diverge even if the lattice is finite and the specius ε is fixed nonzero. If you take the infinite volume limit keeping $\varepsilon > 0$ fixed, the series is finite term by term and asymptotic but not convergent. If you take a continuum "limit" $\varepsilon \searrow 0$ after the infinite volume limit, the individual terms of the series diverge with ε in dimensions $\nu > 1$. The conclusion is that this series will not be useful without extra sophistication. We will aim for this in the rest of the lectures.

5. Integration by Parts with Random Walks

In case it was not clear, the object of the last section was to convince you that (a) perturbation theory is generated by the integration by parts, formula (4.3), and (b) it is of limited use in the form presented. One of the tools with which we shall augment perturbation theory is the Griffiths inequality Theorem 3.1. Another tool is the identity of this chapter, Lemma 5.2. This can be looked on as an integration by parts formula for non-gaussian measures. It will be used to generate an improved (≡ renormalized) perturbation theory with bounds.

For this section we shall use the definitions

$$\begin{aligned} [F] &\equiv \int \exp(\tfrac{1}{2}\underset{\sim}{\phi} J \underset{\sim}{\phi})\, \Pi_x\, g(\phi_x^2) F \\ [F]_{\underset{\sim}{t}} &\equiv \int \exp(\tfrac{1}{2}\underset{\sim}{\phi} J \underset{\sim}{\phi})\, \Pi_x\, g(\phi_x^2 + 2t_x) F \end{aligned} \tag{5.1}$$

where $\underset{\sim}{t} = (t_x)$, $t_x \geq 0\ \forall\, x$. We suppress all reference to the lattice L because it is held fixed in this section. $\tilde{J}$ is changed to J to reflect an additional assumption given below. We <u>assume</u>

$$\begin{aligned} & g(\phi^2)\exp(c\phi^2) \underset{\phi \to \infty}{\longrightarrow} 0 \quad \text{for all } c \geq 0 \\ & J_{xy} = 0 \qquad \text{if} \quad x = y. \end{aligned} \tag{5.2}$$

The second assumption amounts to requiring that any diagonal parts you might like to include in J be absorbed into g.

We begin with a standard fact about random walks and the resolvent of J (which in this form is not as well known as it should be). For λ sufficiently large, depending on J, there is a path representation:

$$(\lambda - J)^{-1}_{xy} = \sum_{\omega:\ x \to y} (\Pi_{s \in \omega} J_s) \lambda^{-|\omega|-1}$$

where: the sum over ω is over all random walks, ω, which start at x and end at y. The steps, s, of ω are from any lattice site to any other (not just nearest neighbors). A step s is a pair of lattice sites; J_s denotes the corresponding matrix element. The length, $|\omega|$, of the random walk is the number of steps in ω. This formula is obtained by considering the expansion:

$$(\lambda - J)^{-1}_{xy} = \lambda^{-1}\delta_{xy} + \lambda^{-1} J_{xy} \lambda^{-1} + \sum_z \lambda^{-1} J_{xz} \lambda^{-1} J_{zy} \lambda^{-1} + \ldots \quad .$$

Set

$$J_\omega \equiv \Pi_{s \in \omega} J_s \tag{5.3}$$

$n(z,\omega) \equiv$ N° at times ω visits site x

then this expansion can be rewritten as

$$(\lambda - J)^{-1}_{xy} = \sum_{\omega:\ x \to y} J_\omega \, \Pi_{z \in L} \, \lambda^{-n(z,\omega)}$$

because $|\omega|+1$ is equal to the sum over $z \in L$ of all the visiting numbers $n(z,\omega)$. An obvious generalization of this formula is:

5.1 Lemma

If $\Lambda \equiv (\lambda_x \delta_{xy})$ is a diagonal matrix with sufficiently large (in absolute value) entries then

$$(\Lambda - J)^{-1}_{xy} = \sum_{\omega:\ x \to y} J_\omega \, \Pi_z \, \lambda_z^{-n(z,\omega)}$$

where the series converges absolutely.

Now we will obtain our non-gaussian integration by parts formula. We start with

$$[\phi_x F] \equiv \int e^{\phi J \phi/2} \, \Pi_{z \in L} \, g(\phi_z^2) \phi_x F \tag{5.4}$$

Substitute into this the fourier inversion

$$g(\phi_z^2) = \int_\Gamma \hat{g}(a_z) e^{-i a_z \phi_z^2} \, da_z \tag{5.5}$$

for all factors $g(\phi_z^2)$. Γ is a contour in the complex plane:

$$\Gamma \equiv \{a \in \mathbb{C}: \text{Im } a = c\}$$

for a constant c chosen large negative. We can choose this contour instead of simply integrating along the real line because of assumption (5.2). It is convenient to give a_z a large negative imaginary part in this way in (5.5) because it will allow us to deal with absolutely convergent ϕ integrals instead of oscillatory ϕ integrals. After making these substitutions we interchange the da, dϕ integrals. Now the $d\phi$ integrals are gaussian and we can integrate by parts using (4.3) with J replaced by $J-\Lambda$ where Λ is the diagonal matrix

$$\Lambda \equiv 2ia_x\delta_{xy} \quad . \tag{5.6}$$

We obtain

$$[\phi_x F] = \sum_y \int_\Gamma d\underset{\sim}{a} \Pi_z \hat{g}(a_z)(\Lambda-J)^{-1}_{xy} \cdot \int d\underset{\sim}{\phi} e^{\underset{\sim}{\phi}(\Lambda-J)\underset{\sim}{\phi}/2} \frac{\partial F}{\partial \phi_y} \tag{5.7}$$

Next we substitute for $(\Lambda-J)^{-1}$ using lemma 5.1. We also interchange $d\underset{\sim}{a}$, $d\underset{\sim}{\phi}$ and the sum over random walks to get

$$[\phi_x F] = \sum_y \sum_{\omega:\, x\to y} J_\omega \int d\underset{\sim}{\phi}\; e^{\underset{\sim}{\phi} J \underset{\sim}{\phi}/2}.$$

$$\cdot \int_\Gamma d\underset{\sim}{a}\; \Pi_z\; \hat{g}(a_z)(2ia_z)^{-n(z,\omega)}\; e^{-\underset{\sim}{\phi}\Lambda\underset{\sim}{\phi}/2} \cdot \frac{\partial F}{\partial \phi_y} \tag{5.8}$$

Notice that the $d\underset{\sim}{a}$ integrals are now independent. They each have the form

$$\int_\Gamma da\; \hat{g}(a)(2ia)^{-n} e^{-ia\phi^2}; \quad n = 0,1,2,\ldots \tag{5.9}$$

and we rewrite each of them as

$$\int_0^\infty d\nu_n(t)\; g(\phi^2+2t)$$

where

$$\begin{aligned} d\nu_n(t) &= \delta(t)\; dt && \text{if } n = 0 \\ &= \frac{t^{n-1}}{(n-1)!}\; dt && \text{if } n = 1,2,\ldots \end{aligned}$$

(This measure is just the laplace transform of $(2ia)^{-n}$. A product has become a convolution.) Since there is one variable $a = a_z$ for each lattice site $z \in L$, we must have a collection of variables $\underset{\sim}{t} \equiv (t_z)$, $z \in L$. Set

$$d\nu_\omega(t) \equiv \Pi_{z\in L}\; d\nu_{n(z,\omega)}\; (t_z) \tag{5.10}$$

After all this, (5.8) becomes

$$[\phi_x F] = \sum_y \sum_{\omega:\, x \to y} J_\omega \int d\nu_\omega(\underset{\sim}{t}) [\frac{\partial F}{\partial \phi_y}]_{\underset{\sim}{t}}$$

We collect this in the following lemma:

5.2 Lemma

Define a product measure on $[0,\infty]^{|L|}$ which depends on a random walk ω by

$$d\nu_\omega(\underset{\sim}{t}) \equiv \Pi_{z \in L}\, d\nu_{n(z,\omega)}(t_z)$$

$$\begin{aligned} d\nu_n(t) &= \delta(t)dt \qquad \text{if } n = 0 \\ &= t^{n-1}/(n-1)!\ dt \quad \text{if } n = 1,2,\ldots \end{aligned} \tag{5.11}$$

then with the assumptions on J, g given in (5.2), the following identity is valid

$$[\phi_x F] = \sum_y \sum_{\omega\, :\, x \to y} J_\omega \int d\nu_\omega(\underset{\sim}{t}) [\frac{\partial F}{\partial \phi_y}]_{\underset{\sim}{t}}$$

where J_ω is defined in (5.3), $[\ ]_{\underset{\sim}{t}}$ in (5.1). F can be any polynomial in $\underset{\sim}{\phi}$.

5.3 Exercise

Show that this identity reduces to gaussian integration by parts (4.3) if $g(\phi^2) = \exp(-a\phi^2/2)$, a sufficiently large positive.

This lemma continues to hold with only cosmetic changes in the proof and statement if one has vector models as in (3.4) with any number of components.

The explicit form of the measure $d\nu_\omega$ will not be important beyond the following properties (1) it depends on ω only through the visiting numbers $\underset{\sim}{n} \equiv (n(z,\omega))$, $z \in L$, thus we can write

$$d\nu_\omega \equiv d\nu_{\underset{\sim}{n}}$$

(2) Suppose $\underset{\sim}{n} = \underset{\sim}{n}_1 + \underset{\sim}{n}_2$ where addition means pointwise addition at each site $z \in L$, then

$$\int d\nu_{\underset{\sim}{n}}(\underset{\sim}{t}) f(\underset{\sim}{t}) = \int d\nu_{\underset{\sim}{n}_1}(\underset{\sim}{t}_1) \int d\nu_{\underset{\sim}{n}_2}(\underset{\sim}{t}_2) f(\underset{\sim}{t}_1 + \underset{\sim}{t}_2) \tag{5.12}$$

(for f arbitrary). This is easy to verify.

5.4 Notation

Set

$$Z_{\underset{\sim}{t}} \equiv [1]_{\underset{\sim}{t}}; \quad Z \equiv [1]_{t=0} = [1]$$

$$<(\cdot)>_{\underset{\sim}{t}} \equiv [(\cdot)]_{\underset{\sim}{t}}/Z_{\underset{\sim}{t}} \tag{5.13}$$

5.5 The two point function

By taking $F = \phi_y$ in lemma 5.2 and dividing through by Z we obtain a representation for the two point function:

$$<\phi_x\phi_y> = \sum_{\omega:\ x\to y} J_\omega \int d\nu_\omega(\underset{\sim}{t})\ \frac{Z_t}{Z} \ . \tag{5.14}$$

Related to this is the following lemma which will be of service later.

5.6 Lemma

If log g is concave then

(a) $\sum_{\omega:\ x\to y,\omega \ni z} \int d\nu_\omega(\underset{\sim}{t})Z_{\underset{\sim}{t}}/Z \leq \sum_{z'} <\phi_x\phi_z>J_{zz'}<\phi_{z'}\phi_y>$

and

(b) $\sum_{\omega:\ x\to y,\ \omega\ni z} \int d\nu_\omega(\underset{\sim}{t})t_z Z_{\underset{\sim}{t}}/Z \leq \sum_z <\phi_x\phi_z><\phi_z\phi_y>$.

Proof of (a): we prove this result under the stronger assumption that log g is a polynomial appropriate to ϕ^4 :

$$\log g(\phi^2) = -a_1\phi^4 - a_2\phi^2 - a_3$$

where $a_1 > 0$, a_2, a_3 are arbitrary. (The result stated above requires a stronger version of the Griffiths inequality than we stated.)

We split ω into three pieces ω_1, ω_2 and a single step zz':

$$\omega = \omega_1 \cup zz' \cup \omega_2$$

ω_1 is the part of ω up to the first time it hits z, zz' is the next step and ω_2 is the rest of ω.

$$\sum_{\omega:\ x\to y,\omega \ni z} J_\omega \int d\nu_\omega(\underset{\sim}{t}) Z_{\underset{\sim}{t}}/Z = \sum_{\omega_1: x\to z} J_{\omega_1} \sum_{z'} J_{zz'} \cdot$$

$$\cdot \sum_{\omega_2:\ z'\to y} J_{\omega_2} \int d\nu_\omega(\underset{\sim}{t}) Z_{\underset{\sim}{t}}/Z.$$

Now we use (5.12) by inserting

$$\int d\nu_\omega(\underset{\sim}{t}) Z_{\underset{\sim}{t}}/Z = \int d\nu_{\omega_1}(\underset{\sim}{t}_1) \int d\nu_{\omega_2}(\underset{\sim}{t}_2) \frac{Z_{\underset{\sim}{t}_1+\underset{\sim}{t}_2}}{Z}$$

obtaining

$$\sum_{z'} J_{zz'} \sum_{\omega_1} J_{\omega_1} \int d\nu_{\omega_1}(\underset{\sim}{t}_1) \frac{Z_{\underset{\sim}{t}_1}}{Z} \cdot \sum_{\omega_2} J_{\omega_2} \int d\nu_{\omega_2}(\underset{\sim}{t}_2) \frac{Z_{\underset{\sim}{t}_1+\underset{\sim}{t}_2}}{Z_{\underset{\sim}{t}_1}}$$

By comparing the quantity preceded by the sum over ω_2 with (5.14) we see that it resums to

$$\sum_{z'} J_{zz'} \sum_{\omega_1} J_{\omega_1} \int d\nu_{\omega_1}(\underset{\sim}{t}_1) \frac{Z_{\underset{\sim}{t}_1}}{Z} \cdot \langle\phi_{z'}\phi_y\rangle_{\underset{\sim}{t}_1} \tag{5.15}$$

By virtue of the hypothesis on log g and the Griffiths inequality

$$\langle\phi_{z'}\phi_y\rangle_{\underset{\sim}{t}_1} \le \langle\phi_{z'}\phi_y\rangle_{\underset{\sim}{t}_1=\underset{\sim}{0}} \tag{5.16}$$

because if t_u is one of the variables in $\underset{\sim}{t}_1$

$$\frac{\partial}{\partial t_u} \langle\phi_{z'}\phi_y\rangle_{\underset{\sim}{t}_1} = -4a_1 \langle\phi_{z'}\phi_y;\phi_u^2\rangle \le 0.$$

On substituting (5.16) into (5.15) the sum over ω_1 also becomes resummable, to $\langle\phi_x\phi_z\rangle$ and we have proven part (a).

Part (b) is proven by a similar argument based on splitting ω into two pieces, up to z, beyond z. The details will be given in [10].

5.7 Remark

The basic idea of turning a non-gaussian integral into a gaussian integral by (5.5) was introduced by Symanzik in [1]. He eliminated all ϕ integrals by evaluating them explicitly and obtained a representation totally in terms of the t's which are "local times" for random walks. Our <u>mixed</u> representation (it has both ϕ's and local times) has some advantages because it can be combined with correlation inequalities.

We discuss both representations in [5].

6. Inequalities

We shall give three applications of the formalism of the last section. The first is a new proof of the gaussian inequalities of Newman [11]. Aside from strengthening the original results, we shall note that these inequalities are merely the first in a sequence of perturbation theoretic inequalities. The next application is a new proof and generalization of "Simons" inequalities [12]. (These inequalities were not solely the invention of Simon. See [13]. However the name does credit to his observations on their usefulness.) Part of the reason for presenting these inequalities is that they are an ingredient in our final application which is to prove Fröhlich's version of Aizenman's inequality. This is also a "perturbation-theoretic" inequality and leads to the results on triviality of ϕ^4 in dimensions greater or equal to four.

All proofs in the rest of these lectures immediately generalize to $N = 2$ vector models (xy models). They would generalize to $N > 2$ non-abelian models if the conjectured inequalities (3.5) were established.

6.1 Notation

As in the last section:

$$[(\cdot)]_{\underset{\sim}{t}} \equiv \int \Pi_x\, g(\phi_x^2 + 2t_x)\, e^{\frac{1}{2}\underset{\sim}{\phi} J \underset{\sim}{\phi}} (\cdot)\, d\underset{\sim}{\phi}$$

$$[(\cdot)] \equiv [(\cdot)]_{\underset{\sim}{t}=\underset{\sim}{0}};\quad Z_{\underset{\sim}{t}} \equiv [1]_{\underset{\sim}{t}},\quad Z = Z_{\underset{\sim}{t}=\underset{\sim}{0}}$$

$$\langle(\cdot)\rangle_{\underset{\sim}{t}} \equiv [(\cdot)]_{\underset{\sim}{t}}/Z_{\underset{\sim}{t}};\quad \langle(\cdot)\rangle \equiv \langle(\cdot)\rangle_{\underset{\sim}{t}=\underset{\sim}{0}}\,.$$

6.2 Hypotheses

Throughout this section we assume log g is concave and J is ferromagnetic, vanishing on the diagonal. These assumptions include ϕ^4 lattice field theories even with mass and charge renormalization. We will only give the proofs for the ϕ^4 lattice field theory case, i.e., log g is a quadratic polynomial (in ϕ^2!) with negative leading coefficient.

The results of this section are all proved on a finite lattice L which is not explicit in the notation. If further conditions on $\log g$ are imposed so that infinite volume limits exist (see (3.2)), then all results transfer to the infinite volume. In particular, if the model is the ϕ^4 lattice field theory with non-zero quartic coupling this infinite volume limit exists.

6.3 Gaussian Inequalities We divide both sides of lemma 5.2 by Z. Assume F is a polynomial with positive coefficients:

$$\begin{aligned}
\langle\phi_x F\rangle &= \sum_y \sum_{\omega:\, x\to y} J_\omega \int d\nu_\omega(\underset{\sim}{t}) \left[\frac{\partial F}{\partial\phi_y}\right]_{\underset{\sim}{t}} /Z \\
&= \sum_y \sum_\omega J_\omega \int d\nu_\omega(\underset{\sim}{t}) \frac{Z_{\underset{\sim}{t}}}{Z} \left\langle\frac{\partial F}{\partial\phi_y}\right\rangle_{\underset{\sim}{t}} \\
&\leq \sum_y \Big(\sum_\omega J_\omega \int d\nu_\omega(\underset{\sim}{t}) \frac{Z_{\underset{\sim}{t}}}{Z}\Big) \left\langle\frac{\partial F}{\partial\phi_y}\right\rangle \\
&= \sum_y \langle\phi_x\phi_y\rangle\left\langle\frac{\partial F}{\partial\phi_y}\right\rangle .
\end{aligned} \tag{6.1}$$

We have used

$$\left\langle\frac{\partial F}{\partial\phi_y}\right\rangle_{\underset{\sim}{t}} \leq \left\langle\frac{\partial F}{\partial\phi_y}\right\rangle \tag{6.2}$$

which is a consequence of the Griffiths inequality, see the proof of lemma 5.6. In the last line we resummed over ω using (5.14). Thus we have proved

6.4 Theorem

If F is a polynomial with positive coefficients

$$\langle\phi_x F\rangle \leq \sum_y \langle\phi_x\phi_y\rangle\left\langle\frac{\partial F}{\partial\phi_y}\right\rangle .$$

We now sketch how these inequalities fit in with perturbation theory. Notice first that by the formula (4.3), integration by parts, the inequality is saturated if $\langle\ \rangle$ is gaussian. The following discussion requires some modifications in the class of diagrams if the expectation is not that of lattice ϕ^4 field theory.

We introduce the graphical notation

$$\langle\phi_1\phi_2\rangle \quad\longleftrightarrow\quad 1\!\sim\!\sim\!\sim\!2$$

where 1,2 are shorthand for x_1, x_2, points in the lattice. In Theorem 6.4 take $F \equiv \phi_2\phi_3\phi_4$ and obtain the following bound on the "4 point function":

$$\langle \phi_{x_1}\phi_{x_2}\phi_{x_3}\phi_{x_4} \rangle \le \begin{matrix} x_1 \sim\!\sim\!\sim x_2 \\ \\ x_3 \sim\!\sim\!\sim x_4 \end{matrix} + \begin{matrix} x_1 \sim\!\sim\!\sim x_3 \\ \\ x_2 \sim\!\sim\!\sim x_4 \end{matrix} + \begin{matrix} x_1 \sim\!\sim\!\sim x_4 \\ \\ x_2 \sim\!\sim\!\sim x_3 \end{matrix} \tag{6.3}$$

(which is known as Lebowitz's inequality [14]). On comparing this with (4.4), (4.5) we see that this is a zeroth order perturbation theory bound and the experts will see that the perturbation theory is renormalized because it involves "interacting propagators".

For the ϕ^4 field theory described above this is merely the beginning of a whole class of "perturbation-theoretic" inequalities. The next one is

$$\langle \phi_{x_1}\phi_{x_2}\phi_{x_3}\phi_{x_4} \rangle \ge \begin{matrix} 1 \sim\!\sim 2 \\ \\ 3 \sim\!\sim 4 \end{matrix} + \begin{matrix} 1 \sim\!\sim 3 \\ \\ 2 \sim\!\sim 4 \end{matrix} + \begin{matrix} 1 \sim\!\sim 4 \\ \\ 2 \sim\!\sim 3 \end{matrix} - 6\lambda \begin{matrix} 1 & & 2 \\ & \times & \\ 4 & & 3 \end{matrix}$$

and the next one is the same except the inequality goes the other way and there is on the right hand side in addition to what is already there the $O(\lambda^2)$ correction

$$+ 18\lambda^2 \left(\begin{matrix} 1 & & 3 \\ & >\!\!\bigcirc\!\!< & \\ 2 & & 4 \end{matrix} + \begin{matrix} 1 & & 2 \\ & >\!\!\bigcirc\!\!< & \\ 3 & & 4 \end{matrix} + \begin{matrix} 1 & & 2 \\ & >\!\!\bigcirc\!\!< & \\ 4 & & 3 \end{matrix} \right) .$$

The experts will recognize this additional term as the next correction, as given by a partly renormalized perturbation theory. (The propagators are renormalized, the vertices still have the bare coupling constant λ.) The graphs of this perturbation theory are all graphs which do not contain any subgraphs connected to the remainder of the graph by two lines. This pattern of alternating lower and upper bounds continues to all orders. For details see [10]. The idea, very briefly, is to write, instead of (6.2)

$$\left\langle \frac{\partial F}{\partial \phi_y} \right\rangle_{\underset{\sim}{t}} = \left\langle \frac{\partial F}{\partial \phi_y} \right\rangle + \int_0^1 d\mu \, \frac{d}{d\mu} \left\langle \frac{\partial F}{\partial \phi_y} \right\rangle_{\mu \underset{\sim}{t}} .$$

The first term yields the right hand side of the gaussian inequality, the second term is a correction. If we are investigating a four point function then $\partial F/\partial\phi$ is a product of two fields so that the second term involves a $<\phi\phi;\phi\phi>_{\mu t}$ correlation (when the $d/d\mu$ derivative is performed -- we are considering a ϕ^4 field theory). This can be estimated by the gaussian inequality applied to $< >_t$. In combination with the path splitting lemma (5.6) this yields these inequalities. Similar results also apply to n point correlations.

These higher order inequalities should be useful in $\nu = 1,2,3$ dimensions. For higher dimensions the "bare coupling" λ should be allowed to depend on the lattice spacing as the continuum limit is taken. This is charge renormalization. This dependence is likely to destroy their usefulness and so we now look for a way around this, which essentially follows Aizenman [4] and Fröhlich [6]. The route I take passes through Simon's inequalities, our next application, which are of independent interest.

6.5 Generalized Simon's Inequalities We begin by investigating the effect of changing the coupling J_{uv} on a given "bond" uv, u,v are sites in L. We investigate the change in the two point function when the coupling is changed from J_{uv} to zero. Let $< >'$ denote the expectation obtained from $< >$ by making this change. Note

$$<\phi_x\phi_y> = <\phi_x\phi_y I>'/<I>' \tag{6.6}$$

where

$$I = \exp \phi_u J_{uv} \phi_v$$

I is a limit of polynomials with positive coefficients. The limit can be taken outside the expectation. Thus we can apply the gaussian inequality to the $< >'$ expectation with F taken to be ϕ_y I:

$$<\phi_x\phi_y I>' \leq <\phi_x\phi_y>'<I>' + <\phi_x\phi_u>' J_{uv} \cdot$$

$$\cdot<\phi_v\phi_y I>' + <\phi_x\phi_v>' J_{uv} <\phi_u\phi_y I> .$$

Dividing through by $<I>'$ yields

$$<\phi_x\phi_y> \leq <\phi_x\phi_y>' + \sum_{zz'=uv,vu} <\phi_x\phi_z>' J_{zz'} <\phi_{z'}\phi_y>$$

where we have used (6.4). Furthermore the Griffiths inequality implies that

$$\langle\phi_x\phi_y\rangle - \langle\phi_x\phi_y\rangle' \geq 0$$

because

$$\frac{\partial}{\partial J_{uv}} \langle\phi_x\phi_y\rangle = \langle\phi_x\phi_y ; \phi_u\phi_v\rangle \geq 0$$

so we have a bound in the other direction also. These considerations extend to the case where one alters the couplings on a set X of bonds instead of just one bond.

6.6 Theorem (Generalized Simon-Lieb inequalities)

Let X be a set of pairs uv of sites in the lattice L. uv and vu are regarded as distinct and X is such that if $uv \in X$ so is vu. $\langle\ \rangle_X$ is defined by setting J_{uv}, for $uv \in X$, equal to zero in the definition of $\langle\ \rangle$. Then for $x,y \in L$

$$0 \leq \langle\phi_x\phi_y\rangle - \langle\phi_x\phi_y\rangle_X \leq \sum_{uv \in X} \langle\phi_x\phi_u\rangle_X$$

$$J_{uv} \langle\phi_v\phi_y\rangle .$$

This type of inequality has been used [12] to obtain good estimates on the decay of two point functions for lattice theories. It is possible to extend these estimates to continuum field theories in less than four dimensions. See [5] for details. To some extent this is a substitute for the more complicated cluster expansion methods; see [3].

6.7 Fröhlich's version of Aizenman's inequality We wish to try and improve on the inequality

$$\langle\phi_1\phi_2\phi_3\phi_4\rangle \geq \begin{matrix} 1 \sim\!\sim 2 \\ 3 \sim\!\sim 4 \end{matrix} + \begin{matrix} 1 \sim\!\sim 3 \\ 2 \sim\!\sim 4 \end{matrix} + \begin{matrix} 1 \sim\!\sim 4 \\ 2 \sim\!\sim 3 \end{matrix} - 6\lambda \begin{matrix} 1 & & 2 \\ & \times & \\ 4 & & 3 \end{matrix}$$

mentioned in application (6.2.1) by eliminating the undesirable "bare" coupling constant λ.

We return to equation (6.1) with $\phi_x F$ taken to be $\phi_1\phi_2\phi_3\phi_4$

$$\langle\phi_1\phi_2\phi_3\phi_4\rangle = \sum_{y=2,3,4} \sum_{\omega:\ 1\to y} J_\omega \int d\nu_\omega(t) \quad \frac{Z_t}{Z} \left\langle \frac{\partial}{\partial\phi_y}(\phi_2\phi_3\phi_4)\right\rangle_t$$

Since decreasing $\underset{\sim}{t}$ gave an upper bound, see (6.2), we now try increasing $\underset{\sim}{t}$ to get a lower bound. We increase the variables t_x to ∞ for all x in the lattice which the random walk ω visits. The effect of this is to change

$$< \frac{\partial}{\partial\phi_y} (\phi_2\phi_3\phi_4)>_{\underset{\sim}{t}} \searrow \; < \frac{\partial}{\partial\phi_y} (\phi_2\phi_3\phi_4)>_\omega$$

where the new expectation $< >_\omega$ is obtained from $< >$ by replacing $g(\phi_x^2)$ by $\delta(\phi_x^2)$ for all x visited by ω $(x \in \omega)$. Thus

$$<\phi_1\phi_2\phi_3\phi_4> \geq \sum_y \sum_{\omega:\ 1\to y} J_\omega (\int d\nu_\omega(\underset{\sim}{t}) \frac{Z_{\underset{\sim}{t}}}{Z}).< \frac{\partial}{\partial\phi_y} (\phi_2\phi_3\phi_4)>_\omega$$

$$= \sum_y \sum_{\omega:\ 1\to y} J_\omega \int d\nu_\omega(\underset{\sim}{t}) \frac{Z_{\underset{\sim}{t}}}{Z} \cdot <\frac{\partial}{\partial\phi_y} (\phi_2\phi_3\phi_4)> + R$$

where

$$R \equiv \sum_y \sum_{\omega:\ 1\to y} J_\omega \int d\nu_\omega(\underset{\sim}{t}) \frac{Z_{\underset{\sim}{t}}}{Z} (< \frac{\partial}{\partial\phi_y} (\phi_2\phi_3\phi_4)>_\omega$$

$$- <\frac{\partial}{\partial\phi_y} (\phi_2\phi_3\phi_4)>) \tag{6.5}$$

Just as in application 6.2.1 we can resum the first term on the right hand side to get

$$<\phi_1\phi_2\phi_3\phi_4> \geq \begin{matrix} 1 \sim 2 \\ 3 \sim 4 \end{matrix} + \begin{matrix} 1 \sim 3 \\ 2 \sim 4 \end{matrix} + \begin{matrix} 1 \sim 4 \\ 2 \sim 3 \end{matrix} + R .$$

Thus the problem is to estimate R.

We write $R = R_2 + R_3 + R_4$ where R_2 corresponds to the possibility that $\partial/\partial\phi_y$ differentiates ϕ_2 etc. in (6.5). Thus

$$R_2 \equiv \sum_{\omega:\ 1\to 2} J_\omega \int d\nu_\omega \frac{Z_{\underset{\sim}{t}}}{Z} (<\phi_3\phi_4>_\omega - <\phi_3\phi_4>) .$$

The estimates for R_2, R_3, R_4 will be the same, so from now on we only consider R_2. To estimate R_2 we need to bound

$$<\phi_3\phi_4>_\omega - <\phi_3\phi_4>$$

(from below). If ω visits either x_3 or x_4 the first term vanishes and the lower bound we are after is the equality

$$<\phi_3\phi_4>_\omega - <\phi_3\phi_4> = -<\phi_3\phi_4>!$$

If ω does not visit x_3 or x_4 we use the generalized Simon inequality, Theorem 6.6. For X in that theorem we take all bonds uv joining sites u visited by ω, $u \in \omega$, to sites v not visited by ω (or vice versa). Then

$$<\phi_3\phi_4>_\omega = <\phi_3\phi_4>_X .$$

This is because in the $< >_\omega$ expectation certain variables ϕ_x, $x \in \omega$, are set to zero by δ functions, but this is the same as setting the couplings J_{xy} with $x \in \omega$, $y \notin \omega$ to zero (recall that these operations take place in both numerator and denominator) as far as its effect on variables ϕ_x with $x \notin \omega$ is concerned. Thus by theorem 6.4

$$<\phi_3\phi_4>_\omega - <\phi_3\phi_4> \geq - \sum_{u,v \in X} <\phi_3\phi_u>_X J_{uv} <\phi_v\phi_u>$$

and by the Griffiths inequality we can remove the X subscript in the right hand side at the cost of a further lower bound, so

$$R_2 \geq -\sum{}'_{\omega:\ 1\to 2} J_\omega \int d\nu_\omega(\underset{\sim}{t}) \frac{Z_{\underset{\sim}{t}}}{Z} \sum_{u,v;v\in\omega} <\phi_3\phi_u> J_{uv} <\phi_v\phi_4> + \cdots$$

where the prime on $\sum$ means omit paths hitting either x_3 or x_4 and ... stands for the compensation for this omission. We can assume v (as opposed to u) $\in \omega$ because $<\phi_3\phi_u>_X$ vanishes if $u \in \omega$. We interchange the two sums.

$$R_2 \geq -\sum_{u,v} \left(\sum_{\omega:\ 1\to 2,\omega\ni v} J_\omega \int d\nu_\omega \frac{Z_{\underset{\sim}{t}}}{Z}\right) <\phi_3\phi_u> J_{uv} <\phi_v\phi_4> + \cdots \quad .$$

We have discarded the prime on $\sum$ at the expense of a further bound. We apply the path splitting lemma 5.6 to the term in round brackets and obtain

$$R_2 \geq -\sum_{u,v,w} <\phi_1\phi_v> J_{vw} <\phi_w\phi_2> <\phi_3\phi_u> J_{uv} <\phi_v\phi_4> + \cdots$$

or in graphical notation

$R_2 \geq -$ $\qquad - \ldots$

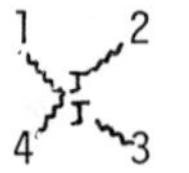

and ... turns out to be

$-$ $\qquad -$

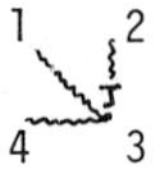

This amounts to a proof of:

6.8 Theorem (Aizenman [4], Frohlich [6])

$$\langle\phi_1\phi_2\phi_3\phi_4\rangle_T \geq -\frac{1}{2} S \left\{ \qquad + \qquad \right\}$$

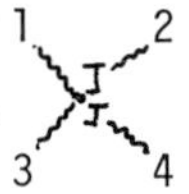

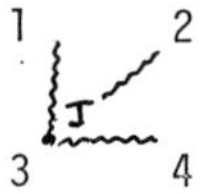

where: $\langle\phi_1\phi_2\phi_3\phi_4\rangle_T \equiv \langle\phi_1\phi_2\phi_3\phi_4\rangle - \langle\phi_1\phi_2\rangle\langle\phi_3\phi_4\rangle$

$$- \langle\phi_1\phi_3\rangle\langle\phi_2\phi_4\rangle - \langle\phi_1\phi_4\rangle\langle\phi_2\phi_3\rangle \qquad (6.6)$$

and S is the unnormalized symmetrization operator (sums over ways of relabeling the ends 1,2,3,4 which lead to distinct graphs).

Notice that this inequality has no bare coupling λ in it. Instead it has J's. This inequality, as we shall see in the next section is complementary to the others in being useful in $d \geq 4$.

The arguments we have presented can be extended to yield higher order inequalities of the same type and also similar inequalities on n point correlation functions.

7. The Continuum Limit of ϕ^4

The continuum limit $\varepsilon \to 0$ is taken after the infinite volume limit where the finite lattice L is increased to an infinite lattice. Thus in this section $\langle\ \rangle$ denotes the infinite volume limit (which exists for ϕ^4 lattice field theories).

For the coupling J we take the translation invariant nearest neighbor coupling of (2.5) appropriate to the lattice ϕ^4 field theory.

As we take the lattice spaceing ε to zero we simultaneously adjust the bare parameters

$$\lambda = \lambda(\varepsilon) > 0, \ a = a(\varepsilon), \ \tilde{Z} = Z(\varepsilon) > 0$$

so that the two point function remains finite; in more detail, so that

$$\text{(i)} \quad \sup_{\varepsilon} \langle \phi_0 \phi_x \rangle^{(\varepsilon)} < \infty \tag{7.1}$$

for x with $0 < |x| \leq 1$. ($|x|$ denotes the length of x.)

$$\text{(ii)} \quad \liminf_{\varepsilon \to 0} \langle \phi_0 \phi_x \rangle^{(\varepsilon)} \neq 0 \tag{7.2}$$

for some x. ε is taken to zero through a sequence such that the corresponding infinite lattices are each refinements of their predecessors and all contain $x = 0$. The final condition on the parameters is that for each ε

$$\text{(iii)} \quad \lim_{|x| \to \infty} \langle \phi_0 \phi_x \rangle^{(\varepsilon)} = 0 \tag{7.3}$$

This says that each of the approximating lattice theories are in their disordered phase.

7.1 Bounds on the two point function

We pause to summarize some results [15], [16], on the two point function

$$\text{(A)} \quad 0 \leq \langle \phi_0 \phi_x \rangle^{(\varepsilon)} \leq \frac{\text{const.}}{\tilde{Z}|x|^{\nu-2}}$$

where const. is independent of $\varepsilon, \lambda, a, \tilde{Z}$.

$$\text{(B)} \quad \langle \phi_0 \phi_x \rangle^{(\varepsilon)} \leq \langle \phi_0 \phi_y \rangle \text{ whenever } |y| \leq \text{const.}|x|$$

where const. is independent of $\varepsilon, \lambda, a, \tilde{Z}$. These results are consequences of Osterwalder-Schrader positivity and correlation inequalities. See [15], [16] for the proofs.

Condition (ii) together with the bound (A) implies that

$$\sup_{\varepsilon} \tilde{Z}(\varepsilon) < \infty . \tag{7.4}$$

7.2 Gaussian Processes

The defining feature of the moments of a gaussian field theory is that for any even n

$$<\phi_{x_1}\phi_{x_2}\cdots\phi_{x_n}> = \sum_\pi \Pi_{ij\,\epsilon\,\pi} <\phi_{x_i}\phi_{x_j}> \tag{7.5}$$

for $x_1,x_2,\ldots,x_n$ arbitrary (distinct) points, where π is summed over all ways of partitioning the set $1,\ldots,n$ into pairs. (The integration by parts formula (4.3) implies this result.) When we say the continuum limit is gaussian we shall mean that the limiting moments obey this identity. [If the limiting correlations obey certain conditions discussed in Professor Osterwalder's lectures then they can be analytically continued, $x \equiv (x^{(0)},x^{(1)},\ldots,x^{(\nu)}) \to (ix^{(0)},x^{(1)},\ldots,x^{(\nu)})$, to become the Wightman functions of a Minkowski field theory. The singularities at coincident points ($x_i = x_j$ for some i,j) can be changed arbitrarily without affecting the resulting Minkowski field theory.]

7.3 Theorem (Aizenman [4], Fröhlich [6])

Any continuum limit constructed in accord with conditions (i) → (iii) in dimension $\nu \geq 5$ is gaussian. The same is true of two component $(\phi^2)^2$ field theories. For dimension $\nu = 4$ the same conclusions hold if (i)-(iii) are supplemented by

$$\text{(iv)}\quad \tilde{Z}(\varepsilon) \to 0 \text{ as } \varepsilon \to 0$$
$$\text{(v)}\quad <\phi_x\phi_y>^{(\varepsilon)} \leq c|x-y|^{-\delta} \quad \text{if } |x-y| \geq 1 \tag{7.6}$$

for some constant c independent of $\varepsilon,\lambda,a,\tilde{Z}$ and some $\delta > 0$ independent of $\varepsilon,\lambda,a,\tilde{Z}$.

Proof. Let us abbreviate notation by writing

$$u^{(\varepsilon)} \equiv <\phi_{x_1}\phi_{x_2}\phi_{x_3}\phi_{x_4}>^{(\varepsilon)} - <\phi_{x_1}\phi_{x_2}>^{(\varepsilon)}<\phi_{x_3}\phi_{x_4}>^{(\varepsilon)}$$
$$- <\phi_{x_1}\phi_{x_3}>^{(\varepsilon)}<\phi_{x_2}\phi_{x_4}>^{(\varepsilon)} - <\phi_{x_1}\phi_{x_4}>^{(\varepsilon)}<\phi_{x_2}\phi_{x_3}>^{(\varepsilon)} \quad . \tag{7.7}$$

We show that if $x_i \neq x_j$, i,j = 1,4, then

$$u^{(\varepsilon)} \to 0 \tag{7.8}$$

This establishes the n = 4 case of (7.5). We omit the (similar) arguments for n > 4. First note that $u^{(\varepsilon)} \leq 0$ by theorem 6.4 . For a bound on the other side we use theorem 6.8 which, when specialized to the case at hand says

$$u_4 \geq -\tfrac{1}{2}\tilde{Z}^2 \varepsilon^{\nu-4} S\{\textstyle\sum_{u,v,w} \varepsilon^{\nu} \langle\phi_{x_1}\phi_u\rangle^{(\varepsilon)}$$

$$\cdot\langle\phi_{x_2}\phi_u\rangle^{(\varepsilon)}\langle\phi_{x_3}\phi_v\rangle^{(\varepsilon)}\langle\phi_{x_4}\phi_w\rangle^{(\varepsilon)} + \ldots\} \tag{7.9}$$

The ... refers to the three line tree graphs in theorem 6.8 . The factors of ε came from the J couplings according to (2.5). u,v,w are constrained by v,w being nearest neighbors to u. Let $\sum^{>}$ denote the sum over u,v,w such that

$$|u-x_i| > \Delta \quad i = 1,\ldots,4 \tag{7.10}$$

where Δ is chosen so that $|x_i-x_j| > 2\Delta$ for i,j = 1,...4. We split the sum in (7.9)

$$\textstyle\sum_{u,v,w} = \sum^{>} + \sum^{1} + \sum^{2} + \sum^{3} + \sum^{4} \tag{7.11}$$

where $\sum^{i}$ sums over u,v,w with $|u-x_i| \leq \Delta$.

Suppose $\nu \geq 5$: in the $\sum^{>}$ contribution we use (A) to bound two of the two point functions and condition (i) and (B) to bound the other two two point functions by constants independent of ε. We obtain a lower bound

$$-3c\varepsilon^{\nu-4} \int d^{\nu}x \; \frac{1}{|x_1-x|^{\nu-2}} \; \frac{1}{|x_2-x|^{\nu-2}} \tag{7.12}$$

where we replaced the Riemann sum by an integral at the cost of a further bound. c is independent of ε. Since $\nu \geq 5$ this lower bound increases to zero as $\varepsilon \searrow 0$.

Now consider the $\sum^{1}$ contribution to (7.9): Again we bound two of the two point functions by (A) and the other two by (i) and (B). One of the two point functions bounded by (A) is chosen to be

$\langle\phi_{x_1}\phi_u\rangle^{(\varepsilon)}$. We obtain the lower bound

$$-3c\varepsilon^{\nu-4}\int_{|x|\leq\Delta} d^\nu x \, \frac{1}{|x_1-x|^{\nu-2}} \, \frac{1}{|x_2-x|^{\nu-2}} \tag{7.13}$$

which goes to zero with ε for $\nu \geq 5$. (Recall $|x_2-x| \geq \Delta$.) In like manner the $\sum^i$ contributions for $i = 2,3,4$ go to zero. The ... part of (7.9) is even more rapidly converging to zero by (B).

$\underline{\nu = 4}$: We refer the reader to [6] for details. The idea is the same with an additional refinement: instead of bounding two two-point functions using (A), use it to bound one and a large fraction of the other:

$$\langle\phi_x\phi_y\rangle = \langle\phi_x\phi_y\rangle^{\alpha}\langle\phi_x\phi_y\rangle^{1-\alpha} \leq \text{const.}\,\frac{1}{|x-y|^{(\nu-2)\alpha}} \tag{7.14}$$

with $\alpha < 1$, $\alpha \simeq 1$. With this procedure, even though the $\varepsilon^{\nu-4}$ factor provides no convergence, because $\nu = 4$, we still have a $\tilde{Z}^{1-\alpha}$. This factor cannot hurt by (7.4) and in fact with hypothesis (iv) gives the requisite convergence.

REFERENCES

[1] Symanzik, K., Euclidean quantum field theory in "Local Quantum Theory", (ed. Jost), New York, London, Academic Press (1969).

[2] See for example the article by E. Brézin in "Methods in Field Theory", Les Houches 1975 Session XXVIII North-Holland Amsterdam-New York-Oxford.

[3] Glimm, J., Jaffe, A., "Quantum Physics", Springer-Verlag, New York, Heidelberg, Berlin, 1981.

[4] Aizenman, M., Phys. Rev. Lett. 47 1 (1981).

[5] Brydges, D.C., Fröhlich, J., Spencer, T., The Random Walk Representation of Classical Spin Systems and Correlation Inequalities, to appear in Comm. Math. Phys.

[6] Fröhlich, J., On the Triviality of $\lambda\phi_d^4$ Theories and the Approach to the Critical Point in d $\underset{(=)}{\geq}$ 4 Dimensions, I.H.E.S. preprint.

[7] Durhuus, B., Fröhlich, J., Comm. Math. Phys. 75, 103 (1980).

[8] See for example, Federbush, P., Battle, G., Phase Cell Cluster Expansion for Euclidean Field Theories, Michigan University preprint.

Gawedski, K., Kupiainen, A., Comm. Math. Phys. 82, 407 (1981).

Fröhlich, J., Spencer, T., The Kosterlitz-Thouless Transition in Two Dimensional Abelian Spin Systems and the Coulomb Gas, I.H.E.S. preprint.

[9] Ginibre, J., Comm. Math. Phys. 16, 310 (1970).

[10] Brydges, D.C., Fröhlich, J., Sokal, A., in preparation.

[11] Newman, C., Z.f. Wahrscheinlichkeitstheorie 33, 75 (1975).

[12] Simon, B., Comm. Math. Phys. 77, 111 (1980). See also the articles by Lieb, Rivasseau and Simon and Aizenman in the same volume.

[13] These inequalities can be subsumed by a "machine" created by Boel, R.J., Kasteleyn, P.W., see Comm. Math. Phys. 61, 191 (1978); 66, 167 (1979); Physica 93A, 503 (1978).

[14] Lebowitz, J.L., Comm. Math. Phys. 28, 313 (1972).

[15] Sokal, A., An Alternate Constructive Approach to the ϕ_3^4 Quantum Field Theory, and a Possible Destructive Approach to ϕ_4^4, Princeton preprint, January 1979.

[16] Schrader, R., Phys. Rev. B15, 2798 (1977).
Messager, A., Miracle-Sole, S., J. Stat. Phys. 17, 245 (1977).
Hegerfeldt, G.C., Comm. Math. Phys. 57, 259 (1977).
Szasz, D., J. Stat. Phys. 19, 453 (1978).

Research supported in part by NSF Grant MCS 79-0249

TOWARD A PROBABILISTIC APPROACH TO QUANTUM FIELD THEORIES WITH FERMI PARTICLES *

by

Francesco Guerra

Istituto Matematico "G.Castelnuovo",

Università di Roma.

* Based on lectures given at the Brasov International Summer School on "Gauge Theories: Fundamental Interactions and Rigorous Results", August 25-September 7, 1981, Poiana Brasov, Romania.

1. INTRODUCTION.

Probabilistic methods in quantum field theory have been extremely useful in the past ten years. In particular the investigation about existence and physical properties of quantum field theory models received great help from the introduction of the probabilistic ideas of Symanzik [20] and Nelson [15] in the frame of the constructive quantum field theory program of Glimm and Jaffe and others [7,21] . On the other hand methods based on the theory of stochastic processes have been extremely successful for numerical investigations on concrete models of physical interest for elementary particle theory, see for example [2,13,16] and references quoted there.

If we take into account the strong connection of probabilistic methods with with the well established Euclidean formulation of relativistic quantum field theory, we realize that probability theory can be considered as a unitary and powerful general frame for the formulation of models of modern theoretical physics, ranging from statistical mechanics, to quantum field theory and elementary particle theory.

Until now a direct application of probabilistic methods has been possible in the case where only Bose fields are present. In the models where Fermi fields partecipate to the interaction then some very powerful techniques of "integrating out" the Fermi degrees of freedom have been developped. In this frame the Fermi structure is transformed into a highly non local effective interaction for Bose fields and only these are directly described through probabilistic methods. This strategy has been very useful both in constructive quantum field theory (see [1,18] and references given there) and for numerical investigation based on Monte Carlo methods (see for example [13]).

On the other hand it would be interesting from a conceptual point of view and also useful (both for general and numerical investigations) to reach a direct probabilistic formulation for fields with Fermi statistics. Clearly this is a formidable task, due to the complications coming from statistics and from half-integer spins. In this connection previous work on the extension of Feynman path integrals to the Fermi case should be mentioned [11] .

In the last few years some steps have been done toward the objective of a complete description of Fermi fields in a probabilistic frame.

The purpose of this note, based on joint work with G.F.De Angelis and D.de Falco [3] , is to provide a coincise and elementary introduction to the proposed general scheme of a probabilistic approach to relativistic quantum field theory with Fermi particles.

For related work we refer to [4] and [5] .

It will be shown that the original ideas of stochastic mechanics, introduced by Nelson [14] some years ago, play a major role. In fact stochastic mechanics (see [8] for a general introduction) can be considered as an alternative (and physically equivalent) formulation of quantum mechanics, based on the theory of stochastic processes and completely equivalent, as far as observable effects are compared, to the standard formulation (see for example [22]) relying on operator theory techniques. The main problem is to show that Fermi fields can be considered as quantum mechanical systems for which the stochastic quantization procedure can be applied.

Our treatment will be extremely simple and paedagogic, stressing the main ideas on the basis of very simple examples.

The paper is organized as follows.

In Section 2 we recall the basic decomposition of the free Fermi-Diracfield in terms of the dynamic variables of a Fermi oscillator. Then a Jordan-Wigner transformation reduces the multi-dimensional Fermi oscillator to a collection of independent one-dimensional oscillators. In Section 3 the elmentary quantum mechanics of the Fermi oscillator will be reviewed. In Section 4 we introduce some basic kinematical considerations related to continuous time random Markov processes taking values on the Z_2 group. In Section 5 the quantum mechanical properties of the Fermi oscillator are formulated in terms of the probabilistic structure described in Section 4 and some additional dynamical assumptions.

Finally Section 6 is dedicated to some conclusions and outlook.

It is with great pleasure that the author would like to thank very much the Organizing Committee, in particular Prof. A.Corciovei and Prof. P.Dita, for the kind hospitality extended to him in Poiana Brasov. Warmful thanks go also to all partecipants of the School for the friendly and stimulating athmosphere created during the two week stay.

2. THE FREE DIRAC FIELD AS A FERMI OSCILLATOR.

We work in the real wave representation of Schwinger 17 according to the general frame described in 6 . Then the Dirac field is represented through selfadjoint operators

$$\psi_a(\vec{x},t) = \psi_a(\vec{x},t)^+ , \tag{1}$$

where the index a describes the inner degrees of freedom related to spin and charge, $a = 1,2,..,8$, while $\vec{x} \in R^3$ and $t \in R$ are space-time labels. The equal time anticommutation relations are given by

$$\{\psi_a(\vec{x},t), \psi_b(\vec{y},t)\} = \delta_{ab}\, \delta(\vec{x}-\vec{y}) . \tag{2}$$

The Wightman two-point function is

$$W_{aa'}(\vec{x},t;\vec{x}',t') = \langle \Omega_0, \psi_a(\vec{x},t)\psi_{a'}(\vec{x}',t')\Omega_0\rangle = \tag{3}$$

$$=(2\pi)^{-3}\int \left(E(\vec{p})+\vec{\alpha}\cdot\vec{p}+M\alpha_5\right)_{aa'} \exp\left[i\vec{p}\cdot(\vec{x}-\vec{x}')-iE(\vec{p})(t-t')\right]\left(2E(\vec{p})\right)^{-1} d\vec{p},$$

where Ω_0 is the vacuum, M is the mass of the particles, $E(p)=\sqrt{\vec{p}^2+M^2}$, $\vec{\alpha} \equiv \{\alpha_1,\alpha_2,\alpha_3\}$ and the 8X8 Hermitian matrices $\alpha_1,\alpha_2,\alpha_3,\alpha_4$ (real) and $\alpha_5,\alpha_6,\alpha_7$ (imaginary) form the Dirac algebra with anticommutation relations

$$\{\alpha_i,\alpha_j\} = 2\,\delta_{ij} \quad , \quad i,j = 1,2,..,7 . \tag{4}$$

It is convenient to discretize the momentum degrees of freedom by enclosing the system in a spatial box. By standard methods we can chose a complete set of real spinors $\{u_j(\vec{x}) \; , \; v_j(\vec{x})\}$,j=1,2,..,such that the following is true. First of all we have the partial wave decomposition for the field

5) $$\psi(\vec{x},t) = \sum_j \left(u_j(\vec{x})\, Q_j(t) + v_j(\vec{x})\, P_j(t) \right),$$

where spinor indices have been suppressed. The operators $\{P_j(t), Q_j(t)\}$ $j = 1,2,\ldots$, describe the full dynamical content of the theory in the Heisenberg representation. They are selfadjoint

6) $$Q_j = Q_j^\dagger \;, \quad P_j = P_j^\dagger \;,$$

and satisfy the anticommutation relations

7) $$Q_i Q_j + Q_j Q_i = 2\delta_{ij} \;, \quad P_i P_j + P_j P_i = 2\delta_{ij} \;, \quad P_i Q_j + Q_j P_i = 0.$$

The dynamical development is given by

8) $$\dot{Q}_j = \omega_j P_j \;, \quad \dot{P}_j = -\omega_j Q_j \;,$$

where $\{\omega_j\}$ are connected with the chosen complete system of spinors. The equations 8) can be considered Heisenberg equations of the type

9) $$i\dot{A} = [A, H]$$

with Hamiltonian

10) $$H = i \sum_j (\omega_j/2)\, Q_j P_j \;.$$

Due to the anticommutation relations 7) the variables P,Q are not very convenient for a direct introduction of probabilistic methods. It is better to perform firstly a Jordan-Wigner transformation

11) $$\{P, Q\} \longleftrightarrow \{\bar{P}, \bar{Q}\} \;,$$

defined recursively by

12) $$\bar{Q}_1 = Q_1 \;, \bar{P}_1 = P_1 \;; \bar{Q}_s = i\,\bar{P}_{s-1} Q_{s-1} Q_s \;, \quad \bar{P}_s = i\,\bar{P}_{s-1} Q_{s-1} P_s \;, \; s = 2,3,\ldots$$

with inverse

13) $$Q_1 = \bar{Q}_1 \;, P_1 = \bar{P}_1 \;; Q_s = i\,P_{s-1} \bar{Q}_{s-1} \bar{Q}_s \;, \quad P_s = i\,P_{s-1} \bar{Q}_{s-1} \bar{P}_s \;, \; s = 2,3,\ldots$$

Then we still have the selfadjointness conditions 6) for $\bar{P},\bar{Q}$ but the anticommutation relations 7) become

14) $$\bar{Q}_i \bar{Q}_j = \bar{Q}_j Q_i \;, \quad \bar{P}_i \bar{P}_j = \bar{P}_j P_i \;, \quad \bar{P}_i \bar{Q}_j = \bar{Q}_j \bar{P}_i \;, \quad i \neq j \;,$$

15) $\bar{Q}_j^2 = 1, \quad \bar{P}_j^2 = 1, \quad \bar{P}_j \bar{Q}_j + \bar{Q}_j \bar{P}_j = 0 .$

Also equations 8),9),10) still hold with $\bar{P},\bar{Q}$ replacing P,Q.

Since variables associated to different degrees of freedom commute, the system splits into independent one-dimensional Fermi oscillators each described by the couple $\bar{P}_j,\bar{Q}_j$ satisfying 6),8),9),10),15). These quantum mechanical systems are very simple and their properties will be recalled in the next Section.

3. QUANTUM MECHANICS OF THE FERMI OSCILLATOR.

The one-dimensional Fermi oscillator is described by dynamical variables P,Q , satisfying the conditions

16) $P = P^{\dagger}, \quad Q = Q^{\dagger}, \quad P^2 = 1, \quad Q^2 = 1, \quad PQ + QP = 0 .$

In the Heisenberg picture these variables evolve according to

17) $\dot{Q}(t) = P(t), \quad \dot{P}(t) = -Q(t),$

with an appropriate choice of the time scale (so that $\omega = 1$).

The Hamiltonian is given by

18) $H = (i/2)\, QP .$

The solution of 17) is

19) $Q(t) = Q \cos t + P \sin t, \quad P(t) = P \cos t - Q \sin t,$

where P and Q in the Schrödinger picture satisfy 16).

We chose to diagonalize the "configuration" variable Q (Schrödinger representation). We introduce the Hilbert space $\mathcal{H} = L^2(Z_2)$ of functions defined on the group $Z_2 = \{-1,1\}$ with normalized measure dx defined by

20) $\int \cdot \, dx = \frac{1}{2} \sum_{x=\pm 1} \cdot .$

Since $\chi_0(x) = 1, \chi_1(x) = x$ are a complete orthonormal basis on Z_2 any function in $\mathcal{H}$ can be represented in the form

21) $\psi = \psi_0 + x\, \psi_1 \quad \text{with} \quad \psi_0, \psi_1 \in \mathbb{C} .$

Scalar products and norms are given by

22) $\langle \psi, \psi' \rangle = \int \psi^*(x)\psi'(x)\,dx = \psi_0^* \psi_0' + \psi_1^* \psi_1' \;, \quad \|\psi\|^2 = |\psi_0|^2 + |\psi_1|^2.$

The action of Q on $\mathcal{H}$ is given by

23) $(Q\psi)(x) = x\,\psi(x)\;.$

Let us introduce the annihilation operator ∇

24) $(\nabla\psi)(x) = \psi_1 = \frac{x}{2}\left(\psi(x) - \psi(-x)\right)$

and its adjoint creation operator $\nabla^+ = Q - \nabla$

25) $(\nabla^+\psi)(x) = \frac{x}{2}\left(\psi(x) + \psi(-x)\right)\,.$

Then we can also write the usual relations

26) $Q = \nabla + \nabla^+ \;, \quad P = i\left(\nabla^+ - \nabla\right)\;, \quad H = \nabla^+\nabla\,.$

27) $(P\psi)(x) = i\,x\,\psi(-x)\;, \quad (H\psi)(x) = \frac{1}{2}\left(\psi(x) - \psi(-x)\right).$

In the Schrödinger picture the wave function $\psi(x,t)$ evolves in time according to

28) $i\,(\partial_t\psi)(x,t) = (H\psi)(x,t) = \frac{1}{2}\left(\psi(x,t) - \psi(-x,t)\right)\,,$

$\psi(x,t) = \psi_0(t) + x\,\psi_1(t)\;, \quad \psi_0(t) \equiv \psi_0\;, \quad \psi_1(t) = e^{-it}\,\psi_1\,.$

It is important to remark that the Hamiltonian semigroup exp(-tH), $t \geqslant 0$, is positivity preserving. In fact from $\psi \geqslant 0$ it follows $\exp(-tH)\psi \geqslant 0$. This is one of the reasons of the possibility of a probabilistic description.

Notice that any state depends only on two parameters, in fact ψ_0 and ψ_1 are related by the normalization condition, moreover an overall phase factor is unessential. It is very well known that the space of pure states of this system is isomorphic to the surface of a sphere in R^3, while mixtures correspond to points in the interior of the sphere.

We can also introduce the Madelung fluid [12,8] associated to this simple system. Let us define

29) $\psi(x,t) = \exp\left[R(x,t) + i\,S(x,t)\right]\,,$

$R(x,t) = R_0(t) + x\,R_1(t)\;, \quad S(x,t) = S_0(t) + x\,S_1(t)\,,$

Then by separating real and imaginary part of the Schrödinger equation 27) we have

$$30)\quad \dot{R}_0 = -\frac{1}{2}\,\mathrm{sh}\,2R_1 \cdot \sin 2S_1 \;,\quad \dot{R}_1 = \frac{1}{2}\,\mathrm{ch}\,2R_1 \cdot \sin 2S_1 \,,$$

$$31)\quad \dot{S}_0 = \frac{1}{2}\left[\cos 2S_1 \cdot \mathrm{ch}\,2R_1 - 1\right] ,\quad \dot{S}_1 = -\frac{1}{2}\cos 2S_1 \cdot \mathrm{sh}\,2R_1 .$$

Introduce the density in configuration space

$$32)\quad \rho(x,t) = |\psi(x,t)|^2 = \exp\left[2R(x,t)\right] = \rho_0(t) + x\,\rho_1(t)$$

and notice that the normalization condition is equivalent to

$$33)\qquad \int \rho(x,t)\,dx = 1$$

is equivalent to

$$34)\quad \rho_0 = e^{2R_0}\,\mathrm{ch}\,2R_1 = 1 \quad ,\quad \rho_1 = \mathrm{tgh}\,2R_1 \;.$$

Therefore the two equations 30) are equivalent to the following continuity condition

$$35)\quad \dot{\rho}_1 = \sin 2S_1 \,/\, \mathrm{ch}\,2R_1 \;.$$

Notice that S_0 has no dynamical meaning, in fact if we solve $30)_2$ and $31)_2$ for R_1 and S_1 then by substitution in $31)_1$ we get S_0 as unessential contribution to a phase factor in the wave function.

In the following we will give an interpretation of $31)_2$ as the analog of the Hamilton-Jacobi equation, expressing a kind of second principle of dynamics in the stochastic frame.

4. MARKOV PROCESSES ON Z_2.

Let us consider a Markov process q(t) taking values on Z_2. If $\rho(x,t)$ is the normalized density at time t,

$$36)\qquad \rho(x,t) = 1 + x\,\rho_1(t) \,,$$

then the expectations are written as

$$37)\quad E\big(F(q(t),t)\big) = \int F(x,t)\,\rho(x,t)\,dx = F_0(t) + F_1(t)\,\rho_1(t)$$

for a generic time dependent function on Z_2

38) $$F(x,t) = F_0(t) + x F_1(t).$$

We can introduce the mean forward and backward derivatives of the process

39) $$p_{(\pm)}(x,t) \equiv (D_{(\pm)} q)(t) \equiv p_0^{(\pm)}(t) + x p_1^{(\pm)}(t) =$$
$$= \pm \lim_{\Delta t \to 0^+} (\Delta t)^{-1} E\left(q(t \pm \Delta t) - q(t) / q(t) = x\right),$$

where E(/) are conditional expectations.

In general for generic functions of the type 38) we define

40) $$(D_{(\pm)} F)(x,t) = \pm \lim_{\Delta t \to 0^+} (\Delta t)^{-1} E\left(F(q(t \pm \Delta t), t \pm \Delta t) - F(q(t),t) / q(t) = x\right)$$

and we find from 39)

41) $$(D_{(\pm)} F)(x,t) = (\partial_t F)(x,t) + p_{(\pm)}(x,t)(\nabla F)(x,t).$$

The derivatives $p_{(\pm)}$ in 39) can be easily expressed in terms of the transition probability p(x',t';x,t) from x at time t to x' at time t', $t \leq t'$. In fact we have

42) $$p_{(+)}(x,t) = \lim_{\Delta t \to 0^+} (\Delta t)^{-1} \sum_{x' = \pm 1} (x' - x) p(x', t + \Delta t; x, t) =$$
$$= -2x \lim_{\Delta t \to 0^+} (\Delta t)^{-1} p(-x, t + \Delta t; x, t).$$

Since $p(x',t';x,t) \geq 0$ we have the positivity conditions

43) $$x p_{(+)}(x,t) \leq 0 \qquad \text{or} \qquad p_1^{(+)}(t) \leq |p_0^{(+)}(t)|,$$

which can be also written as

44) $$p_{(+)}(1,t) \leq 0, \quad p_{(+)}(-1,t) \geq 0$$

with immediate physical interpretation.

We can also define

45) $$D = (D_{(+)} + D_{(-)})/2 = \partial_t + p\nabla, \quad \delta D = (D_{(+)} - D_{(-)})/2 = \delta p \nabla,$$

$$p(x,t) = (p_{(+)} + p_{(-)})/2 = p_0(t) + x p_1(t), \quad \delta p(x,t) = (p_{(+)} - p_{(-)})/2 = (\delta p_0)(t) + x \delta p_1(t).$$

Trough standard methods [3,8,14] we easily find the following kinematical conditions

46) $$E\left((\delta p)(q(t),t)\right) = 0, \quad \delta p_0 + p_1 \delta p_1 = 0$$

$$E\left(q(t)\,p(q(t),t)\right) = 0 \quad , \quad p_1 + \rho_1 p_0 = 0 \; ,$$
$$\dot{\rho}_1(t) = E\left(p(q(t),t)\right) \quad , \quad \dot{\rho}_1 = p_0 + \rho_1 p_1 \; .$$

The positivity condition 43),44) can also be written as

47) $$\delta p_1 \leq -|p_0| \; .$$

Equations 46) show that we can assume $\rho_1, \dot{\rho}_1, \delta p_1$ as basic dynamical variables, while $\delta p_0, p_1, p_0$ can be expressed as

48) $$\delta p_0 = -\rho_1 \delta p_1 \quad , \quad p_1 = -\rho_1 p_0 \quad , \quad p_0 = \dot{\rho}_1 / (1-\rho_1^2) \; .$$

5. STOCHASTIC MECHANICS OF THE FERMI OSCILLATOR.

According to the basic strategy of stochastic mechanics [14,8] we now try to associate to each quantum state of the system, as described in Section 3, some stochastic process q(t) satisfying the general kinematical conditions of Section 4 and some additional dynamical assumptions. The association must reproduce fixed time quantum averages in terms of stochastic expectations of the type 37).

First of all let us recall that the dynamical content of the quantum description is carried by the equations 35),31)$_2$. Taking an additional time derivative in 35) and exploiting 32) we see that ρ_1 must satisfy

49) $$\ddot{\rho}_1 + \rho_1 = 0 \quad ,$$

whose general solution is

50) $$\rho_1(t) = A\cos(t-t_0) \quad \text{with} \quad 0 \leq A \leq 1 \; .$$

Therefore in this simple case the equation 49) describes the essential dynamical content of the theory, while 35) can be taken as a definition of S_1 (as a matter of fact we have two different states associated to each solution of 49)).

Let us now consider a Markov process q(t) satisfying the kinematical conditions of Section 4. In analogy with the dynamical equation 17) we assume that q(t) satisfies also the averaged equation

51) $\frac{d}{dt} E(p(q(t),t)) = - E(q(t))$.

Notice that $p_{\pm}$ could be equivalently substituted to p in 51) without any modification because $p_{\pm} = p \pm \delta p$ and 46) holds.

But the dynamical assumption 51) turns out to be equivalent to 49), if we take into account $46)_3$ and the obvious $E(q(t)) = \rho_1(t)$. Therefore, by collecting all results , we have the following statement: any Markov process on Z_2 satisfying the kinematical conditions of Section 4 and the dynamical assumption 51) will reproduce at each fixed time stochastic expectations identical to the quantum mechanical ones.

While in usual stochastic mechanics Nelson's assumption of universal Brownian motion [14] specifies completely the process, here no such assumption is available for the moment, therefore δp has been left completely unspecified, but for the condition $46)_1$,47).

In the paper [3] we have explored the possibility of replacing 51) with the stronger

52) $\frac{1}{2}\left(D_{(+)} P_{(-)} + D_{(-)} P_{(+)} \right) = - q$

which is the analog of the smoothed form of the second principle of dynamics, introduced by Nelson [14] . Clearly 52) implies 51) but is stronger because it also determines δp_0 and δp_1 . In [3] it is shown that 52) gives very good results for states near the ground state.

Let us describe in detail the process associated to the ground state Ω_0 . It is very simple to derive from 52) in this case the following expressions

53) $\varphi = 1$, $R = S = 0$, $\rho = 1$, $\rho_1 = 0$,

$p = 0$, $\delta p_0 = 0$, $(\delta p)(x,t) = p_{(+)}(x,t) = - p_{(-)}(x,t) = -x$.

Therefore for the stochastic process associated to the ground state we have

54) $D_{(\pm)} q = \mp q$,

and the correlation functions are easily expressed in the form

55) $$E\left(q(t_1)\ldots q(t_n)\right) = e^{-(t_n - t_{n-1})} \ldots e^{-(t_2 - t_1)}$$

for n even and $t_1 \leq t_2 \leq \ldots \leq t_n$, while they vanish for n odd.

As in more general cases [10], also here we find the rather surprising result that the correlation functions of the ground state process can be indirectly obtained through the analytic continuation of the Wightman functions and vice versa. In fact the Wightman functions for the Fermi oscillator are easily found from 19) in the form

56) $$W(t_1, \ldots, t_n) = \langle \Omega_0, Q(t_1)\ldots Q(t_n)\Omega_0 \rangle = e^{-it_1} e^{it_2} \ldots e^{it_n}$$

for n even, while they vanish for n odd.

On the other hand we also have the standard possibility (see [9]) of deriving the whole quantum mechanical content of the theory from the knowledge of the ground state process. In fact let us assume that (Q, Σ, μ) is the probability space of the process q(t). Then we can verify that all Hilbert spaces $L^2(Q, \Sigma_t, \mu)$, where Σ_t is the σ-algebra generated by the process at time t, are isomorphic to the physical Hilbert space. By introducing the embedding operators j_t defined by

57) $$j_t : \mathcal{H} \longrightarrow L^2(Q, \Sigma, \mu),$$
$$\psi_0 + x\psi_1 \longrightarrow \psi_0 + q(t)\psi_1,$$

we verify that the Hamiltonian semigroup is correctly given by the standard transfer matrix equation

58) $$\langle \psi', e^{-tH}\psi \rangle = E\left((j_0\psi')^*(j_t\psi)\right),$$

in analogy with the case of Bose fields [9].

6. CONCLUSIONS AND OUTLOOK.

In the previous Sections we have seen that also quantum mechanical systems with Fermi statistics can be in principle described in a stochastic frame. The right strategy is to find complete sets of observables which can be promoted to stochastic processes, such that the Hamiltonian semigroup can be obtained through the transfer matrix.

Since the transfer matrix is obviously positivity preserving we have that, in the chosen representation, the Hamiltonian semigroup must be positivity preserving. It was found that the Jordan-Wigner transformed "configuration" variables Q are possible candidates for the right variables.

In [5] a probabilistic framework of the type described here is extended to realistic (strongly cut off) quantum field theories containing Bose-Fermi interactions.

Clearly the situation is rather satisfactory from a conceptual point of view and the goal of obtaining a unified direct probabilistic description of Bose and Fermi fields seems very near. But additional work is necessary in order to show that the proposed probabilistic scheme is really useful for the mathematical control of the ultra-violet and infrared cut off removal or for numerical investigations.

REFERENCES.

[1] G.A.Battle and L.Rosen, Journal of Statistical Physics,22,123(1980).

[2] M.Creutz, L.Jacobs and C.Rebbi, Phys.Rev.Lett.,42,1390(1979).

[3] G.F.De Angelis,D.de Falco and F.Guerra, Phys.Rev.,D23,1747(1981).

[4] G.F.De Angelis and G.Jona-Lasinio, A Stochastic Description of Spin $\frac{1}{2}$ Particle in a Magnetic Field, Marseille Preprint CPT-81/P 1334, November 1981.

[5] D.de Falco, S.De Martino and S.De Siena, Probabilistic Ideas for Two Dimensional Interacting Fermi Fields on a Lattice, Salerno Preprint, October 1981.

[6] D.de Falco and F.Guerra, J.Math.Phys.,21,1111(1980).

[7] J.Glimm and A.Jaffe, Quantum Physics, A Functional Integral Point of View, Springer Verlag,Berlin,1981.

[8] F.Guerra, Structural aspects of Stochastic Mechanics and Stochastic Field Theory, Phys.Rep.C, to appear.

[9] F.Guerra,L.Rosen and B.Simon, Ann.Math. 101, 111 (1975).

[10] F.Guerra and P.Ruggiero,Phys.Rev.Lett. 31, 1022 (1973).

[11] J.R.Klauder, Ann. Phys. 11, 123 (1960).

[12] E.Madelung, Zeits.f.Physik, 40, 322 (1926).

[13] E.Marinari,G.Parisi and C.Rebbi, Monte Carlo Simulation of the Massive Schwinger Model, CERN preprint, TH 3080, May 1981.

[14] E.Nelson, Dynamical Theories of Brownian Motion, Princeton University Press, 1967.

[15] E.Nelson, Probability Theory and Euclidean Field Theory, in 21 .

[16] G.Parisi and Wu Yong-shi, Perturbation Theory without Gauge Fixing, ASITP Preprint, 1980, to appear on Scientia Sinica.

[17] J.Schwinger, Phys.Rev. 115, 721 (1959).

[18] E.Seiler, Commun.Math.Phys. 42, 163 (1965).

[19] B.Simon, The $P(\phi)_2$ Euclidean (Quantum) Field Theory, Princeton University Press, 1974.

[20] K.Symanzik, Euclidean Quantum Field Theory, in Local Quantum Theory, R.Jost, ed., Academic Press, 1969.

[21] G.Velo and A.Wightman, eds, Constructive Quantum Field Theory, Springer Verlag, Berlin, 1973.

[22] J.Von Neumann, Mathematical Foundation of Quantum Mechanics, Princeton University Press, 1955.

IV. Related Topics

GEOMETRIC QUANTISATION

D.J. SIMMS
School of Mathematics
Trinity College, Dublin 2

Introduction

In these lectures I want to describe the differential geometric approach to canonical quantisation which is largely due to B. Kostant [1] and J.M. Souriau [4] and is generally known as geometric quantisation. Many of the ideas originate in the work of Kirillov and others in the representation theory of Lie groups. The ideas are also very much in the spirit of the symplectic approach to quantisation pioneered by I. Segal.

The constructions are concerned with a smooth manifold M representing the phase space of a classical dynamical system. For systems with a finite number of degrees of freedom, M is a finite dimensional manifold. For field theories we have to work with an infinite dimensional manifold of some suitable type. We confine ourselves to the finite dimensional case except briefly in the final section.

We assume that a non-degenerate closed differential 2-form ω is given on M. Thus $d\omega = 0$, and $X \lrcorner \omega = 0$ if and only if $X = 0$. Here $X \lrcorner \omega$ denotes the contraction of the vector field X with the 2-form ω. Such a 2-form ω is called a symplectic form and M is then called a symplectic manifold. Each smooth function H on a symplectic manifold M defines a vector field X_H on M called the Hamiltonian vector field associated to H, defined by the equation

$$dH + X_H \lrcorner \omega = 0.$$

If H and H' are smooth functions on M, we define their Poisson bracket $\{H,H'\}$ to be the function

$$\{H,H'\} = X_H(H') = \langle dH', X_H \rangle = \omega(X_H, X_{H'}).$$

With respect to the Poisson bracket the smooth functions on M form a Lie algebra and $H \to X_H$ is a Lie algebra homomorphism into the algebra of vector fields on M with Lie bracket.

The motivation for these concepts comes from the classical theory of Hamiltonian dynamics. Here we start with a configuration space Q and construct the corresponding cotangent bundle $M = T^*Q$ whose points are the covariant vectors on Q. Then M is called the classical <u>phase space</u> corresponding to the configuration space Q. Each local coordinate system $q^1, \dots, q^n$ on Q defines an associated coordinate system $q^1, \dots, q^n, p_1, \dots, p_n$ on M. The differential 1-form $\Sigma p_i dq^i$ is independent of the initial choice of coordinates $q^1, \dots, q^n$ and hence is the local expression for a canonical differential 1-form β defined globally on M. The 2-form $\omega = d\beta$ whose local expression is $\Sigma dp_i \wedge dq^i$ is a canonical symplectic form on M. We have

$$X_H = \Sigma\left(\frac{\partial H}{\partial p_i}\frac{\partial}{\partial q^i} - \frac{\partial H}{\partial q^i}\frac{\partial}{\partial p_i}\right)$$

and hence the integral curves of X_H are the solutions of the Hamiltonian equations of motion

$$\frac{dq^i}{dt} = \frac{\partial H}{\partial p_i}, \quad \frac{dp_i}{dt} = -\frac{\partial H}{\partial q^i}$$

The aim of geometric quantisation is to study the geometry associated with the symplectic form ω, and to use it to derive the kinematics and the dynamics of the corresponding quantum mechanical system.

Symplectic potentials

The phase space M under consideration may not be associated with a configuration space Q. For example, the natural phase space associated with a spin degree of freedom is the 2-sphere S^2 with some suitable multiple of the area element as the symplectic form. Since S^2 is compact it cannot be the cotangent bundle of any configuration space Q.

However, a theorem of Darboux ensures that on any symplectic manifold M we can always choose coordinates $p_1, \dots, q^n$ in the neighbourhood of any point of M so that

$$\omega = \Sigma\, dp_i \wedge dq^i$$

Such coordinates are called <u>canonical coordinates</u> on the symplectic manifold M.

A 1-form θ defined on an open subset of M is called a <u>symplectic potential</u> of $d\theta = \omega$. Any choice of canonical coordinates $p_1, \dots, p_n, q^1, \dots, q^n$ gives us a symplectic potential $\Sigma\, p_i dq^i$. In particular, if M is a phase space derived from a configuration space then M carries a canonical and globally defined symplectic potential β whose local expression is $\Sigma\, p_i\, dq^i$ for any choice of configuration coordinates $q^1, \ldots, q^n$ and corresponding momentum coordinates $p_1, \ldots, p_n$. For a given Hamiltonian H, this choice of symplectic potential leads to the usual Lagrangian

$$\Sigma\, p_j \frac{dq^i}{dt} - H = X_H \lrcorner\, \beta - H.$$

For a general symplectic manifold with Hamiltonian H we call $X_H \lrcorner\, \theta - H$ the Lagrangian associated with the symplectic potential; it is a function with the same domain as θ.

The quantum line bundle

The basic idea of geometric quantisation is that the quantum wave functions are associated with a U(1) gauge theory whose field strength is $\frac{2\pi\omega}{h}$ where ω is the symplectic form and h is Planck's constant.

More precisely, the wave functions are sections of a complex line bundle L having the phase space M as base, U(1) as structure group, and carrying a connection whose curvature form is $\frac{2\pi\omega}{h}$. Such a line bundle is called a quantum line bundle. In the usual language of gauge theory this may be equivalently expressed as follows: a choice of symplectic potential θ with domain an open subset of M is called a choice of gauge; the wave functions are represented in the gauge θ by complex valued functions on the domain of θ; each vector field X on the phase space defines a covariant derivative ∇_X acting on the wave functions, which is represented by the operator $X - \frac{2\pi i}{h} X \lrcorner\, \theta$ in the gauge θ; on change of symplectic potential by a gauge transformation $\theta \to \theta + du$, the representative ϕ of a wave function transforms as

$$\phi \to \left(\exp \frac{2\pi i u}{h}\right) \phi.$$

The existence of such a line bundle L with connection having $\frac{2\pi\omega}{h}$ as curvature form imposes a topological condition on ω. The de Rham cohomology class of the 2-form $\frac{\omega}{h}$ must be an integral valued element of $H^2(M,R)$. In other words, the integral of ω over each closed surface in M must be an integral multiple of h. This is a quantisation condition on the phase space M, and we say that M is quantisable if the condition is satisfied. It is indirectly related to the old Bohr-Sommerfeld quantisation rule.

When M is quantisable and simply connected there is, up to isomorphism, exactly one line bundle with connection having $\frac{2\pi\omega}{h}$ as curvature form. For a non-simply connected quantisable space, the set of distinct line bundles with connection having $\frac{2\pi\omega}{h}$ as curvature form is parametrised by the character group of the fundamental group of M.

When ω is an exact form, as for example in the case when M arises from a configuration space and $\omega = d\beta$, the quantisation condition is trivially satisfied since then ω represents the zero de Rham cohomology class. In this case if θ is any globally defined symplectic potential, $\omega = d\theta$, we can define a connection on the product line bundle $L = M \times C$ by identifying smooth sections of L with complex valued functions on M, in the usual way, and then defining the covariant derivative by

$$\nabla_X = X - \frac{2\pi i}{h} X \lrcorner \theta$$

for each vector field X on M. This connection has $\frac{2\pi\omega}{h}$ as curvature form, and thus L is a quantum line bundle.

As an example, consider a particle with charge e in a region Q in R^3 with a magnetic field given by a 2-form B. Here the phase space is $M = T^*Q$ and the motion of the particle can be derived from the free Hamiltonian and the <u>charged symplectic form</u> $\omega = dp \wedge dq + e\,\sigma^*B$ where σ is the projection of M onto Q. Thus the Lorentz force on the particle is incorporated into the symplectic form instead of into the Hamiltonian. The quantisation condition on ω requires that the integral of eB over each closed surface in Q be an integral multiple of h. This gives Dirac's result on charge quantisation. If B is derived from a vector potential A, $B = dA$, then ω is exact with a globally defined symplectic potential $pdq + e\,\sigma^*A$. In the Bohm-Aharonov experiment the region Q is the complement of a cylinder in R^3. There is a magnetic field in R^3 which is zero in Q, while inside the cylinder it is non-zero and has a constant orientation parallel to the cylinder. Thus on R^3 the magnetic field B equals dA where A is closed on Q and the integral of A on a path in Q around the cylinder is non-zero. In this case ω is equal to $dp \wedge dq$ on the phase space $M = T^*Q$ and the two globally defined symplectic potentials pdq and $pdq + e\sigma^*A$ on M give inequivalent connection on the product line bundle $L = M \times C$. The connection defined by $pdq + e\sigma^*A$ in the correct one, according to Bohm-Aharonov.

If L is a complex line bundle over M we shall denote by $L^\times$ the open subset consisting of all non-zero elements of L. Let L have structure group $U(1)$ and let a connection be given on L. Then we may regard $L^\times$ as the principal bundle of L when we extend the structure group from $U(1)$ to the group $C^\times$ of non-zero complex numbers. The connection therefore defines a connection 1-form α (say) on $L^\times$. Thus $d\alpha = \frac{2\pi}{h} pr^* \omega$ where pr denotes the projection of $L^\times$ onto the base space M. When L is the product bundle, and when the connection is that defined by a globally defined symplectic potential θ, then

$$\alpha = \frac{2\pi}{h} pr^* \theta + i \frac{dz}{z}$$

where z is the function on $L^\times = M \times C^\times$ given by a complex linear co-ordinate on the fibre $C^\times$.

For each smooth function H on M we lift the Hamiltonian vector field X_H on M to a vector field V_H on $L^\times$ normalised by the condition

$$V_H \lrcorner \alpha = \frac{2\pi}{h} pr^* H.$$

This ensures that V_H preserves α since then the Lie derivative of α along V_H is

$$V_H \lrcorner d\alpha + d(V_H \lrcorner \alpha) = V_H \lrcorner \frac{2\pi}{h} pr^* \omega + d(\frac{2\pi}{h} pr^* H)$$

$$= \frac{2\pi}{h} pr^* (X_H \lrcorner \omega + dH) = 0.$$

The vector field V_H is called the <u>quantum vector field</u> generated by H. It gives the quantum operator associated to H, as will be explained.

To see how the quantum vector field V_H acts on the quantum wave functions, we first note that the wave functions are sections of the complex line bundle L and that each section ψ may be regarded as a complex valued function ψ on the principal bundle $L^\times$. More precisely, the value of ψ on $v \in L^\times$ is defined to be the component of ψ in the basis v:

$$\psi(m) = \psi(v)v$$

if v belongs to the fibre of $L^\times$ at $m \in M$. In this way we have a natural operation of the vector field V_H on the sections of L. As an operator on the sections of L we have

$$V_H = \nabla_{X_H} + \frac{2\pi i}{h} H.$$

The quantum vector field V_H may be regarded as a differentiated form of the Dirac amplitude $\exp(\frac{2\pi}{ih}\text{ action})$. To see this, suppose that in time t the flow of the vector field V_H carries a section of L whose representative in the θ-gauge is ϕ into a section whose representative is ϕ_t. Then

$$\phi_t(m_o) = \exp\left[\frac{2\pi}{ih} \int_o^t (X_H \lrcorner \theta - H)\, ds\right] \phi(m_t)$$

where the integral is taken along the integral curve $s \to m_s$ of X_H from m_o to m_t. The integrand $X_H \lrcorner \theta - H$ is the Lagrangian associated with the Hamiltonian H and the gauge θ. The vector field V_H itself, and its action on the sections of L, are gauge invariant concepts.

The way in which quantum mechanical states are respresented by wave functions will depend on what representation we are working with, for example a configuration space representation or a momentum space representation, and in any given representation not all the sections of L will be wave functions. So to examine the action of V_H on the wave functions we must consider which sections are wave functions in a particular representation.

Polarisations and representations

To fix a 'representation' in the sense of Dirac we choose a 'maximal commuting set of classical observables'. For example, if the system is based on a configuration space Q then any set of coordinates $q^1, \dots, q^n$ on Q will Poisson commute, $\{q^j, q^k\} = 0$, and we will be lead to the configuration space representation.

If $H_1, \dots, H_n$ are independent Poisson commuting functions on an open set W in M then

$$\omega(X_{H_j}, X_{H_k}) = \{H_j, H_k\} = 0$$

and $X_{H_1}, \dots, X_{H_n}$ span, over W, a sub-bundle of the complex tangent bundle of M. Here we are concerned only with functions on an open subset of M. To get a suitable global notion, we define a polarisation of M to be a sub-bundle F of the complex tangent bundle of M such that

i) $\omega(X,Y) = 0$ for any two sections X and Y of F
ii) the sections of F are closed under the Lie bracket
iii) the fibre dimension of F is equal to half the dimension of M.

A polarisation is thus the global geometric notion which corresponds to the idea of a 'maximal commuting set of classical observables'.

In the case when M is the cotangent bundle of a configuration space Q and $q^1, \dots, q^n$ are configuration coordinates, we have

$X_{q^j} = \frac{\partial}{\partial p_j}$. Thus the polarisation corresponding to the configuration coordinates is generated by vectors tangent to the cotangent fibres, and hence is defined globally and is independent of the choice of coordinates on Q. This is an example where the commuting observables are real valued functions on M, leading to real tangent vectors spanning F. In this case we call F a real polarisation.

In the case of the phase space for a spin degree of freedom, the 2-sphere S^2, there cannot be a real polarisation since there is no non-singular real vector field on the sphere. However if we introduce complex analytic coordinates on S^2 as the Riemann sphere then we get a globally defined polarisation which is spanned by $\frac{\partial}{\partial \bar{z}}$ on the domain of a complex coordinate z.

Having chosen a polarisation F, we define the wave functions in the F-representation to be the smooth sections of L which are covariant constant along sections of F. Thus the F-wave functions are represented in the gauge θ by complex valued functions ϕ on the domain of θ which satisfy the condition

$$(X - \frac{2\pi i}{h} X \lrcorner \theta)\, \phi = 0$$

for each section X of F.

When H is a classical observable whose Hamiltonian vector field X_H preserves the polarisation F, then the quantum vector field V_H will leave invariant the space of wave functions of the F-representation. The operator $\frac{h}{2\pi i} V_H$ acting on the F-wave functions is called the quantised operator $\hat{H}$ which represents the classical observable H in the F-representation. Thus

$$\hat{H} = \frac{h}{2\pi i} \nabla_{X_H} + H$$

as an operator on the F-wave functions.

To illustrate these ideas, suppose p and q are canonical coordinates on a phase space M with $\omega = dp \wedge dq$ and let q be chosen as the maximal commuting observable. Then the corresponding polarisation F is generated by the vector field $X_q = \frac{\partial}{\partial p}$. In the gauge pdq we find that the wave functions in the F-representation are complex valued functions of q alone. Moreover we find that the quantised operators representing p and q are given by the usual Schrödinger representation

$$\hat{q} = q \quad , \quad \hat{p} = \frac{h}{2\pi i} \frac{\partial}{\partial q} .$$

If on the same phase space we choose the complex coordinate $z = p + iq$ as the commuting observable then the polarisation is generated by $\frac{\partial}{\partial \bar{z}}$ and the wave functions in the gauge $-\frac{i}{2}\bar{z}dz$ form the Fock space of holomorphic functions of z. The quantised operators representing z and $\bar{z}$ are the creation and annihilation operators

$$\hat{z} = z \quad , \quad \hat{\bar{z}} = \frac{h}{\pi}\frac{\partial}{\partial z}$$

For a particle in a magnetic field we can use the charged symplectic form on R^3, $\omega = \Sigma dp_j \wedge dq^j + e\,\sigma^* B$, the functions q^1, q^2, q^3 as commuting set, and the gauge $pdq + e\,\sigma^* A$ where A is a vector potential. We get

$$\hat{q}^j = q^j \;,\; \hat{p}_j = \frac{h}{2\pi i}\frac{\partial}{\partial q_j} - eA_j$$

acting on functions of q along; the usual minimal coupling operators, where $A_j = \Sigma A_j dq^j$.

Inner products

An important question is the definition of a Hilbert space inner product on the space of wave functions of a given representation. There is also the question of relating different representations, just as the Fourier transform relates the configuration space and momentum space representations.

These questions are tackled in two steps. The first step is to note that for a system based on a configuration space Q, the space of wave functions is $L^2(Q)$. If Q carries no natural measure then a natural way to define $L^2(Q)$ is to regard its elements as half densities (transforming as the square root of the absolute value of the Jacobian) or half-forms (transforming as the square root of the Jacobian). The space $L^2(Q)$ corresponds to the wave functions relative to the polarisation generated by the configuration space coordinates. For a general polarisation F we are led to redefine the F-wave functions so that they are half-forms based on the normal bundle of F and with values in the line bundle L. The second step is to define a pairing between wave functions relative to polarisation F and F' respectively, to produce a density which can be integrated over the quotient of M by the real foliation defined by the intersection of F and the complex conjugate of F'. This pairing, due to Kostant, Blattner and Sternberg, gives (formally at least) an inner product on the F-wave functions and an integral transform from the F-wave functions to the F'-wave functions.

The Fourier transform and the Bargmann transform are examples of these ideas, and a basic motivation for them. Let $M = R^2$ and $\omega = dp \wedge dq$ with the polarisations F, F' and F" generated by the functions q, p and $z = p + iq$ respectively. In the gauge pdq the F-wave functions have the form $\phi(q)(dq)^{\frac{1}{2}}$. In the gauge-qdp the F'-wave functions have the form $\psi(p)(dp)^{\frac{1}{2}}$. Since $pdq = -qdp + d(pq)$, the F'-wave functions have the form

$$(\exp \frac{2\pi i}{h} pq)\ \psi(p)(dp)^{\frac{1}{2}}$$

in the gauge pdq. A pairing of F- and F'-wave functions leads to the expression

$$\frac{1}{h} \int (\exp \frac{-2\pi i}{h} pq)\ \phi(q)\overline{\psi(p)}\ dp \wedge dq$$

which leads to the Fourier transform between the configuration space and momentum space representations.

In the gauge $-\frac{i}{2}\bar{z}\,dz$ the F" -wave functions are of the form $f(z)\ (dz)^{\frac{1}{2}}$ where f is a holomorphic function. Since $pdq = -\frac{i}{2}\bar{z}\,dz + \frac{1}{4}\,d\,(2pq + iz\bar{z})$, the F" -wave functions have the form

$$(\exp \frac{\pi i}{2h} (2pq + iz\bar{z}))f(z)(dz)^{\frac{1}{2}}$$

in the gauge pdq. A pairing of F- and F" -wave functions leads to the expression

$$\frac{1}{2^{\frac{1}{4}}h} \int (\exp \frac{\pi i}{2h}\ (2pq + iz\bar{z}))\phi(q)\overline{f(z)}dp \wedge dq$$

which leads to the Bargmann transform between the configuration space and Fock space representations. See Woodhouse [5] pages 168-170.

The pairing of two Fock space representations leads to the Bogoliubov transformation [6].

The Schrödinger operator and the Feynman path integral

If the flow of the Hamiltonian vector field X_H generated by a classical observable H does not preserve a given polarisation F, then the action of the quantum vector field V_H on the F-wave functions must be obtained by using the pairing. The basic idea is that if in time t the flow of X_H carries F into a new polarisation F_t, then the flow of V_H will carry an F-wave function ϕ into an F_t-wave function ϕ_t. We then transform ϕ_t back into an F-wave function $T(t)\phi$ using the pairing. The quantised operator $\hat{H}$ representing H in the F-representation is then formally defined

$$\hat{H}\,\phi = \frac{h}{2\pi i}\,\frac{d}{dt}\,T(t)\phi\Big|_{t=0}$$

All this is rather formal. Its justification is that for a system based on a configuration space Q and Hamiltonian of the form

$$H = \tfrac{1}{2}\Sigma g^{jk} p_j p_k + V(q)$$

where g_{jk} is a Riemannian metric on Q, and with the polarisation generated by the configuration space coordinates, a calculation due to Śniatycki [3] page 134 gives

$$\hat{H} = -\frac{h^2}{8\pi^2}\left(\Delta - \frac{R}{6}\right) + V$$

where Δ is the Laplace-Beltrami operator and R is the scalar curvature of g_{jk}.

The operator T(t) is given by the flow of V_H, which involves the Dirac factor $\exp\left(\frac{2\,i}{h}\text{ action}\right)$, followed by an integral transform derived from the pairing. Thus T(t) is an integral transform with the Dirac factor appearing in the integrand. The quantum evolution $\phi(t) = (\exp \frac{2\,it}{h}\hat{H})\,\phi(0)$ with initial wave function $\phi(o)$ is, conceptually, tangential to $t \to T(t)\phi(0)$ at $t = 0$. Thus, formally, $\phi(t)$ is given by a limit of iterations

$$\phi(t) = \lim T(t_N - t_{N-1}) \ldots\ldots T(t_2 - t_1)T(t_1)\phi(0)$$

where the limit is taken in some sense over partitions $0 \leq t_1 \leq \ldots\ldots \leq t_N = t$ of the interval [0,t]. This has a natural interpretation as a Feynman path integral.

Symmetry groups

If the phase space M admits a Lie group G of symmetries then it is appropriate to choose a G-invariant polarisation of M for the representation.

For example, let M be the phase space of an elementary classical relativistic system, in the sense that the Poincaré group G acts transitively on M preserving ω. Then the moment map, discussed by Sternberg in his lectures, identifies M with an orbit of G in the dual $\mathcal{G}$ of its Lie algebra $\mathcal{G}$. The usual Casimir polynomials P^2 and W^2 are smooth functions on $\mathcal{G}^*$ and are constant on M. Suppose

$P^2 = m^2 > 0$ and $W^2 = -s^2m^2 < 0$ on M. Then M is diffeomorphic to $R^6 \times S^2$ and is quantisable if and only if $4\pi s$ is an integral multiple of h. Thus the spin is quantised. Since G preserves ω we have a representation of the Lie algebra $\mathcal{G}$ in the Poisson bracket algebra of smooth functions on M. If we use a G-invariant polarisation F to represent functions on M by operators on the F-wave functions we obtain a representation of $\mathcal{G}$ by operators. This is the usual Wigner (m,s) representation.

Constraints

In the classical mechanics, if M is a phase space with symplectic form ω, then a constraint is introduced by choosing a constraint submanifold N. If the restriction of ω to N has constant rank then we consider the quotient $\tilde{M}$ of N by the foliation given by the null-spaces of the restriction. The quotient $\tilde{M}$, with its induced symplectic form $\tilde{\omega}$, is called the reduced phase space. The reduction at the quantum level is carried out as follows. We start with a quantum line bundle L over M, giving the quantum mechanics of the unconstrained system. We then require of the constraint submanifold N that parallel transport in L be trivial on each leaf of the foliation of N. This enables us to pass from L to a quotient line bundle $\tilde{L}$ over $\tilde{M}$. This gives the quantum mechanics of the constrained system.

Field theory

A careful treatment of some of the preceding ideas in the context of a quantum field interacting with an external classical potential may be found in [6].

Finally, a point to bear in mind is that when fermions are to be quantised as well as bosons, then the phase space M should be taken as a graded-symplectic manifold. All other notions such as the quantum line bundle should also be taken in the graded sense, see [2].

References

[1] Kostant B. Quantization and unitary representations, Lecture Notes in Mathematics 170, 87-208, Springer, Berlin 1970.

[2] Kostant B. Graded manifolds, graded Lie theory, and prequantization, Lecture Notes in Mathematics 570, 177-306, Springer, Berlin 1977.

[3] Śniatycki J. Geometric Quantization and Quantum Mechanics, Springer, Berlin 1980.

[4] Souriau J.-M. Structure des Systèmes Dynamiques, Dunod, Paris 1970.

[5] Woodhouse N. Geometric Quantization, Oxford University Press 1980.

[6] Woodhouse N. Geometric quantization and the Bogoliubov transformation. Proc. Royal Society of London 378, 119-139, 1981.

PATH INTEGRALS OVER PHASE SPACE,
THEIR DEFINITION AND SIMPLE PROPERTIES *

Jan Tarski

International Centre for Theoretical Physics, Trieste, Italy,
and
Institut für Theoretische Physik, Technische Universität Clausthal,
3392 Clausthal-Zellerfeld, Federal Republic of Germany.

Abstract

Path integrals over phase space are defined in two ways. Some properties of these integrals are established. These properties concern the technique of integration and the quantization rule $i^{-1}\partial_q \leftrightarrow p$.

1. Introduction

Let us recall two familiar representations of the Schrödinger Green's function in terms of path integrals:

$$G(t,q;0,q') = N_a \int_{\substack{\eta(0)=q' \\ \eta(t)=q}} \mathcal{D}(\eta) \exp\left\{ i\int_0^t d\tau \left[\tfrac{1}{2} m \dot{\eta}^2(\tau) - V(\eta(\tau)) \right] \right\} \tag{1.1a}$$

$$= N_b \int_{\substack{\eta(0)=q' \\ \eta(t)=q}} \mathcal{D}(\eta)\mathcal{D}(p) \exp\left\{ i\int_0^t d\tau \left[p(\tau)\dot{\eta}(\tau) - (2m)^{-1} p^2(\tau) - V(\eta(\tau)) \right] \right\}. \tag{1.1b}$$

---

* This article should be regarded as a research article rather than as lecture notes. It emphasizes new results and includes only a part of the material on path integrals that was presented at the Poiana Braşov School. A seminar based on the content of this article was also given by the author at a conference in Clausthal in July 1981.

The last two terms constitute the Hamiltonian,

$$H(p,q) = p^2/2m + V(q). \tag{1.1c}$$

The function G refers to the time interval $[0,t]$, and N_a, N_b are two normalizing constants, which depend on the definition of the integral that one adopts. The paths $\eta(\tau)$, $p(\tau)$ are over R^n, and in writing $\dot{\eta}^2$, $p\eta$, and $\dot{p}^2$, we presupposed scalar products in R^n.

The form (1.1b), which is the main concern of this article, was already given in the early work of Feynman on path integrals [1]. The form (1.1a) is historically somewhat older. One sees that the two are equivalent (at least at the heuristic level) by doing the p-integration in (1.1b).

The form (1.1b) has gained some importance because of its usefulness for investigating constrained systems, in particular gauge theories [2]. Moreover, this form was discussed in a number of foundational studies, e.g. in [3] and [4]. (Cf. also the review article [5].)

The form (1.1b) can be generalized somewhat. Let us denote the propagator G by $\langle t,q|0,q'\rangle$, and then the expectation of an operator-valued, time-ordered function $(f(p,x))_+$ can be written as

$$\langle t,q|(f(p(\cdot),x(\cdot)))_+|0,q'\rangle = N_b \int_{\substack{\eta(0)=q' \\ \eta(t)=q}} \mathcal{D}(\eta)\mathcal{D}(p)$$

$$\times \exp\left[i\int_0^t d\tau(p\dot{\eta} - p^2/2m - V)\right] f(p(\cdot),\eta(\cdot)). \tag{1.2}$$

We now sketch a possible definition of such a path integral, as a background for the present article. We follow [3] and [5].

We subdivide the time interval $[0,t]$ by the division points

$$t_j = jt/(2n+2), \qquad j = 0, 1, \ldots, 2n+2, \tag{1.3}$$

and we introduce the independent variables

$$p(t_1),\ \eta(t_2),\ p(t_3),\ \dots,\ \eta(t_{2n}),\ p(t_{2n+1}), \tag{1.4}$$

while $\eta(t_0) = q'$ and $\eta(t_{2n+2}) = q$ are fixed. We approximate the integrand in (1.2) by a suitable function of the variables in (1.4) (and of q', q), and we replace m by $m-i\delta$ with $\delta > 0$. Finally, we approximate the functional integral by the (suitably normalized) multiple integral

$$N_n \int\cdots\int_{-\infty}^{\infty} dp(t_1)\, d\eta(t_2) \cdots dp(t_{2n+1}) [\cdots], \tag{1.5}$$

and we integrate.

The functional integral then results from passing to the limits $n \to \infty$, $\delta \searrow 0$. Various options are possible, with regard to approximating the integrand and with regard to taking the limits. (Moreover, some modification that would ensure translational invariance might be desirable, cf. Secs.2 & 3.)

The definition just sketched appears to fulfill various desiderata for the integral. However, we know of no attempt to develop a rigorous theory of integration on the basis of such a definition.

In this connection we may note,that on the whole, very little rigorous work has been done on the path integral over phase space. We cite in particular [6], and there are some brief comments in [7] and [8]. The present article examines the integral in question from a broader perspective than was done in [6]-[8].

The three succeeding sections discuss in turn alternative definitions of the phase-space integral, some elementary invariance and integrability properties, and an application to the quantization problem. One proof is relegated to an appendix. Another appendix discusses nontangential limits, and rectifies a shortcoming of Ref.8.

The author is indebted to Professor H.D. Doebner for hospitality at T.U. Clausthal, and to Professor Abdus Salam, the International Atomic Energy Agency and UNESCO for hospitality at the International Centre for Theoretical Physics, Trieste.

2. Two definitions

The two definitions that follow are adaptations from [8] and [9], respectively. The first of these depends on the Hilbert space structure associated with the paths $\eta(\tau)$, $p(\tau)$. In the present case we have a direct sum of two real Hilbert spaces, $\mathcal{H}_\eta$ and $\mathcal{H}_p$, defined, respectively, by the scalar products

$$\langle \dot{\eta}, \dot{\eta} \rangle = \int_0^t d\tau \sum_{j=1}^n (d\eta^j/d\tau)^2, \quad \langle p, p \rangle = \int_0^t d\tau \sum_{j=1}^n (p^j)^2, \tag{2.1}$$

and for $\mathcal{H}_\eta$, we impose also the condition $\eta(0) = 0$.

We observe that there is a natural correspondence relating $\mathcal{H}_\eta$ and $\mathcal{H}_p$, i.e. an isomorphism, defined by: $\eta \leftrightarrow p$ if and only if $\dot{\eta}(\tau) = p(\tau)$ almost everywhere. In view of this isomorphism, we will also write $\int d\tau p\dot{\eta}$ as $\langle p, \dot{\eta} \rangle$, etc.

For the abstract discussion that follows, we will assume two real Hilbert spaces and a given isomorphism between them. These will be denoted by $\mathcal{H}_1$, $\mathcal{H}_2$, or by $\mathcal{H}_\xi$, $\mathcal{H}_\zeta$, etc., and we will denote the scalar products which are encountered as $\langle \xi, \xi \rangle$, $\langle \zeta, \zeta \rangle$, also $\langle \xi, \zeta \rangle$ etc. Furthermore, we fix a mass parameter κ which satisfies $\operatorname{Im} \kappa \geq 0$, $\kappa \neq 0$. (In the familiar applications $\kappa = m > 0$.)

We start with the case $\dim \mathcal{H}_1 = \dim \mathcal{H}_2 = k < \infty$. Let $\alpha \in \mathcal{H}_1$, $\alpha_0 \in \mathcal{H}_2$, and let b, b_0 satisfy $\operatorname{Re} b$, $\operatorname{Re} b_0 > 0$. We set

$$I^{b,\alpha;\, b_0,\alpha_0}(F) = \{[b(b_0 + i/\kappa) + 1]^{\frac{1}{2}}/2\pi\}^k \int d^k u\, d^k v \exp(-\tfrac{1}{2} b \langle u-\alpha, u-\alpha\rangle)$$

$$\times \exp(-\tfrac{1}{2} b_0 \langle v-\alpha_0, v-\alpha_0\rangle)\, e^{i\langle u,v\rangle}\, e^{-i\langle v,v\rangle/2\kappa}\, F(u,v). \tag{2.2}$$

[The coefficient $\{[\ldots\}^k$ is adjusted so that $I^{b,0;b_0,0}(1) = 1$.] We then take the nontangential double limit as $b, b_0 \to 0$, and require that this limit be independent of α, α_0. If such a limit exists, it is the desired phase-space integral of F, and will be denoted by $I_{ps}(P)$.

In specifying the nontangential double limit $b, b_0 \to 0$, we presupposed here an extension of the standard result [10] on uniqueness of such a limit if it exists, from one to two complex variables. The desired extension can be easily established, by utilizing Vitali's theorem for two variables (proved in effect in [8], Sec.5) and by adopting the proof of [10] from one variable to two. See also Appx.B.

If $\dim \mathcal{H}_\xi = \dim \mathcal{H}_\zeta = \infty$, we employ sequences of projections, as in [8]. We introduce the set $\mathcal{P}$ of finite-dimensional orthogonal projections P on $\mathcal{H}_\xi$, and in view of the given isomorphism, the action of such a P extends to $\mathcal{H}_\xi \dotplus \mathcal{H}_\zeta$. Consider next increasing sequences of such projections, and let

$$\hat{\mathcal{Q}} = \{\{P_j\} : P_j \in \mathcal{P},\ P_{j+1} \geq P_j \text{ for } \forall j,\ \lim_{k\to\infty} P_k = 1\}. \tag{2.3}$$

For $P \in \mathcal{P}$, we set

$$I_P^{\,b,\alpha;\, b_0,\alpha_0}(F(\cdot,\cdot)) = I^{b,P\alpha;\, b_0,P\alpha_0}(F(P\cdot,P\cdot)) \tag{2.4a}$$

and for $\Pi = \{P_j\} \in \hat{\mathcal{Q}}$

$$I_{\Pi}^{b,\alpha;b_0,\alpha_0}(F) = \lim_{j\to\infty} I_{P_j}^{b,\alpha;b_0,\alpha_0}(F), \tag{2.4b}$$

$$I_{\Pi}(F) = \lim_{b,b_0\to 0} I_{\Pi}^{b,\alpha;b_0,\alpha_0}(F), \tag{2.4c}$$

where the last limit is to be nontangential, and to be independent of α, α_0.

Since the limit $I_{\Pi}^{b,\alpha;b_0,\alpha_0}(F)$ might not be analytic in b, b_0, the existence of a unique nontangential limit may raise additional problems.

Definition 1. Let $\mathfrak{A}$ be a determining family of sequences Π (cf.[8]). If the $I_{\Pi}(P)$'s are equal for $\forall\, \Pi \in \mathfrak{A}$, we call the common value the phase-space integral of F, and we use the notations

$$I_{ps}(F) = \int \mathcal{D}(\xi)\,\mathcal{D}(\zeta)\, e^{i\langle\xi,\zeta\rangle} e^{-i\langle\zeta,\zeta\rangle/2\kappa} F(\xi,\zeta). \tag{2.5}$$

We rewrite this equation for the quantities considered originally:

$$I_{ps}(F) = \int_{\gamma(0)=0} \mathcal{D}(\gamma)\,\mathcal{D}(p)\, e^{i\langle p,\dot{\gamma}\rangle} e^{-i\langle p,p\rangle/2m} F(p,\gamma). \tag{2.6}$$

Determining families of sequences of projections were introduced as a device that enables us to eliminate some sequences of $\hat{\mathfrak{A}}$ which may be inconvenient, and to retain agreement of the $I_{\Pi}(F)$'s for a sufficiently large family of Π's. We refer to [8] for the definition of a determining family. The following kind of example will suffice for our needs (here $P \in \mathcal{P}$):

$$\mathfrak{A}(P) = \{\{P_j\} \in \hat{\mathfrak{A}} :\ P_k \geq P \text{ for } \forall k\}. \tag{2.7}$$

We may note that, in the short discussion of this integral in [8], projections were considered which act in independent ways on $\mathcal{H}_{\xi}$ and on $\mathcal{H}_{\zeta}$.

It is not clear if the broader definition of loc. cit. has any advantages. However, it is more difficult to handle.

For the second definition of the phase-space integral we introduce, following [9]:

$$J(b,\alpha;b_0,\alpha_0;b_1;F) = \int \mathcal{D}(\xi)\mathcal{D}(\zeta)\exp(-\tfrac{1}{2}b\langle\xi-\alpha,\xi-\alpha\rangle)$$
$$\times\exp(-\tfrac{1}{2}b_0\langle\zeta-\alpha_0,\zeta-\alpha_0\rangle)\exp(i\langle\xi,\zeta\rangle)\exp(-\tfrac{1}{2}b_1\langle\zeta,\zeta\rangle)F(\xi,\zeta). \tag{2.8}$$

This is to be Gaussian integral over $\mathcal{H}_\xi \dotplus \mathcal{H}_\zeta$, but the occurrence of $\langle\xi,\zeta\rangle$ in the exponent is a complication. One way to define this integral is by doing first the ζ-integration, and then (if F is a "well-behaving" function) we obtain the factor $\exp[-\frac{1}{2}(b_0+b)^{-1}\langle\xi,\zeta\rangle]$. Then for the ξ-integration we will have the variance $(b+(b_0+b_1)^{-1})^{-1}$. Alternatively, the ξ-integral could be done first.

For definiteness, let us require that the two evaluations be equal.

Definition 2. Suppose that J is analytic in b, b_0, b_1 in a region which includes the set

$$\{\mathrm{Re}\ b > 0,\quad \mathrm{Re}\ b_0 > 0,\quad \mathrm{Re}\ b_1 \geq 0\}. \tag{2.9}$$

If the nontangential double limit of $J(b,\alpha;b_0,\alpha_0;i/\kappa;F)$ as b, $b_0 \to 0$ exists and is independent of α, α_0, we call this limit the phase-space integral in the sense of analytic continuation, and we denote it by $J_{ps}(F)$.

3. Some direct consequences

We start with an assertion about invariance principles of the two integrals I_{ps}, J_{ps}. The following proposition can be proved readily (cf. Proposition 4 of [8]):

Proposition 3. Let $\beta \in \mathcal{H}_\xi$, $\beta_0 \in \mathcal{H}_\zeta$. Let R be an orthogonal transformation on $\mathcal{H}_\xi$ and on $\mathcal{H}_\zeta$, acting jointly on the two spaces in such a way that this action commutes with the assumed isomorphism. (I.e., so that $\langle \xi,\zeta \rangle = \langle R\xi, R\zeta \rangle$.) Let F be integrable for I_{ps}. Then

$$I_{ps}(F) = \int \mathcal{D}(\xi)\,\mathcal{D}(\zeta) \exp\left(i\langle \xi+\beta, \zeta+\beta_0\rangle\right)$$

$$\times \exp\left(-i\langle \zeta+\beta_0, \zeta+\beta_0\rangle/2\kappa\right) F(\xi+\beta, \zeta+\beta_0) \tag{3.1a}$$

$$= \int \mathcal{D}(\xi)\,\mathcal{D}(\zeta)\, e^{i\langle \xi,\zeta\rangle} e^{-i\langle \zeta,\zeta\rangle/2\kappa} F(R\xi, R\zeta)\,. \tag{3.1b}$$

If I_{ps} converges with reference to the determining family $\mathfrak{a}$, the right-hand side of (3.1a) likewise converges with reference to $\mathfrak{a}$, and the right-hand side of (3.1b) with reference to the rotated family, as is implied by $\{P_j\} \to \{R^{-1} P_j R\}$. Furthermore, if $J_{ps}(F)$ exists, then the analogue to (3.1a) holds, with the integral on the right-hand-side converging in the sense of analytic continuation.

We did not consider the rotation invariance of J_{ps}, since the requisite Gaussian measures require an extension of the Hilbert space, and rotational invariance of such extensions would require a separate discussion.

The next proposition is of a different kind, and in effect it extends slightly a result obtained in [11].

Proposition 4. Let $\xi_1,\ldots,\xi_n \in \mathcal{H}_\xi$ and $\zeta_1,\ldots,\zeta_\ell \in \mathcal{H}_\zeta$ be arbitrary, and let μ be a measure on $\mathcal{H}_\xi$ such that

$$\int d|\mu|(\xi')(1+|\langle \xi_1, \xi'\rangle|)\cdots(1+|\langle \xi_n, \xi'\rangle|)(1+|\langle \zeta_1, \xi'\rangle|)$$

$$\times \cdots (1+|\langle \zeta_\ell, \xi'\rangle|) < \infty\,. \tag{3.2}$$

(Here $|\mu|$ is the absolute variation of μ. The scalar products $\langle \zeta_j, \xi' \rangle$ presuppose the given ismorphism.) Then

$$F(\xi,\zeta) := (-i)^{n+\ell} \langle \xi_1, \xi \rangle \ldots \langle \xi_n, \xi \rangle \langle \zeta_1, \zeta \rangle \ldots \langle \zeta_\ell, \zeta \rangle \int d\mu(\xi') e^{i\langle \xi, \xi' \rangle} \tag{3.3a}$$

is integrable, with reference to the Definitions 1 and 2. (For I_{ps}, one has convergence for the maximal family $\hat{\mathcal{Q}}$.) The resulting phase-space integral is

$$I_{ps}(F) = J_{ps}(F) = (-1)^{n+\ell} \int d\mu(\xi') \langle \xi_1, \delta/\delta\xi' \rangle \ldots \langle \xi_n, \delta/\delta\xi' \rangle$$
$$\times [\langle \zeta_1, \delta/\delta\zeta' \rangle \ldots \langle \zeta_\ell, \delta/\delta\zeta' \rangle e^{\langle \xi', \xi' \rangle / 2i\kappa} e^{-i\langle \xi', \zeta' \rangle}]_{\zeta'=0} . \tag{3.3b}$$

The differential operators like $D_{\xi_1} = \langle \xi_1, \delta/\delta\xi' \rangle$ are in the sense of Gâteaux. However here their action is algebraic, and trivial. We should also note that F is the Fourier transform of the distribution

$$d\mu_D(\xi',\zeta') := D_{\xi_1} \cdots D_{\xi_n} D_{\zeta_1} \cdots D_{\zeta_\ell} d[\mu(\xi') \delta_0(\zeta')] , \tag{3.4}$$

where $\delta_0(\zeta')$ is the δ-measure at 0 of $\mathcal{H}_\zeta$. Equations (3.3) can be expressed in terms of μ_D as follows:

$$\int \mathcal{D}(\xi) \mathcal{D}(\zeta) e^{i\langle \zeta, \xi \rangle} e^{-i\langle \zeta, \zeta \rangle / 2\kappa} \int d\mu_D(\xi', \zeta') e^{i\langle \xi, \xi' \rangle} e^{i\langle \zeta, \zeta' \rangle}$$
$$= \int d\mu_D(\xi', \zeta') \int \mathcal{D}(\xi) \mathcal{D}(\zeta) e^{i\langle \zeta, \xi \rangle} e^{-i\langle \zeta, \zeta \rangle / 2\kappa} e^{i\langle \xi, \xi' \rangle} e^{i\langle \zeta, \zeta' \rangle} . \tag{3.5}$$

The last functional integral yields just $e^{\langle \xi', \xi' \rangle / 2i\kappa} e^{-i\langle \xi', \zeta' \rangle}$, cf. (3.3b).

Suppose now that $\ell = 0$ (with the empty product being unity). Then F depends only on ξ, and the evaluation of $I_{ps}(F)$ agrees with that given in [11], in terms of an integral over $\mathcal{H}_\xi$.* We have therefore:

Corollary 5. Let $\ell = 0$ in Proposition 4, so that F is independent of ζ . Then,

$$I_{ps}(F) = J_{ps}(F) = \int \mathcal{D}(\xi) e^{\frac{1}{2} i\kappa\langle\xi,\xi\rangle} F(\xi) . \tag{3.6}$$

(The last integral converges as I(F) with reference to $\hat{\mathfrak{a}}$, and also in J(F).)

Let us return to Proposition 4. We prove it in Appx.A. The choice of the measure $\delta_0(\zeta')$ for $\mathcal{H}_\zeta$ simplifies the manipulations, but we believe that (3.5) would hold also if a more general measure $\nu(\xi',\xi')$ is allowed in (3.4).

We also believe that all the evaluations of [8] and [9] can be extended to integrals over phase space, at least in the trivial way shown in (3.6). I.e., let $F(\xi,\zeta)$ be defined as equal to the old $F(\xi)$, independently of ζ , and we expect that the phase-space integral should reduce to the original integral over $\mathcal{H}_\xi$. One can also say: The phase-space integral should be equal to the corresponding iterated integral $\int \mathfrak{D}(\xi)\, F(\xi)\, [\int \mathfrak{D}(\zeta)\ldots]$.

However, Proposition 4 extends a Feynman-type integral over $\mathcal{H}_\xi$ in a non-completely-trivial way,and various evaluations of [8] and [9] should allow such extensions as well.

4. Quantization

The path-integral over phase space suggests an approach to the problem of quantization, i.e. to the problem of assigning quantum-mechanical operators to classical entities. We give first a short heuristic analysis, following [3],

* In Eq.(4b) of [11], $(-i)^n$ should be replaced by $(-1)^n$.

and will then follow with some rigorous considerations. For brevity in writing we consider a particle on R^1.

The action for a system under discussion can be written as

$$S = \int_{q'}^{q} d\eta\, p - \int_0^t d\tau\, H , \tag{4.1}$$

so that

$$i^{-1}\partial_q e^{iS} = e^{iS} p(t) , \qquad -i^{-1}\partial_t e^{iS} = e^{iS} H(t) . \tag{4.2}$$

If we now construct a functional integral and interchange operations, then we obtain

$$i^{-1}\partial_q G(t,q;0,q') = N \int_{\substack{\eta(0)=q' \\ \eta(t)=q}} \mathcal{D}(\eta)\mathcal{D}(p) e^{iS} p(t) , \tag{4.3a}$$

$$-i^{-1}\partial_t G(t,q;0,q') = N \int_{\substack{\eta(0)=q' \\ \eta(t)=q}} \mathcal{D}(\eta)\mathcal{D}(p) e^{iS} H(t) . \tag{4.3b}$$

The last two equations provide a basis for the correspondence

$$i^{-1}\partial_q \leftrightarrow p(t) , \qquad -i^{-1}\partial_t \leftrightarrow H(t) . \tag{4.4}$$

We emphasize that this argument is only heuristic. E.g., $\int d\tau\, H$ depends on q through $V(\eta(t))$.

We should like to make the following observation about the integrals in (4.3). Noncommutativity of p and q should have its counterpart in path integrals, and $p(t)$, $H(t)$ might not be meaningful. Therefore, we have to be prepared to replace in these integrals $p(t)$, $H(t)$ by $p(t-\varepsilon)$, $H(t-\varepsilon)$ where $\varepsilon \searrow 0$ afterwards, or to introduce other modifications.

A need for modification in fact appears when we scrutinize (4.3) from the point of view of the Hilbert spaces $\mathcal{H}_\eta$, $\mathcal{H}_p$, in preparation for a rigorous analysis. The quantity $q = \eta(t)$ can be written as a scalar product, and so is meaningful in the context of $\mathcal{H}_\eta$. Indeed if we set

$$\theta_T(\tau) = \min(T,\tau), \qquad \text{so that} \quad \theta_t(\tau) = \tau, \tag{4.5}$$

then

$$\langle \dot\eta, \dot\theta_T \rangle = \int_0^T d\tau\, \dot\eta(\tau) = \eta(T), \qquad \langle \dot\eta, \dot\theta_t \rangle = \eta(t). \tag{4.6}$$

(For R^n, we have to utilize also scalar products in R^n, e.g. $\underset{\sim}{e}\cdot\underset{\sim}{\eta}(t)$.)

Consequently, $\partial_q = \partial_{\eta(t)}$ also has a meaning in the context of $\mathcal{H}_\eta$, namely, as the Gâteaux derivative $\langle \dot\theta_t, \delta/\delta\dot\eta \rangle$. On the other hand, $p(t)$ and $H(t)$, occurring in (1.3), do not have such a meaning. In view of this circumstance we expect that the usual **p-q** symmetry will have to be altered. The following lemma illustrates this point.

Lemma 6. Let the potential be of the form

$$V(x) = \int_{-\infty}^{\infty} d\nu(y)\, e^{ixy}, \tag{4.7a}$$

where ν satisfies

$$\int_{-\infty}^{\infty} d|\nu|(y)\,(1+|y|) < \infty . \tag{4.7b}$$

Then, for small t,

$$i^{-1}\partial_q G(t,q;0,0) = \int_{\eta(0)=0} \mathcal{D}(\eta)\,\mathcal{D}(p)\, e^{iS}\,\delta(\eta(t)-q)$$
$$\times\, t^{-1}\int_0^t d\tau\, p(\tau) \;+\; O(t^2). \tag{4.8}$$

The phase-space integral converges as I_{ps} or as J_{ps}. In the former case, the reference family is $\mathcal{Q}(\bar{P})$, where $\bar{P}$ projects onto the subspace of linear functions in $\mathcal{H}_\eta$, or, of constant functions in $\mathcal{H}_p$. Cf. (2.6) and (4.5).

By requiring that all projections P_k satisfy $P_k \geqslant \bar{P}$, we can handle the δ-function in (4.7) in a simple way. We can also say: convergence of the phase-space integral could be deduced from Proposition 4, except that this proposition does not envisage δ-functions. We therefore compensate for the δ-function by requiring $P_k \geqslant \bar{P}$.

Outline of proof: The possibility of expressing $\exp(-i \int d\tau\, V)$ as the Fourier transform of a measure on $\mathcal{H}_\eta$ which is sufficiently strongly bounded is discussed fully in [11]. Convergence of the phase-space integral in (4.8) therefore follows from the above remarks. To prove the lemma now requires, in essence, justification of the interchange of operations, and the estimate of $i^{-1}\partial_q$. The validity of interchange, like integrability, follows from the boundedness of the measure. We turn therefore to the action of $i^{-1}\partial_q$.

After removing the δ-function we obtain an intergal for G as in (1.1b). The quantity $q = \eta(t)$ occurs there in $\langle p,\dot{\eta}\rangle$ and in $\int d\tau\, V$. The decomposition of $\langle p,\dot{\eta}\rangle$ into contributions from the spaces $\bar{P}\,\mathcal{H}_\eta$ and $(1-\bar{P})\,\mathcal{H}_\eta$ takes the following form:

$$\langle p,\dot{\eta}\rangle = (t^{-\frac{1}{2}}q)\left[t^{-\frac{1}{2}}\int_0^t d\tau\, p(\tau)\right] + (\text{terms independent of } q), \tag{4.9}$$

since on $(1-\bar{P})\,\mathcal{H}_\eta$, $\eta(t) = 0$. Applying $i^{-1}\partial_q$ to $e^{i\langle p,\dot{\eta}\rangle}$ will therefore yield the phase-space integral in (4.8).

Moreover, in view of (4.6), the action of $i^{-1}\partial_q$ on the potential factor yields

$$i^{-1}\langle\dot\theta_t, \delta/\delta\dot\eta\rangle \exp\left[-i\int_0^t d\tau\, V(\langle\dot\eta,\dot\theta_\tau\rangle)\right] = -\int_0^t d\sigma\, V'(\langle\dot\eta,\dot\theta_\sigma\rangle)\langle\dot\theta_t,\dot\theta_\sigma\rangle$$

$$\times \exp\left[-i\int_0^t d\tau\, V(\langle\dot\eta,\dot\theta_\tau\rangle)\right] = -\int_0^t d\sigma\, V'(\eta(\sigma))\sigma \exp\left(-i\int_0^t d\tau\, V\right) . \tag{4.10}$$

The condition (4.7b) implies that V' is bounded, while $0 \leq \sigma \leq t$. Since the path-integral reduces to a measure-theoretic one, we conclude that $\int d\sigma\, V'\sigma \sim \mathcal{O}(t^2)$. (□).

By employing a composition law for G, it might be possible to prove that $i^{-1}\partial_q G$ can be estimated by:

$$\int_{\eta(0)=0} \mathcal{D}(\eta)\mathcal{D}(p)\, e^{iS}\, \delta(\eta(t)-q)\, \varepsilon^{-1}\int_{t-\varepsilon}^{t} d\tau\, p(\tau) + \mathcal{O}(\varepsilon^2) . \tag{4.11}$$

We do not pursue here the problem of justifying this estimate.

Let us still return to (4.3b), where we have $H(t)$ in the (heuristic) integral. We note that $p^2(t)$, even if averaged over time as in (4.8), would not be an integrable functional. Therefore, an analysis such as preceding would not apply.

Appendix A: Proof of Proposition 4

The following evaluation will be basic for us:

$$\int\mathcal{D}(\chi)\, e^{-\frac{1}{2}\langle\chi, A\chi\rangle} e^{i\langle\chi,\varphi\rangle} = \exp\left(-\tfrac{1}{2}\langle\varphi, A^{-1}\varphi\rangle\right) . \tag{A.1}$$

At the heuristic level, A could be any number, real or complex, or , more generally, a symmetric operator. The implied integration is over a suitable real space of the χ's, but φ could be real, or complex. If A and/or φ are complex, then the scalar product must be understood as bilinear, or symmetric (but not hermitian). The assumed normalization in (A.1) is such that for $\varphi = 0$, the integral yields unity.

As Eq. (3.5) shows, the proof requires in essence, the justification of the interchange of order of operations. The derivatives are essentially algebraic here, and can be ignored. (The polynomial growth which they imply can be accomodated, in view of the hypothesis on the measure.) Integration over ζ' is also trivial. Only integration over ξ' needs some discussion.

We consider first Definition 2, and we evaluate $J(b,\alpha;b_0,\alpha_0;b_1;F)$ by a two-fold application of (A.1), with $F(\xi,\zeta) = \exp(i\langle\xi,\xi'\rangle)$, independently of ζ. In this case the evaluation can be readily justified, and we obtain

$$J(b,\alpha;b_0,\alpha_0;b_1;F) = \exp\left(-\tfrac{1}{2}b\langle\alpha,\alpha\rangle - \tfrac{1}{2}b_0\langle\alpha_0,\alpha_0\rangle + \tfrac{1}{2}b_0^2(b_0+b_1)^{-1}\langle\alpha_0,\alpha_0\rangle\right)$$

$$\times \exp\left\{-\tfrac{1}{2}[b+(b_0+b_1)^{-1}]^{-1}\langle\xi'+\psi, \xi'+\psi\rangle\right\}, \tag{A.2a}$$

where

$$\psi = -ib\alpha + b_0(b_0+b_1)^{-1}\alpha_0 . \tag{A.2b}$$

The evaluation (A.2) clearly has the desired analyticity properties, and analytic continuation in b_1 to i/κ is immediate. In order to justify the passage to the limit $b,b_0 \to 0$ inside the ξ'-integral, it suffices to show that this J is bounded (as a function of ξ'). Note that the exponential involving $\langle\alpha,\alpha\rangle$ and $\langle\alpha_0,\alpha_0\rangle$ is irrelevant.

If $\operatorname{Im}\kappa > 0$, then we have a damped Gaussian which remains damped also as $b,b_0 \to 0$. Similarly, if $\operatorname{Re} b$, $\operatorname{Re} b_0 > 0$ then we have a damped Gaussian. In all these cases J is a bounded function of ξ'. Therefore it suffices for us to consider the case κ real ($\neq 0$) and $|b|$, $|b_0| < \delta$ for some sufficiently small δ.

We observe that if $\operatorname{Re} b$, $\operatorname{Re} b_0 > 0$ then

$$\operatorname{Re}\left[b + (b_0 + i/\kappa)^{-1}\right]^{-1} > 0 , \tag{A.3}$$

but we need a more sensitive bound. We use the criterion for continuous differentiability of a function (cf. e.g. [12]), in the following form:

$$|f(b,b_0) - [f(0) + b\partial_b f(0) + b_0 \partial_{b_0} f(0)]| \leq \varepsilon(b,b_0)(|b| + |b_0|) \,, \qquad \text{(A.4)}$$

where $\varepsilon \to 0$ as $b, b_0 \to 0$. Take $f = [b + (b_0 + i/\kappa)^{-1}]^{-1}$, and since $\operatorname{Re} f(0) = 0$,

$$-\varepsilon(b,b_0)(|b| + |b_0|) + \operatorname{Re}[b\partial_b f(0) + b_0 \partial_{b_0} f(0)] \leq \operatorname{Re} f(b,b_0) \,. \qquad \text{(A.5)}$$

The two derivatives in question are

$$\partial_b f(0) = \kappa^{-2} \,, \qquad \partial_{b_0} f(0) = 1 \,. \qquad \text{(A.6)}$$

The hypothesis of a nontangential limit allows us to restrict b, b_0 to values which satisfy

$$\gamma |b| \leq \operatorname{Re} b \,, \qquad \gamma_0 |b_0| \leq \operatorname{Re} b_0 \qquad \text{(A.7)}$$

for some $\gamma, \gamma_0 > 0$ (and of course < 1). Since ε can be made arbitrarily small by restricting b, b_0 to sufficiently small values, we may conclude:

$$a(|b| + |b_0|) \leq \operatorname{Re} f(b,b_0) \qquad \text{(A.8)}$$

provided $|b|$, $|b_0| < \delta$, where a and δ are some quantities > 0. The last inequality implies

$$|\exp\{-\tfrac{1}{2}[b + (b_0 + i/\kappa)^{-1}]^{-1}\langle \xi', \xi' \rangle\}| \leq$$
$$\leq \exp[-\tfrac{1}{2} a(|b| + |b_0|) \|\xi'\|^2] \,. \qquad \text{(A.9)}$$

The integral J also contains a factor of the form $\exp[g(b,b_0)\langle\xi',\psi\rangle]$ which could be exponentially growing. However, by employing an argument similar to the previous one and by restricting $|b|$, $|b_0|$ sufficiently, we can bound this factor by

$$\exp\left[a(|b|+|b_0|)\,\|\xi'\|\,\|\psi'\|\right], \tag{A.10}$$

for some new vector ψ'. The product of the bounds in (A.9) and (A.10) is obviously bounded by a constant, also as $b, b_0 \to 0$. This completes the proof of integrability for Definition 2.

For Definition 1, we observe that the bound just obtained will necessarily apply to the evaluations determined by projections P_j. Therefore, we are able to take $\lim(j \to \infty)$ inside the μ-integral. The limit as $b, b_0 \to 0$ can then be taken, as before. □

APPENDIX B: The nontangential limit

In this article, as well as in [8] [9], nontangential limits were assumed in the definitions. One reason for considering such limits rather than simply $b \searrow 0$ (along R^1_+) comes from path integrals over phase space. Consider Eq.(2.2). If we set there $\alpha = \alpha_0 = 0$, if F is independent of v, and if we do the v-integration, then we obtain

$$\{[b+(b_0+i/\kappa)^{-1}]/2\pi\}^{\frac{1}{2}k}\int d^k u\, \exp\{-\tfrac{1}{2}[b+(b_0+i/\kappa)^{-1}]\langle u,u\rangle\}\,F(u)\ . \tag{B.1}$$

We see that we can restrict the limits to those along the real axes by taking first $b_0 \searrow 0$ and then $b \searrow 0$. However, it seems useful to allow also for other ways of approaching $b = b_0 = 0$, and the coefficient $[b + \ldots]$ will then vary along complex directions.

If we have a function $f(b)$ defined in the right-half-plane $C^1_>$, then the following conditions are sufficient for a (unique) nontangential limit at $b = 0$: (1) the existence of the limit along any nontangential curve, (2) analyticity in $C^1_>$, (3) boundedness in $N \cap C^1_>$, where N is any neighbourhood of 0.

In [8] [9] the third condition, that of boundedness, was not adequately incorporated into the discussion. However, the proofs of loc. cit. do not require revision, since the necessary boundedness, or the nontangential limits themselves, were established as needed.

Note:

After completing this article, the author became aware of some other rigorous work on path integrals over phase space, as follows:

13. В.П. Маcлob u А.М. Чебo apeb, Теор. Мат. Физ. 28, 291 (1976); translation: V.P. Maslov and A.M. Chebotarev, Teor. Mat. Fiz. 28. 793 (1976).

14. J.R. Klauder and I. Daubechies, "Measures for path integrals", preprint.

References

[1] R.P. Feynman, Phys. Rev. 84, 108 (1951), especially Appx.B.

[2] Л.Д. Фаддеев, Теор. Мат. Физ. 1, 3 (1969); translation: L.D. Faddeev, Teor. Mat. Fiz. 1, 1 (1969).

[3] A. Katz, Classical Mechanics, Quantum Mechanics, Field Theory (Academic Press, New York and London, 1965), especially Secs.18-19.

[4] G. Rosen, Formulations of Classical and Quantum Dynamical Theory (Academic Press, New York and London, 1969).

[5] C. Garrod, Rev. Mod. Phys. 38, 483 (1966).

[6] K. Gawędzki, Repts. Math. Phys. 6, 327 (1974).

[7] J. Tarski, in Complex Analysis and Its Applications, Vol.III (IAEA, Vienna, 1976), p.193.

[8] J. Tarski, in Feynman Path Integrals - Proc. Marseille 1978, Eds. S. Albeverio, Ph. Combe, R. Hoegh-Krohn, G. Rideau, M. Sirugue-Collin, M. Sirugne and R. Stora (Springer-Verlag, Berlin-Heidelberg-New York, 1979; Lecture Notes in Physics 106), p.254.

[9] J. Tarski, in Functional Integration: Theory and Applications (Proc. Louvain-la-Nevue, 1979), Eds. J-P. Antoine and E. Tirapegui (Plenum Press, New York, 1980), p.143.

[10] I.I. Priwalow, Randeigenschaften Analytischer Funktionen, 2nd edition (VEB Deutscher Verlag der Wissenschaften, Berlin, 1956), especially pp.18-19.

[11] H.P. Berg and J. Tarski, J. Phys. A: Math. Gen. 14, 2207 (1981).

[12] S. Lang, A Second Course in Calculus, 2nd edition (Addison-Wesley Publ., Co., Reading, Mass., 1968), especially p.368.

Progress in Mathematics

Edited by J. Coates and S. Helgason

Progress in Physics

Edited by A. Jaffe and D. Ruelle

- A collection of research-oriented monographs, reports, notes arising from lectures or seminars
- Quickly published concurrent with research
- Easily accessible through international distribution facilities
- Reasonably priced
- Reporting research developments combining original results with an expository treatment of the particular subject area
- A contribution to the international scientific community: for colleagues and for graduate students who are seeking current information and directions in their graduate and post-graduate work.

Manuscripts

Manuscripts should be no less than 100 and preferably no more than 500 pages in length.

They are reproduced by a photographic process and therefore must be typed with extreme care. Symbols not on the typewriter should be inserted by hand in indelible black ink. Corrections to the typescript should be made by pasting in the new text or painting out errors with white correction fluid.

The typescript is reduced slightly (75%) in size during reproduction; best results will not be obtained unless the text on any one page is kept within the overall limit of 6x9½ in (16x24 cm). On request, the publisher will supply special paper with the typing area outlined.

Manuscripts should be sent to the editors or directly to:
Birkhäuser Boston, Inc., P.O. Box 2007, Cambridge, Massachusetts 02139

PROGRESS IN MATHEMATICS

Already published

PM 1 Quadratic Forms in Infinite-Dimensional Vector Spaces
Herbert Gross
ISBN 3-7643-1111-8, 431 pages, paperback

PM 2 Singularités des systèmes différentiels de Gauss-Manin
Frédéric Pham
ISBN 3-7643-3002-3, 339 pages, paperback

PM 3 Vector Bundles on Complex Projective Spaces
C. Okonek, M. Schneider, H. Spindler
ISBN 3-7643-3000-7, 389 pages, paperback

PM 4 Complex Approximation, Proceedings, Quebec, Canada, July 3-8, 1978
Edited by Bernard Aupetit
ISBN 3-7643-3004-X, 128 pages, paperback

PM 5 The Radon Transform
Sigurdur Helgason
ISBN 3-7643-3006-6, 202 pages, paperback

PM 6 The Weil Representation, Maslov Index and Theta Series
Gérard Lion, Michèle Vergne
ISBN 3-7643-3007-4, 345 pages, paperback

PM 7 Vector Bundles and Differential Equations
Proceedings, Nice, France, June 12-17, 1979
Edited by André Hirschowitz
ISBN 3-7643-3022-8, 255 pages, paperback

PM 8 Dynamical Systems, C.I.M.E. Lectures, Bressanone, Italy, June 1978
John Guckenheimer, Jürgen Moser, Sheldon E. Newhouse
ISBN 3-7643-3024-4, 300 pages, paperback

PM 9 Linear Algebraic Groups
T. A. Springer
ISBN 3-7643-3029-5, 304 pages, hardcover

PM10 Ergodic Theory and Dynamical Systems I
A. Katok
ISBN 3-7643-3036-8, 352 pages, hardcover

PM11 18th Scandinavian Congress of Mathematicians, Aarhus, Denmark, 1980
Edited by Erik Balslev
ISBN 3-7643-3034-6, 528 pages, hardcover

PM12 Séminaire de Théorie des Nombres, Paris 1979-80
Edited by Marie-José Bertin
ISBN 3-7643-3035-X, 408 pages, hardcover

PM13 Topics in Harmonic Analysis on Homogeneous Spaces
Sigurdur Helgason
ISBN 3-7643-3051-1, 142 pages, hardcover

PM14 Manifolds and Lie Groups, Papers in Honor of Yozô Matsushima
Edited by J. Hano, A. Marimoto, S. Murakami, K. Okamoto, and H. Ozeki
ISBN 3-7643-3053-8, 480 pages, hardcover

PM15 Representations of Real Reductive Lie Groups
David A. Vogan, Jr.
ISBN 3-7643-3037-6, 771 pages, hardcover

PM16 Rational Homotopy Theory and Differential Forms
Phillip A. Griffiths, John W. Morgan
ISBN 3-7643-3041-4, 264 pages, hardcover

PM17 Triangular Products of Group Representations and their Applications
S.M. Vovsi
ISBN 3-7643-3062-7, 150 pages, hardcover

PM18 Géométrie Analytique Rigide et Applications
Jean Fresnel, Marius van der Put
ISBN 3-7643-3069-4, 215 pages, hardcover

PM19 Periods of Hilbert Modular Surfaces
Takayuki Oda
ISBN 3-7643-3084-8, 144 pages, hardcover

PM20 Arithmetic on Modular Curves
Glenn Stevens
ISBN 3-7643-3088-0, 214 pages, hardcover

PM21 Ergodic Theory and Dynamical Systems II
A. Katok, editor
ISBN 3-7643-3096-1, 215 pages, hardcover

PM22 Séminaire de Théorie des Nombres, Paris 1980-81
Marie-José Bertin, editor
ISBN 3-7643-3066-X hardcover

PM23 Adeles and Algebraic Groups
A. Weil
ISBN 3-7643-3092-9, 136 pages, hardcover

PROGRESS IN PHYSICS

Already published

PPh1 Iterated Maps on the Interval as Dynamical Systems
Pierre Collet and Jean-Pierre Eckmann
ISBN 3-7643-3026-0, 256 pages, hardcover

PPh2 Vortices and Monopoles, Structure of Static Gauge Theories
Arthur Jaffe and Clifford Taubes
ISBN 3-7643-3025-2, 275 pages, hardcover

PPh3 Mathematics and Physics
Yu. I. Manin
ISBN 3-7643-3027-9, 111 pages, hardcover

PPh4 Lectures on Lepton Nucleon Scattering and Quantum Chromodynamics
W.B. Atwood, J.D. Bjorken, S.J. Brodsky, and R. Stroynowski
ISBN 3-7643-3079-1, 587 pages, hardcover

PPh5 Gauge Theories: Fundamental Interactions and Rigorous Results
P. Dita, V. Georgescu, R. Purice, editors
ISBN 3-7643-3095-3, 375 pages, hardcover